K

20 023 662 17

281|96017543
15|8|97 *

UNIVERSITY OF NORTHUMBRIA
AT NEWCASTLE LIBRARY
WITHDRAWN

Phytochemistry and Agriculture

PROCEEDINGS OF THE PHYTOCHEMICAL SOCIETY OF EUROPE

23. **The Genetic Manipulation of Plants and its Application to Agriculture**
    Edited by P.J. Lea and G.R. Stewart

24. **Membranes and Compartmentations in the Regulation of Plant Functions**
    Edited by A.M. Boudet and others

25. **The Biochemistry of Plant Phenolics**
    Edited by C.F. Van Sumere and P.J. Lea

26. **Plant Products and the New Technology**
    Edited by K.W. Fuller and J.R. Gallon

27. **Biologically Active Natural Products**
    Edited by K. Hostettmann and P.J. Lea

28. **Biochemistry of the Algae and Cyanobacteria**
    Edited by L.J. Rogers and J.R. Gallon

29. **Bryophytes: Their Chemistry and Chemical Taxonomy**
    Edited by H.D. Zinsmeister and R. Mues

30. **Secondary Plant Products from Tissue Culture**
    Edited by B.V. Charlwood and M.J.C. Rhodes

31. **Ecological Chemistry and Biochemistry of Plant Terpenoids**
    Edited by J.B. Harborne and F.A. Tomas-Barberan

32. **Biochemistry and Molecular Biology of Plant–Pathogen Interactions**
    Edited by C.J. Smith

33. **Nitrogen Metabolism of Plants**
    Edited by K. Mengel and D.J. Pilbeam

34. **Phytochemistry and Agriculture**
    Edited by T.A. van Beek and H. Breteler

PROCEEDINGS OF THE
PHYTOCHEMICAL SOCIETY OF EUROPE

Phytochemistry and Agriculture

Edited by

T.A. VAN BEEK

Phytochemical Section, Department of Organic Chemistry, Wageningen Agricultural University, Dreijenplein 8, 6703 HB Wageningen, The Netherlands

and

H. BRETELER

Agricultural Research Department (DLO), PO Box 59, 6700 AB Wageningen, The Netherlands

CLARENDON PRESS · OXFORD

1993

Oxford University Press, Walton Street, Oxford OX2 6DP
Oxford New York Toronto
Delhi Bombay Calcutta Madras Karachi
Kuala Lumpur Singapore Hong Kong Tokyo
Nairobi Dar es Salaam Cape Town
Melbourne Auckland Madrid
and associated companies in
Berlin Ibadan

Oxford is a trade mark of Oxford University Press

Published in the United States
by Oxford University Press Inc., New York

© The Phytochemical Society of Europe, 1993

All rights reserved. No part of this publication may be reproduced, stored in a retrieval system, or transmitted, in any form or by any means, without the prior permission in writing of Oxford University Press. Within the UK, exceptions are allowed in respect of any fair dealing for the purpose of research or private study, or criticism or review, as permitted under the Copyright, Designs and Patents Act, 1988, or in the case of reprographic reproduction in accordance with the terms of licences issued by the Copyright Licensing Agency. Enquiries concerning reproduction outside those terms and in other countries should be sent to the Rights Department, Oxford University Press, at the address above.

This book is sold subject to the condition that it shall not, by way of trade or otherwise, be lent, re-sold, hired out, or otherwise circulated without the publisher's prior consent in any form of binding or cover other than that in which it is published and without a similar condition including this condition being imposed on the subsequent purchaser.

A catalogue record for this book is available from the British Library

Library of Congress Cataloging in Publication Data
Phytochemistry and agriculture / edited by T.A. van Beek and H. Breteler.
p. cm.—(Proceedings of the Phytochemical Society of Europe; 34)
Based on papers given at a symposium held April 22–24, 1992, in Wageningen, The Netherlands.
Includes bibliographical references and index.
1. Botanical pesticides—Congresses. 2. Plants, Protection of—Congresses.
3. Botanical chemistry—Congresses. 4. Pesticidal plants—Congresses.
I. Beek, Teris André van. II. Breteler, H. III. Series.
SB951.145.B68P49 1993 630'.2'54—dc20 93–4235
ISBN 0–19–857762–1

Typeset by Apex Products, Singapore
Printed in Great Britain by
St Edmundsbury Press, Bury St Edmunds

UNIVERSITY OF
NORTHUMBRIA AT NEWCASTLE
SS No.
2002366217 581.19 PHY

Preface

This volume contains an introduction and 18 chapters based on reviews presented by invited speakers at the International Symposium, entitled *Phytochemistry and Agriculture* held in Wageningen, The Netherlands on 22–24 April 1992. More than 150 scientists from 22 countries from all five continents attended this meeting and made it a truly international scientific happening.

This was the first meeting of the Phytochemical Society of Europe (PSE) ever held in The Netherlands, in spite of the fact that there are many active phytochemists and PSE members in The Netherlands. Probably the best known phytochemical group is the one at the Department of Pharmacognosy of the State University of Leiden, where Prof. Hegnauer, Prof. Baerheim Svendsen, and Prof. Verpoorte successively earned themselves a good reputation in the fields of chemotaxonomy, essential oils, alkaloids, and plant cell biotechnology. But also in Wageningen at many of the institutes of the Agricultural Research Department and at the Agricultural University various groups are working on phytochemical problems in such different fields as ecology, food technology, crop protection, plant cell biotechnology, agrification, toxicology, and organic chemistry. This is evidenced by the large percentage of the 48 displayed posters originating from Wageningen. Because of the broad agricultural interest and the international character the Board of the Agricultural University declared this symposium *the* distinctive symposium of the University in 1992 (the third symposium in a series of Wageningen Conferences).

The present meeting was organized by the Phytochemical Section of the Department of Organic Chemistry of Wageningen Agricultural University under auspices of the PSE and in co-operation with the Royal Netherlands Chemical Society. We hope that after this meeting other scientists in The Netherlands have become enthusiastic about this type of symposia and that a second PSE meeting in The Netherlands will be organized within a few years.

We are indebted to the officers of the PSE for their assistance, particularly to Dr B.V. Charlwood who could be consulted at all times, and Dr P.J. Hylands for supervising the printing of the first and second circulars. Further we wish to thank the staff of the International Agricultural Centre where the symposium took place, the scientific advisory committee, and the various chairmen who all did a marvellous job. Finally, we thank the plenary lecturers who prepared their manuscripts for this volume in the series 'Proceedings of the PSE'. It will hopefully turn out to be an often consulted hard proof of this stimulating meeting.

Wageningen
April 1993

T.A. van Beek
H. Breteler

Sponsorship

The Organizing Committee gratefully acknowledges the following non-profit institutions and commercial organizations for their financial guarantee or support of the symposium *Phytochemistry and Agriculture* held in Wageningen, 22–24 April 1992.

Non-profit institutions

Landbouw Universiteit Wageningen
Commission of the European Communities (ECLAIR)
Koninklijke Academie van Wetenschappen
Ministerie van VROM, Dir.-Gen. Milieubeheer
Koninklijke Nederlandse Chemische Vereniging
Stichting Biotechnologie Nederland
Cluster Biomoleculaire Wetenschappen der LUW
Ministerie van LNV, Kabinet
Phytochemical Society of Europe
Stichting 'Fonds Landbouw Export Bureau 1916/1918'

Commercial organizations

E. Merck Nederland
Hoechst Holland
Denka International
Unilever
Hewlett Packard Nederland
Varian Nederland
Bruker Spectrospin
Elsevier Science Publishers
Fisons Instruments
Meyvis en Co.
Chrompack
JEOL (Europe)
Schering AAgrunol
AVEBE
Mettler-Toledo
Shimadzu Benelux
J. Wiley & Sons
Royal Society of Chemistry UK
Chapman & Hall
Royal Sluis

Contents

Contributors

M. Bokel: Institut für Chemie, Universität Hohenheim, 70593 Stuttgart, Germany

Dirk Bosch: CPRO-DLO, PO Box 16, 6700 AA Wageningen, The Netherlands

Alle Bruggink: DSM Andeno B.V., PO Box 81, 5900 AB Venlo, The Netherlands

Barry V. Charlwood: Division of Life Sciences, King's College London, University of London, Campden Hill Road, London W8 7AH, UK

Marianne Delarue: Laboratoire de Biologie de la Rhizosphère, Institut National de la Recherche Agronomique, F-78026 Versailles Cedex, France

Arlette Goldmann: Laboratoire de Biologie de la Rhizosphère, Institut National de la Recherche Agronomique, F-78026 Versailles Cedex, France

Edwin Haslam: Department of Chemistry, University of Sheffield, Sheffield S3 7HF, UK

Bhupinder P.S. Khambay: Department of Insecticides & Fungicides, AFRC Institute of Arable Crops Research, Rothamsted Experimental Station, Harpenden, Herts, AL5 2JQ, UK

W. Kraus: Institut für Chemie, Universität Hohenheim, 70593 Stuttgart, Germany

Enno Krebbers: Plant Genetic Systems N.V., Joseph Plateaustraat 22, B-9000 Gent, Belgium

C. Lavaud: Laboratoire de Pharmacognosie (URA CNRS 492), Faculté de Pharmacie, Université de Reims Champagne-Ardenne, 51, rue Cognacq-Jay, 51096 Reims Cedex, France

Loic Lecoeur: Laboratoire de Biologie de la Rhizosphère, Institut National de la Recherche Agronomique, F-78026 Versailles Cedex, France

Monique Maille: Laboratoire de Biologie de la Rhizosphère, Institut National de la Recherche Agronomique, F-78026 Versailles Cedex, France

G. Massiot: Laboratoire de Pharmacognosie (URA CNRS 492), Faculté de Pharmacie, Université de Reims Champagne-Ardenne, 51, rue Cognacq-Jay, 51096 Reims Cedex, France

Brigitte Message: Laboratoire de Biologie de la Rhizosphère, Institut National de la Recherche Agronomique, F-78026 Versailles Cedex, France

Joseph N.M. Mol: Department of Genetics, Faculty of Biology, Free University, De Boelelaan 1087, 1081 HV Amsterdam, The Netherlands

Russell J. Molyneux: Western Regional Research Center, Agricultural Research Service/USDA, 800 Buchanan Street, Albany, CA 94710, USA

Adolf Nahrstedt: Institut für Pharmazeutische Biologie und Phytochemie, Westfälische Wilhelms-Universität, Hittorfstraße 56, D-48149 Münster, Germany

Norval O'Connor: Department of Insecticides & Fungicides, AFRC Institute of Arable Crops Research, Rothamsted Experimental Station, Harpenden, Herts, AL5 2JQ, UK

J. O'Reilly: Cara Partners, Little Island, Co. Cork, Ireland

John A. Pickett: Department of Biological & Ecological Chemistry, AFRC Institute of Arable Crops Research, Rothamsted Experimental Station, Harpenden, Herts, AL5 2JQ, UK

Peter W. Price: Department of Biological Sciences, Northern Arizona University, Flagstaff, AZ 86011-5640, USA

L.M. Schoonhoven: Department of Entomology, Wageningen Agricultural University, PO Box 8031, 6700 EH Wageningen, The Netherlands

M. Schwinger: Institut für Chemie, Universität Hohenheim, 70593 Stuttgart, Germany

Asfaq Sheak: Department of Drug Administration, Thapathali, Kathmandu, Nepal

R. Soellner: Central Research and Development, Bayer AG, 51273 Leverkusen, Germany

R. Steffens: Agricultural Division, Bayer AG, 51273 Leverkusen, Germany

Sten Stymne: Department of Plant Physiology, Swedish University of Agricultural Sciences, PO Box 7047, S-750 07 Uppsala, Sweden

David Tepfer: Laboratoire de Biologie de la Rhizosphère, Institut National de la Recherche Agronomique, F-78026 Versailles Cedex, France

Joel Vandekerckhove: University of Gent, Department of Physiological Chemistry, Ledeganckstraat 35, B-9000 Gent, Belgium

B. Vogler: Institut für Chemie, Universität Hohenheim, 70593 Stuttgart, Germany

U. Wachendorff: Agricultural Division, Bayer AG, 51273 Leverkusen, Germany

Lester J. Wadhams: Department of Biological & Ecological Chemistry, AFRC Institute of Arable Crops Research, Rothamsted Experimental Station, Harpenden, Herts, AL5 2JQ, UK

D. Wendisch: Central Research and Development, Bayer AG, 51273 Leverkusen, Germany

Michael Wink: Institut für Pharmazeutische Biologie, Ruprecht-Karls-Universität Heidelberg, Im Neuenheimer Feld 364, 6920 Heidelberg, Germany

Christine M. Woodcock: Department of Biological & Ecological Chemistry, AFRC Institute of Arable Crops Research, Rothamsted Experimental Station, Harpenden, Herts, AL5 2JQ, UK

Introduction

T. A. VAN BEEK and H. BRETELER

Nobody will argue about the tremendous importance of agriculture and forestry. It furnishes mankind with vegetable food and feed for his domestic animals. Further, plants supply many other kinds of products which are important for normal life or which make life more agreeable such as firewood for cooking and heating, medicines for healing, dyes for painting, pesticides for protection of crops, toxins for hunting or fishing, spices for flavouring, volatile oils for soaps, and stimulants like coffee and tea. In the industrialized western world often the crude plants or extracts are no longer used for these purposes, instead the pure, active principles or still more active synthetic replacements are used. Thus in many products used in every day life agriculture and phytochemistry have played, at some stage, an important role. It was therefore easy for the Department of Organic Chemistry of Wageningen Agricultural University, when it was approached to organize the Spring 1992 meeting of the PSE, to decide what the topic of the first PSE meeting in *the* agricultural centre of The Netherlands should be. '*Phytochemistry and Agriculture*' seemed an obvious choice, even more so because this topic had not been reviewed at any previous PSE symposium.

It was decided at an early stage to approach this theme broadly and not to limit the symposium to naturally occurring agrochemicals and crop protection only. The latter subject was to be only one of the three themes of the symposium. The three themes that were selected are:

(1) phytochemicals and crop protection;

(2) adverse effects of phytochemicals in food and feed;

(3) production of phytochemicals by field crops and tissue culture.

The first theme had been the subject of four recent symposia in the USA (Cutler 1988; Arnason *et al.* 1989; Hedin 1991; Nigg and Seigler 1992). Apart from paying attention to the beneficial role of phytochemicals in the development of, or application as, crop protectants, attention was given to the fact that some phytochemicals in our food and feed are causing unwanted side effects and to the question of how to cope with these. This formed the second theme. In the last theme of this symposium 'Production of phytochemicals' emphasis was given to the breeding, cultivation, extraction, and usage of various kinds of useful plant constituents. Both modern and classical methods were discussed. In the remainder of this introduction the different topics will be highlighted and the contents of the different chapters will be put in the perspective of the topics covered by the symposium.

To protect our food crops and other plants from attacks by herbivores, various infectious diseases, and competition by weeds many pesticides nowadays are used, e.g. herbicides, insecticides, nematicides, and fungicides. In spite of these measures it is estimated that 10–35 per cent of all crops are lost either before harvest on the field or during storage. Especially due to the current intensive cultivation practices, which are often dictated by an economy of scale, pests can have a devastating effect in the large monocultures. Although most of the synthetic pesticides are nowadays highly potent compounds many of them possess several disadvantages, like lack of selectivity, build-up of resistance, toxicity to man, and accumulation in the environment. The classical example of lack of bio-degradability is DDT, initially thought to be the final solution for almost all insect pests. The development of resistance is one of the biggest problems because one has either to resort to other agents or to use more during each application thus adding to the problem of pollution. To stop this pollution process the use of certain pesticides is therefore diminished. For instance in the Netherlands many pesticides can no longer be used for all crops, some will be totally banned, and the overall usage of pesticides has to diminish by at least 50 per cent before the year 2000. On the other hand the world population is still growing and more and more food will be needed. Other approaches to achieve this goal are therefore called for and an economic form of Integrated Pest Management appears to give the optimal result.

Pesticides will still be needed however, and if some are or will be forbidden new ones have to be found. One group of compounds which surely deserves attention, either to be used as such or to serve as model compounds, are the native constituents of plants which have proven themselves time after time. This idea is certainly not new but only temporarily forgotten due to the instant successes of many of the synthetic pesticides. L. M. Schoonhoven in the opening chapter most appropriately cites Abbott (1887) who more than 100 years ago wrote: 'It has been recently suggested that many of the chemical compounds may serve the plant as means of defence against animals, and when we camphorize our furniture and poison our flower-beds, we are only imitating and reinventing what the plants practised before the existence of man.'

If one closely watches all the ecological systems in nature, leads for the development of new crop protection agents abound. Until quite recently only compounds which *directly* influenced a pest species were used as models. This group includes for instance plant-derived insecticides, insect hormones or hormone mimics, repellants, neuropeptides or pheromones, phytoalexins, antifeedants, and plant growth regulators. For some groups this has been a highly successful approach, e.g. the pyrethroids. However Schoonhoven gives several examples of the many intricate interactions between insects and plants and clearly shows that we have only just started to understand the many factors involved. The examples clearly show that for a further successful development and application of natural pesticides a thorough knowledge of the biology and chemistry of the system under study is a *conditio-sine-qua-non*. It is becoming more and more clear that many compounds play a role in interactions involving more than just two species and in future this concept may prove to be useful for the development of *indirectly* acting crop protection agents.

One group of natural directly-acting pesticides mentioned above are the antifeedants. By the term 'antifeedants' plant compounds are meant which interact with taste receptors of herbivores in such a way that the process of consuming is strongly repressed or stopped. An intrinsic advantage of antifeedants is that by their nature they prevent damage, instead of killing the insect after it has damaged the plant as in the case of ordinary insecticides. Although the concept is simple many practical problems have so far prevented large-scale usage of antifeedants for the protection of crops. Problems include availability of the antifeedant, low activity, activity against too few pest insects, necessity of frequent application, and toxicity towards non-target organisms including humans. The most successful antifeedants until now are a group of tetranortriterpenoids from the neem tree (*Azadirachta indica*). A crude aqueous extract prepared from the abundant seeds of this large tropical tree is now used by small farmers in India, the Philippines, and other tropical countries for the protection of various crops. The crude extract is further registered as a pesticide in the USA (Morgosan-O®) for use on non-food plants.

The most active constituent in this extract is azadirachtin, a highly complex natural product which has not yet been synthesized. Its structure has only recently been elucidated by the group of Kraus, after two incorrect structures had first been published. It seemed therefore appropriate that W. Kraus *et al.* would deal with the occurrence and antifeedant activity of many of these triterpenoids which have been isolated from several Meliaceae genera, e.g. *Azadirachta*, *Melia*, and *Khaya* (Chapter 2). From this work which is a nice blend of phytochemistry, synthesis, and pharmacology it follows that subtle changes in a molecule can profoundly influence the antifeedant activity, and that several terpenes are as active as azadirachtin itself. However little is known of these novel compounds with regard to toxic effects on insects and mammals. One of the strong assets of azadirachtin is that, in addition to being a excellent antifeedant, it is toxic to insects but virtually non-toxic and tasteless to humans.

Notwithstanding the introduction of novel approaches such as the antifeedants mentioned above, the development of newer traditional insecticides will remain necessary to safeguard our crops. About half of the insecticides (mainly pyrethroids and carbamates) used today are originally derived from plant sources and B. Khambay and N. O'Connor in Chapter 3 make it clear that higher plants are still being considered a promising source and are being screened for novel insecticidal lead structures. So far only a small percentage of all plants have actually been screened and in addition to this some of the earlier studies are not up to the current standards with regard to the bioassays used. After finding a promising new lead structure, nowadays structure–activity relationship (SAR) approaches play an important role in the further development of new products. SAR studies are also important for the on-going development of a well-established group of compounds such as that of the pyrethroids. Photostable synthetic pyrethroids which are more than 1000 times as potent as the original lead compound pyrethrin I now exist, and further SAR studies may produce still other useful pyrethroids and thus prolong the use of this group. The authors further show that several other basic insecticidal compounds from plants, many of which are generally considered as obsolete, are worth investigating anew with today's sophisticated techniques, e.g. certain coumarins and lignans, rotenoids, and veratrum alkaloids. Apart from higher plants nowadays microorganisms and animals are also screened for insecticides. From bacteria both peptidic (e.g. *Bacillus thuringiensis*) and non-peptidic (e.g. avermectin) insecticides are known. Efforts in the development of crop protection agents from animals are concentrated on extremely active insect neuropeptides and some spider venoms.

Even though it may not be fully justified on the basis of scientific evidence, in future there may be a shift from purely toxic insecticides, such as mentioned above, to more regulatory insecticides which solely affect the behaviour of the insect in a, for us, favourable way. As regulatory compounds functioning within the organism may be difficult targets, e.g. extremely active neuropeptides which fail to penetrate the insect, regulators which act externally on the organism such as allelochemicals and pheromones seem to have a brighter future. J. A. Pickett *et al.* in Chapter 4 show that pheromones on their own are not capable of producing a long-lasting effect on insect populations. Behaviour-controlling phytochemicals, and semiochemicals in general, will in future find use with push-pull, or 'stimulo-deterrent diversionary strategies'

(SDDS). In such strategies insects are repelled from the crop by anti-feedants or deterrent pheromones and at the same time attracted by sex pheromones to trap-crops. On these trap-crops either a selectively acting insecticide or a pathogenic biological control agent is placed. However for a proper application of this strategy it is—again—necessary to know which ecologically relevant compounds are involved. The authors of this chapter discuss various techniques for the identification of such compounds, among others gas chromatography coupled on-line with an insect antenna as detector. They also pay attention to the cheap(er) production of some regulatory compounds starting from extracted phytochemicals.

In recent years the study of plant–bacterium relationships has become a major research topic as witnessed by the many papers published on gene transfer by *Agrobacterium tumefaciens* and *A. rhizogenes* and *Rhizobium* strains involved in symbiotic nitrogen fixation by legumes. Secondary metabolites play an important role in the maintenance of these relationships. It has recently become clear that other secondary metabolites are implicated in specific nutritional relationships between bacteria and their host-plants. Some plant substances are selectively absorbed (and at the same time degraded) by soil bacteria and thus influence the dynamics of microbial populations. In Chapter 5 the discovery and ecological significance of an exciting new group of secondary metabolites named calystegins are discussed by D. Tepfer and co-workers (Goldmann *et al.*). These substances are produced by only a few plant families and are selectively used by *Rhizobium meliloti*. Although at present these compounds play no role in the protection or improvement of crops this may change in the future. It may be possible to introduce specific catabolic genes in a beneficial soil organism. This may increase the survival rate of these symbiotic soil microorganisms which in turn may be beneficial for the plant. Or alternatively, plants could be genetically manipulated for the release of novel metabolites into the rhizosphere causing new interactions. Eventually such new types of interactions could be beneficial for agriculture.

As already mentioned by Schoonhoven (Chapter 1) and Pickett *et al.* (Chapter 4) many ecological interactions are much more complex than was imagined only a few years ago. For instance it is probably more rule than exception that upon attack by herbivores plants produce chemicals (synomones) which attract enemies of the initial invader. This 'bodyguard', 'indirect defence', or 'tritrophic' concept was put forward for the first time by Price *et al.* (1980). Price in the final chapter of the 'Phytochemicals and crop protection' theme shows how this idea may be put to work in the field. Enhancement of natural extrinsic defences through natural enemies of herbivores is one of the few long-term solutions to diminish the use of externally applied pesticides and resistance problems. In the field attention should be paid both to the plant *and* the predator or parasite. The production of synomones ('bottom-up' approaches) by the first trophic level should be maximized while at the same time the production of intrinsic defences should not be neglected. On the predator/parasite side mass release and use of regulatory chemicals inducing, for instance, host-seeking should be used (top-down approaches).

The second section 'Adverse effects of phytochemicals in food and feed' again clearly illustrates that the popular belief 'synthetic is toxic and natural is healthy' is a severe misconception. In this context one does not just have to think of the highly toxic strychnine from *Strychnos* species or aflatoxins from certain poorly stored food stuffs to show that this belief is wrong. One the largest staple crops in the world is *Manihot esculenta* (cassava, manioc, tapioca). This food plant has the advantage that it can grow under rather poor conditions. The tubers are rich in starch and serve as an energy supply for an estimated half billion people in the tropics. Unfortunately the bark of the tubers contains two toxic cyanogenic glycosides. If no proper detoxification procedure is carried out, long-term consumption leads to chronic cyanide intoxication. However as Nahrstedt shows in Chapter 7 breeding for lower cyanogenic glycoside levels may not be the final answer to all problems as some low cyanogenic strains suffer from enhanced attack by herbivores. The short term solution seems to lie in adapted processing procedures or the controlled usage of insecticides.

One of the larger environmental problems in the Netherlands is the production of dung by the ever increasing intensive animal production. At this moment there is a clear excess and although much research has been devoted in recent years to a technological solution of this problem, e.g. biogas plants and processing of manure, no clear-cut solution is expected in the near future. Apart from diminishing the number of animals the only possibility to *decrease* the dung production is to improve the digestibility of the poultry or cattle feed. Then correspondingly less will be needed. One of the best protein sources for animal feed under temperate climatic conditions is alfalfa. Yields of proteins for this crop are in the range of 2 tons per hectare, two to three times more than, for instance, with soya. Unfortunately, alfalfa extracts contain certain 'digestibility-inhibiting substances' which limit their use. G. Massiot and C. Lavaud in Chapter 8 describe the multidisciplinary research which led to the elucidation of at least some of the responsible compounds. A combination of chemical and highly advanced spectroscopic methods, and accurate bioassays was necessary to unravel the structures of the complex saponins present in alfalfa.

A clear correlation was established between at least one saponin and growth of test animals. The next step will be a selection of different alfalfa varieties for low saponin levels while at the same time maintaining good agricultural properties. Alternatively a genetic modification of alfalfa with the aim of reducing or changing the saponin content could be pursued.

A problem that is somewhat unfamiliar to cattle farmers in western Europe with their fine grazing lands or intensive indoor cattle systems is the death of livestock by the consumption of poisonous plants. In the Western USA, however, this is a very serious economic problem. Annual losses are estimated at US $340 million. For individual farmers losses may be catastrophic and lead to bankruptcy. Since the beginning of this century the United States Department of Agriculture (USDA) has been investigating the disease symptoms, the plants, and the constituents causing the diseases, and possible countermeasures or solutions for the problem. The most notorious plant is locoweed (*Astragalus* or *Oxytropis* species), in some cases 25 000 pieces have been lost locally. Other species causing problems are *Halogeton* species, death camas (*Zygadenus paniculatus*), English yew (*Taxus baccata*), tansy ragwort and groundsel (*Senecio* species), Western false hellebore (*Veratrum californicum*), poison hemlock (*Conium maculatum*), tree tobacco (*Nicotiana glauca*), lupines (*Lupinus* species), and many more. Sometimes the effect is not a direct poisoning but malformation of the offspring. R. J. Molyneux in Chapter 9 discusses these problems. He shows that in spite of a considerable body of knowledge regarding the majority of toxic plant species inhabiting rangelands, serious livestock losses still occur and that this is primarily caused by a lack of understanding of the phytochemistry involved. One lacks the ability to predict a safe level of exposure. The toxin content of the plants varies greatly depending on the species, stage of growth, and numerous environmental factors. His group has developed an idealized sequence of research stages which, if successfully accomplished, will lead to management recommendations. The four steps are:

(1) determination of the etiology of poisoning;
(2) isolation and characterization of toxins;
(3) development of analytical methods for quantitation of toxins in plants and animals;
(4) devising practical methods of managing the range lands and herds so that no harmful doses are ingested.

This approach works well for the locoweed problem.

The last nine chapters belong to the section 'Production of phytochemicals by field crops and tissue culture'. The bulk of all food and non-food plants are still traditionally cultivated and harvested. Food plants supply us with carbohydrates, fats and oils, and proteins while non-food plants are sources of textiles like cotton, paper, flavours, fragrances, and all kind of specialty chemicals. The possible large-scale cultivation and use of plants for the production of non-food products (agrification) is receiving more and more attention in the highly industrialized countries. The first four chapters of this section deal with such plants and products ('Production of phytochemicals by field crops'). The opening chapter (10) 'Production and application of phytochemicals from an agricultural perspective' by M. Wink serves as a link between the crop protection theme (Chapters 1 to 6) and this theme because he deals mainly with plants that contain insecticides. The next three chapters are concerned with plants containing polyphenols, the cultivation and improvement of the medicinal tree *Ginkgo biloba*, and the production of essential oil bearing and medicinal plants in a developing country. Attention is also given to the type, extraction, and usage of the commercially interesting constituents. The last four chapters of this section ('Production of phytochemicals by plant biotechnology') focus on new breeding and production systems. In these chapters either modern biotechnology plays a pivotal role or its impact on future development is addressed. Plants producing oils and fats, colours, aroma compounds, and proteins are reviewed.

Wink points out that from a variety of defence mechanisms against their predators, plants in general have developed a preference for chemical defence systems. The systems comprise compounds as numerous as the number of plant species, viz. at least a quarter of a million. This wealth of compounds not only serves the survival of the species producing them, but may also be used as pharmaceuticals, which can be produced in fermentors or in the field.

However the use of phytochemicals in crop protection is still underexposed, and for various reasons a survey of what has been done and what still can be expected is timely. Data from Wink's own group convincingly demonstrate that, for example, quinolizidine alkaloids from lupines and other legumes have a number of functions and biological activities in a plant. The ecological aspects of compounds should be kept in mind when cultivating or breeding crops for various purposes. Wink indicates that in some agronomic strategies valuable compounds have been lost to make crops seemingly more attractive, e.g. the removal of certain alkaloids to decrease bitterness. The disappearance of the chemical must then be compensated for by using man-made pesticides. Wink also makes clear that the development of new natural pesticides is only one aspect of the changes that are required to develop new and safe agricultural practices. All these aspects should be integrated and phytochemistry and chemical ecology play a pivotal role in this whole process.

A class of phytochemicals with interesting properties and which are extracted on a large scale from plants are the polyphenols. Natural polyphenols or vegetable tannins as they are also called, give complexation reactions with many other compounds and their commercial usage is related to this intrinsic ability. They are used for the tanning of leather, in foodstuffs and beverages, and in stomach pills. Plants use these compounds themselves for chemical defense and for pigmentation purposes. E. Haslam gives in Chapter 11 an account of the different classes of polyphenols and illustrates with several examples how they work at a molecular level. The complexation processes may be reversible or irreversible and various co-substrates may be involved. Irreversible polyphenol complexation is much more common than one would think at first. Not only the tanning of leather, but also the browning processes in fruits, tea and coffee fermentation, humic acid formation in soils, necrosis of plant tissues to avoid further damage by infections or predators, and sclerotization in insects are examples of irreversible polyphenol complexations. Sometimes as in the case of *Sorghum bicolor*, an important foodcrop in Africa, tannins constitute an important intrinsic chemical defense against pathogens and herbivores. However a too high concentration makes this cereal unpalatable so that some kind of compromise has to be found between possible plant damage and a too high concentration of antinutritional compounds. A better understanding of these processes could lead to more useful varieties.

The last few years have seen a tremendous increase in the sales of extracts of the medicinal plant *Ginkgo biloba*. In 1989 Tebonin®, one of several available ginkgo extracts, was the most prescribed drug in Germany with a sale of 219 million DM. The ginkgo tree, which was called 'a living fossil' by Darwin because of its long unchanged habitat and unique botanical position, was used for healing purposes by the Chinese several thousand years ago. In the 1960s and 1970s some chemical and pharmacological research was carried out, but only after Schwabe in Germany discovered that some unique diterpene trilactones (trivial name: ginkgolides) possessed strong and selective anti-platelet-activating-factor activity did the patented extract become popular (Hasler *et al.* 1990). It is now one of the few available drugs against problems related with old age and early dementia, i.e. extracts diminish forgetfulness and increase concentration and mental agility. The ginkgo extracts achieve this by an increase of the cerebral bloodflow. They also stimulate the peripheral bloodflow and are of use for intermittent claudication and similar problems. In a recent Ph.D. thesis the supposedly beneficial properties of many phytotherapeuticals were very critically assessed and according to the high standards which are nowadays used for the registration of new drugs only two were active: *Ginkgo biloba* extracts and garlic preparations (Kleijnen 1991). To obtain a sufficient supply of high quality ginkgo leaves, many steps had to be taken by the producers. J. O'Reilly in Chapter 12 gives a fine review of the cultivation, extraction, constituents, and pharmacological activity of the ginkgo tree. Particularly interesting is the account of the set-up of the ginkgo orchards. On, in total, 660 ha 17 million ginkgo trees have been planted from which the leaves are mechanically harvested every autumn.

Experiences of a very different kind are related by A. Sheak in Chapter 13. As the managing director of a commercial company trading in herbs, extracts, and essential oils from its foundation in 1981 until 1991 he, more than anyone else, is able to tell about the great difficulties of such an enterprise in a poor developing country without much infrastructure such as Nepal. His story shows, however, that it is not *a priori* impossible and that projects in developing countries *can* be highly successful. From its foundation the turnover of the company increased rapidly and in 1990 it was already 50 times higher than in 1981. It is becoming one of the larger firms in Nepal. Its vegetable products are used both for the local paint, dye, soap, and drug industries thus saving valuable hard currency, and for exportation (mainly essential oils) thus earning hard currency. One of the most difficult tasks was the penetration of the Japanese, European, and American markets. Both contacts and reliable delivery and production of standardized high quality products for a competitive price are necessary. This takes a lot of time, training, and investment. Cultivation practices for exotic plants had to be learned, extraction and distillation facilities had to be prepared, and a quality control laboratory had to be erected. Especially problematic was the transportation in this almost roadless hilly country. The project had also interesting social sides. Many previously landless farmers and their families obtained a piece of company land where they are now responsible for looking after the crops. They get a fixed, relatively high price for their harvests.

S. Stymne reviews in Chapter 14 plant storage lipids, commonly known as vegetable oils or fats which are the second most valuable commodity in world trade and have a global annual production exceeding 50 million metric tons. There is still a market for new oils with interesting composition and properties. Introduction of new vegetable oils could be achieved in a number of ways, e.g. domestication of wild species, mutation breeding, or alteration of oil qualities by genetic engineering. Many oils contain 'unusual' fatty acids and may be useful for specific purposes, e.g. erucic acid, γ-linolenic acid, ricinoleic acid, vernolic acid, and petroselinic acid. This last acid is for instance a potential raw material for the production of adipic acid. Our ability to manipulate plants genetically has opened new horizons for oil-containing crops. The economic potential of introducing foreign fatty acids into oil-crops by interspecies gene-transfer is enormous. Stymne's paper highlights the target enzymes, whose modification is nowadays under investigation in various laboratories. With the increased knowledge about fatty acids and storage lipid biosynthesis it may become possible to produce some special 'tailor-made' oils for the production of lubricants, surfactants, varnishes, and polymers. The ecological consequences both for plants and man are expected to be small. In the meantime the first report of the alteration of the fatty acid composition, and thus the quality, of seed oil by genetic engineering has been published (Knutzon *et al.* 1992).

The modification of pigmentation patterns and flower colours (Chapter 15) as achieved by the group of J. N. M. Mol is an example of research on the regulation of basic synthetic processes which also attracts the attention of practical floriculturists. By using sense and anti-sense constructs of genes involved in flavonoid biosynthetic pathways from a number of phylogenetically remote species like maize (Gramineae) and petunia (Solanaceae) it has now become possible to engineer flower colour in a predetermined way. Blue roses, carnations, and tulips may become commercially available in due time. This breakthrough may give an enormous impetus for the design of other new crop varieties by the same techniques. One of the traits that may be tackled is the taste and aroma of edible vegetable products. This is discussed in the next chapter by B. V. Charlwood.

His account in Chapter 16 on recent advances in the production of aroma compounds *in vitro* shows a number of possibilities and stumbleblocks. With emphasis on the production of lower isoprenoids (especially the monoterpenoids) in cell, organ, and tissue culture the ecological function and biochemistry of some aroma compounds are examined in order to produce marketable commodities in the long run. In undifferentiated systems (monoterpenoid containing) essential oils seldom accumulate. The explanation for this phenomenon at the level of enzyme activity and compartmentation of products is given. At a higher degree of tissue organization such as in genetically transformed root and shoot cultures significant accumulation of aroma compounds or their precursors occurs. Experiments with these cultures may be a necessary interphase before the introduction of foreign 'aroma genes' into intact plants can be attempted. Interestingly, the knowledge of plant defence systems (phytoalexins induced by elicitors) may be used here for other purposes than crop protection.

In the last chapter (17) of this section, E. Krebbers *et al.* expand the possibilities of the new gene technologies by considering the production of foreign proteins and peptides in higher plants. These proteinaceous compounds encompass human proteins and proteins used in medicine, such as immunoglobulin. The relatively low cost of growing plants and the availability of technology for harvesting and processing of crops are an incentive to study this type of heterologous gene expression.

Although yields of correctly expressed target proteins are still disappointingly low and the products are prone to modifications that make them inactive, some successes are expected soon. Extrapolation of expression levels attained in *Arabidopsis thaliana* and in rape indicate that the current state of the technology is the production of proteins at the kg/ha level. The authors systematically indicate which problems are likely to be solved and which hurdles may be more difficult to overcome. Apart from science and technology, Krebbers *et al.* draw attention to legislatory issues related to growing transgenic plants in the field and the uncertainty about the requirement of clinical trials even when shown that the produced proteins or peptides are chemically identical to those already tested and approved. This legislatory aspect and consumer acceptance of products of the new gene technology have a bearing on the economic prospect of this approach. The choice between various production systems for chemicals with an established market is elaborated further in the last chapter of this book.

Probably many phytochemists, pharmacognosists, natural product chemists, and biologists active in the plant cell biotechnology field secretly *hope* that once in their life they may create a breakthrough and come up with a new drug, a novel pesticide, or a commercially interesting production system. The journals are full of new promising natural products but the harsh reality is that almost no products or production systems *become* a commercial success. It is already 30 years since the last two drugs developed from higher plants, vincristine and vinblastine, entered the market, and also plant cell biotechnology has not realized the bright future predicted some 10 years ago. It is clear that the introduction of new products and production routes is an *evolutionary* and not a *revolutionary* process. This is illustrated in the last chapter by A. Bruggink who is both an R & D director of a company producing fine chemicals and a professor in industrial fine chemistry. He shows by analysing the production processes of different fine chemicals that technology changes in industry occur much slower than anticipated from scientific developments and that these changes can be rationalized and are to some extent even predictable. The changes are always driven by *economic* reasons. This facet especially tends to be forgotten now and then by university researchers. Further there is never an ever-winning product or technology; every product has a certain lifetime and at any given time and place its own optimal production process.

It is of course impossible to consider all aspects and review all plants, products, and technologies which have played, play, or will play a role in such a broad field as 'Phytochemistry and Agriculture' in 18 chapters. Some interesting topics such as the development of herbicides from plant allelopathics, the possible role of phytoalexins in crop protection, or the production of alkaloids such as taxol have not been discussed at all. The editors nevertheless hope that the concepts and research technologies which have been discussed give a good idea about what is going on in this field and will increase the general awareness about the positive and negative properties of phytochemicals and the plants containing them in current and future agriculture.

References

Abbott, H. C. de S. (1887). Comparative chemistry of higher and lower plants. *Am. Nat.*, **21**, 800–10.

Arnason, J. T., Philogène, B. J. R., and Morand, P. (ed.) (1989). *Insecticides of plant origin*. ACS symposium series 449, American Chemical Society, Washington.

Cutler, H. G. (ed.) (1988). *Biologically active natural products—potential use in agriculture*. ACS symposium series 380, American Chemical Society, Washington.

Hasler, A., Meier, B., and Sticher, O. (1990). *Ginkgo biloba*. Botanische, analytische und pharmakologische Aspekte. *Schweiz. Apoth. Ztg.*, **128**, 342–7.

Hedin, P. A. (ed.) (1991). *Naturally occurring pest bioregulators*. ACS symposium series 387, American Chemical Society, Washington.

Kleijnen, J. (1991). Food supplements and their efficacy. Ph.D. thesis. Maastricht.

Knutzon, D. S., Thompson, G. A., Radke, S. E., Johnson, W. B., Knauf, V. C., and Kridl, J. C. (1992). Modification of *Brassica* seed oil by anti-sense expression of a stearoyl-acyl carrier protein desaturase gene. *Proc. Natl. Acad. Sci., USA*, **89**, 2624–8.

Nigg, H. N. and Seigler, D. (ed.) (1992). *Phytochemical resources for medicine and agriculture*. Plenum Press, New York.

Price, P. W., Gross, C. E., McPheron, B. A., Thompson, J. N., and Weis, A. E. (1980). Interactions among three trophic levels: Influence of plants on interactions between insect herbivores and natural enemies. *Annu. Rev. Ecol. Syst.*, **11**, 41–65.

1. Insects and phytochemicals—nature's economy

L. M. SCHOONHOVEN

Department of Entomology, Agricultural University, P.O. Box 8031, 6700 EH Wageningen, The Netherlands

Introduction

Terrestrial plant life, though restricted to less than one third of the Earth's surface, absorbs more than twice as much solar energy as aquatic plant life. As a result terrestrial plants transformed what was once a forbidding desert planet, indistinguishable from Mars, into a luxuriant garden, in which insects, birds, and mammals could flourish. Although its uniform green hue suggests a basic homogeneity of the living plant world, its external appearance hides a wealth of chemicals. The variety of these phytochemicals is larger than chemists even in the near past could imagine and remains difficult to appreciate. Upon this seemingly endless number of organic compounds, which are constantly being transformed and replenished, many herbivores and parasitic organisms thrive and enrich the living world with their presence. The animal kingdom, in size and shape just as diverse as the plant world, includes many herbivores, ranging from plant lice to elephants. Among the herbivores insects occupy a paramount position, as exemplified by their biomass. Total body weight of social insects alone in tropical rainforests is seven times that of vertebrates (Fig. 1.1) (Holden 1989). However, biomass is not a representative trait for their role in a biotic community, because turnover rates, which include feeding, growth, reproduction and decay, are in insects much higher than in vertebrates. Therefore their role in the dynamics of energy and nutrient cycles is much greater than their tiny sizes would suggest. It is estimated that roughly 10 per cent of plant primary production in natural ecosystems is consumed by insects. (In cultivated plants the losses are, despite intense pest control measures, higher and under western conditions amount to 13 per cent.) Actually, in view of the fact that herbivorous insects generally possess short generation cycles, high reproductive rates, and, by being winged, can easily exploit new food sources, it is surprising that, despite the omnipresence of green plants, insects under natural conditions rarely become a threat to the survival of a given plant species. That is, resistant plants are the rule

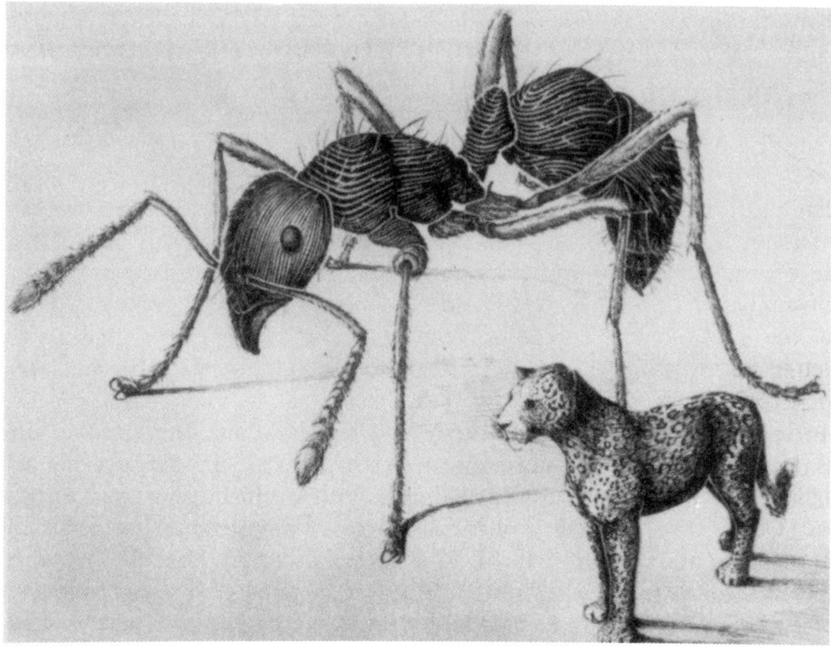

Fig. 1.1. In this drawing the ant represents the biomass of all ant populations and the leopard the biomass of the total land vertebrate population in the Brazilian Amazon. The ants outweigh the vertebrates by about 4 to 1 (from Holden 1989, drawing by K. Brown-Wing).

rather than the exception, and very susceptible plants are actually rare. Biologists as well as applied entomologists are interested in learning the nature of the protective measures which plants have developed against insect attack, in order to further decrease insect damage to crop plants.

How insects recognize host-plants

There is little doubt that insects devouring plants are in many ways influenced by what they eat. Their food not only affects their digestive processes, but also their ecological interactions (Strong *et al.* 1984) and even their evolutionary development (Ehrlich and Raven 1964; Jermy 1976). Knowledge that some plant species contain substances that repel insects can be traced to classical times (Smith and Secoy 1981), but the notion that such compounds provide the plants producing them protection to herbivores, has arisen more recently: 'It has been recently suggested that many of the chemical compounds may serve the plant as means of defence against animals, and when we camphorize our furniture and poison our flower-beds, we are only imitating and re-inventing what the plants practised before the existence of man.' (Abbott 1887). On the other hand from time immemorial it was recognized that most insects, guided by their unerring 'botanical instinct' feed on certain plant species only (Schoonhoven 1990). The Dutch botanist E. Verschaffelt first proved the decisive role of specific secondary plant substances in host-plant recognition by insects. When he offered larvae of the cabbage white butterfly the leaves of various non-host-plants wetted with a solution of sinigrin, the glucoside from black mustard, they were eagerly eaten as if they were host plants (Verschaffelt 1910).

It took some time to realize that the concept of an insect recognizing its host-plant by the presence or absence of only one single chemical signal is a too simplistic view. The chemical image of a plant is richer and based upon several if not many compounds. When we want to know in some detail which chemical stimuli determine host selection behaviour, a rewarding approach is to measure to which compounds the sensory system reacts to and to analyse in which way the quality and concentration of the chemical is translated into sensory messages. With the aid of modem electronic equipment neural signals can be recorded in individual olfactory or taste receptors. Thus the neural code, on which the brain decides what to eat or what not to eat, may be tapped into and analysed. From the results obtained with such electrophysiological methods it appears that food discrimination in the cabbage worm is based upon a double-factor. First host-specific compounds, i.e. glucosinolates, stimulate specialized taste neurones, which may release in the brain a command to start or maintain feeding behaviour (Fig. 1.2). If, however, the caterpillar accidentally takes a bite from some non-host-plant, taste cells may be activated which serve like smoke detectors and signal the presence of distasteful (deterrent) compounds (Schoonhoven and Blom 1988). Thus, leaves from the ginkgo tree, a kind of tree which is notoriously resistant to insect attack, are unacceptable to hungry cabbage worms, even after being treated with a strong sinigrin solution. Chemical analysis, combined with feeding tests have shown the presence of several feeding deterrents, including the ginkgolides, each of which strongly stimulates the deterrent cell of cabbage butterfly larvae, thereby markedly reducing or even preventing the intake of food (Yan *et al.* 1990).

For a long time the relationships between insects and plants were thought to fit a relatively simple pattern: host-plants contain one or a few related secondary plant substances that stimulate feeding and/or oviposition behaviour in insects specialized on them. Insects refrain from

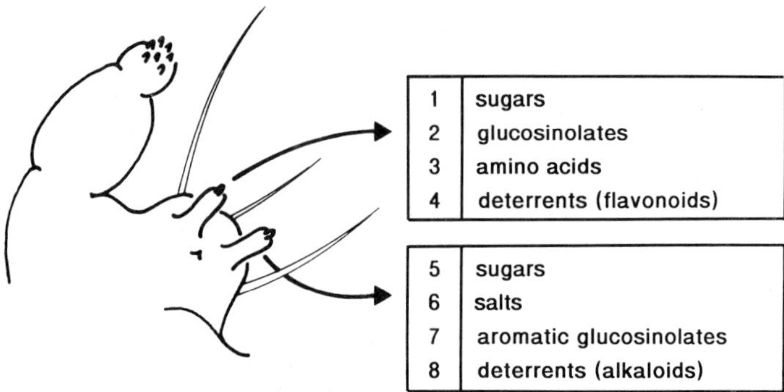

Fig. 1.2. The two taste hairs on each maxilla of the cabbage worm each contain four taste neurons. Some of them respond to low concentrations of host-plant specific chemicals (glucosinolates), others to various deterrent compounds.

feeding on non-host-plants because of the presence of some deterrent substances, which activate specialized deterrent receptors.

As is often the case in nature, reality is far more intricate and subtle than was first thought. It now appears that the plant stimuli governing insect behaviour are usually very complex; sometimes a complex profile is more critical than individual compounds. Another complication is that the insect's sensory reactions may vary considerably depending on internal factors, such as physiological condition and previous experiences. Thus, the chemoreceptory system does not show a fixed and invariable sensitivity, but rather appears to be modulated by the chemical composition of the food (Blaney *et al.* 1986). As a result the taste of potato leaves to a caterpillar which was grown on potato foliage, is quite different from that experienced by a caterpillar grown on tomato leaves. Their chemoreceptors are adapted to the different food types, and one could say that the chemoreceptors (which are primary nerve cells) possess a memory function. Concomitantly the insect's food preferences have been altered.

There is increasing evidence that not single chemicals but specific combinations are representative for certain plant taxa. Thus the odour emanated by potato plants, which is strongly attractant to Colorado potato beetles, consists of a number of general plant volatiles. Specificity in this case cannot be attributed to the compounds themselves, but lies in the ratio in which they occur in the headspace of the potato plant. None of these components, when tested singly, is attractive for the beetles. Indeed, if the level of one component in the natural potato plant odour is increased, the blend does not typify the scent of potato

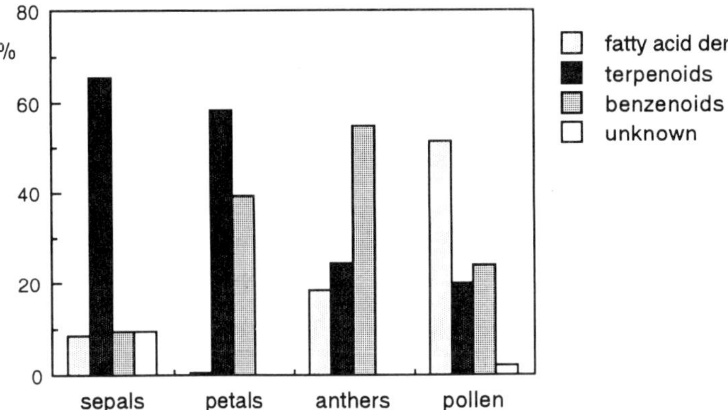

Fig. 1.3. Relative odour composition of different parts of *Rosa rugosa* flowers. Volatile components have been categorized into three different chemical groups (data from Dobson *et al.* 1990, with permission).

plants anymore, and, accordingly, Colorado potato beetles do not respond (Visser 1986).

The fragrances of flowers, even more than foliage, are composed of complex and delicate mixtures of volatiles. The floral fragrance of sunflowers, for example, consists of at least 144 constituents. Honeybees recognize sunflower odour on the basis of a certain combination of 28 components out of the complete bouquet (Pham-Delegue *et al.* 1990). The subtlety which characterizes chemical profiles of flowers becomes still more evident when variations in space and time are taken into account. Flowers often possess complex spatial patterns of odours, due to the fact that different parts contribute differently to the whole-flower fragrance (Fig. 1.3). This spatial arrangement most likely serves to lead the way for insects such that their pollination effectiveness is optimized (Dobson *et al.* 1990). Honeybees indeed show different behavioural reponses to different regions of flowers emitting diverse fragrances (Lex 1954). Moreover, the composition of flower odours may vary with time of the day, as well as with the developmental stage of the blossom. Thus, the fragrance of clover flowers alters when pollination is completed. This change is so evident that even we can smell the difference between a fertilized and an unfertilized flower.

Plant chemicals affect insect endocrinology

Secondary plant substances not only govern behavioural responses of herbivorous insects, but may also affect physiological processes, such as their morphogenesis, retardation or acceleration of reproduction, or the induction of diapause. Although behavioural responses have been studied in considerably more detail than physiological reactions, several cases in which physiological responses are triggered by host-plant specific phytochemicals are known. Two examples may be mentioned.

An American geometrid moth (*Nemoria arizonaria*) which lives on oak, produces two generations per year. Although at hatching the larvae of both generations are completely identical, larvae of the spring generation, which feed on oak catkins, develop a rich yellow coloured integument, densely rugose in texture and with many papillae. These morphs are virtually indistinguishable from the catkins on which they live. The appearance of the summer generation, feeding on leaves, is quite different. They look greenish-grey, remain frozen for long periods and show 'twig mimicry' typical of geometrid larvae (Fig. 1.4). The caterpillars of this species can be raised on an artificial diet. When the diet is made up free of tannins, thus imitating the situation in catkins which contain

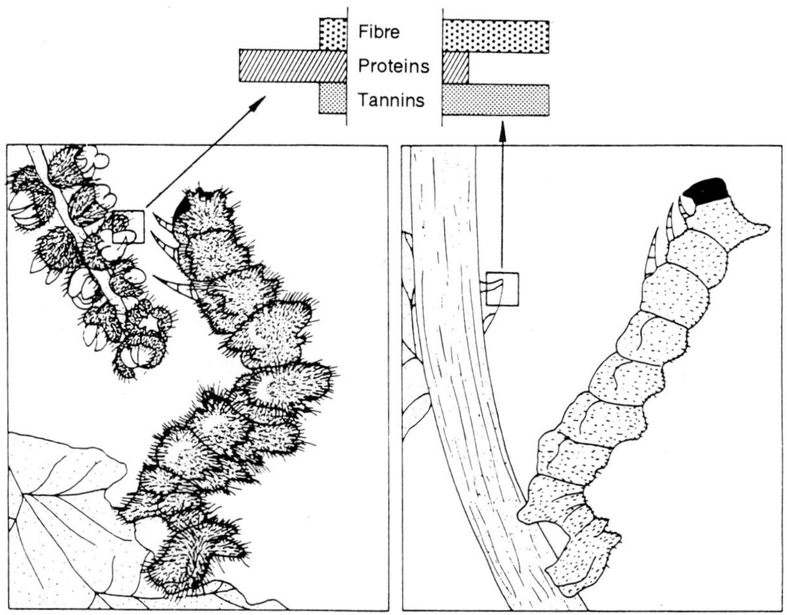

Fig. 1.4. Different morphs of the larvae of *Nemoria arizonaria* which were raised on oak catkins (*left*) or oak summer leaves (*right*). Food composition (not to scale) is shown for catkins (*top left*) and summer leaves (*top right*) (from Dettner 1989, after Greene 1989).

only small amounts of tannins, larvae develop into spring morphs. When, on the other hand, the diet is supplemented with tannin levels which occur during the summer in host leaves, summer morphs are obtained. Apparently the presence or absence of tannin in the food triggers some neuro-endocrine mechanism, regulating this cryptic polymorphism (Greene 1989).

Another type of physiological response in which host-plant chemicals have a signal function concerns reproductive behaviour. Adult females of several insect species delay the production of sex pheromones and egg development until they have been exposed to volatiles or contact chemicals of their host-plants. In this way they prevent the development of 'egg-pressure' when host-plants are not yet available for oviposition. Thus females of the corn earworm (*Helicoverpa zea*) only start the production of male-attracting pheromone and perform their typical calling behaviour after they have smelled for some time the volatiles from their host-plant, for example, those produced by immature ears of corn. Chemical analysis revealed that in this case phenylacetaldehyde and ethylene are responsible for stimulating a neuro-endocrine cascade of reactions, which eventually leads to pheromone production (Raina *et al.* 1991). These examples suffice to show that insects may use various types of chemical signals, not only to recognize suitable host plants, but also to synchronize their life cycles with that of their hosts.

Nature: a web of complex relations

So far only the effects of a plant's secondary metabolites upon an insect herbivore have been discussed. This, of course, represents only a fragment of the natural system, taken out for the sake of analytical convenience. In reality, as sensed already by Darwin (1859), 'plants and animals, most remote in the scale of nature, are bound together by a web of complex relations'. Plants and herbivorous insects, together with their natural enemies, interact with several other species in often bewilderingly complex food webs. When trying to determine the role of secondary compounds in this more extended natural context, clearly a whole new dimension with regard to their ecological function is introduced. Herbivores extracting their energy and body materials from green plants, fall in their turn victim to parasitoids and predators. These natural enemies base their foraging decisions on information gained not only from the organisms they prey upon, that is herbivores belonging to the second trophic level, but also from their host plants, i.e. the first trophic level (Vet and Dicke 1992). Indeed, several studies have shown that insect parasitoids are attracted to chemical stimuli released by the plants upon which their hosts may be found (see Chapter 6). In some instances the plant volatiles have been identified, as in the case of cotton odour, which attracts an ichneumonid parasitoid *Campoletis sonorensis*. Females of this insect use a blend of terpenoid volatiles, as occurring in cotton, as a cue to locate their host habitat (Williams *et al.* 1988).

A more efficient strategy for a parasitoid would be to fly preferentially to insect-damaged plants, rather than to plants which are devoid of hosts. Indeed several parasitoid species have been found to be strongly attracted to damaged host-plants. Thus, the braconid *Cotesia marginiventris*, which among others parasitizes beet army worms, flies in a choice situation to damaged corn seedlings rather than to undamaged controls. Due to the damage inflicted by the herbivore the seedlings appear to release a blend of 11 volatiles, including some so-called 'green leaf odours' (aldehydes and alcohols) and some terpenoids (Turlings *et al.* 1991). Because several terpenoids are known to deter feeding, such plant responses may function as direct defences against their attackers, but also serve as signals to attract natural enemies of the herbivore. This functional relationship may show some further refinements. Some parasitoid species do not show an innate response to the volatiles emitted by damaged plants, but are only attracted after they have learned, through previous experiences, to associate these odours with the occurrence of a potential host. Thus, naive females of the parasitoid *Leptopilina heterotoma* are repelled by the odour of *cis*-3-hexen-1-ol. However, when the females had been exposed only once to the novel odour while ovipositing, this compound appeared from then on to be strongly attractive (Fig. 1.5) (Vet and Groenewold 1990).

The idea that plants, once they are attacked, can do nothing to impede further onset from herbivores has recently been challenged by the finding that some plant species, in addition to activating a kind of immune response, produce signal substances which not only attract natural enemies of the attacker, but also elicit defence responses in nearby conspecific plants. Such an alarm reaction has been observed in cotton plants which were infested by spider-mites. Uninfested plants located in an airstream downwind of mite-infested plants, after some time become less attractive to spider-mites than plants positioned upwind. The infested cotton plants produce certain volatiles, which activate the defence mechanism in neighbouring plants, thereby reducing their susceptibility to spider-mite infestation. Moreover, the same volatiles are attractive to predatory mites, which, guided by these plants signals, have an increased chance of finding their prey and thus reducing infestation levels (Takabayashi *et al.* 1991). But, at a closer look, the picture shows some finer details. Recent findings indicate that volatiles produced through the interaction of a herbivore and its host-plant may vary qualitatively and quantitatively

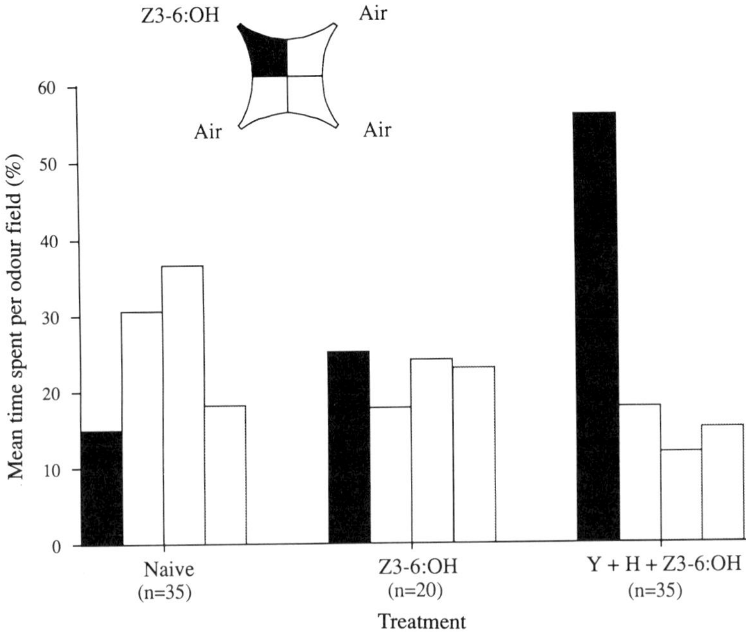

Fig. 1.5. Response (mean percentage time spent per odour field) of *Leptopilina heterotoma* females with different previous experiences to the odour of *cis*-3-hexen-1-ol (*Z*3–6:OH) produced in one arm of a four-armed air-flow olfactometer (inset shows one quadrant containing the odour, the other three quadrants produced control air only). Responses are shown of naive insects (*left*), insects which have been habituated to *Z*3–6:OH during an earlier exposure for 1 h to this compound (*middle*), and insects attracted to the *Z*3–6:OH after a 1 h exposure to host larvae in the presence of this compound (*right*) (from Vet and Groenewold 1990, with permission).

with herbivore species, as well as plant species and even plant variety. Since natural enemies are affected by this variation, it adds to the complexity of the environment in which the various participants of a food web operate (Vet and Dicke 1992)

Tritrophic relationships, in which phytochemicals play a prominent role, feature only portions of intricate community structures. Food web complexity is exemplified by the often large numbers of its constituent species. A study set up to control bark beetles by collecting them *en masse* in traps baited with attractants, clearly demonstrated the existence of interactions between species participating in a particular food web, based on using the same chemical signals. Sticky traps baited with the pheromone of the southern pine beetle in combination with host-tree volatiles attracted not only many southern pine beetles, but also considerable numbers of insects belonging to a total of 28 other species in beetle-infested forests. The accompanying species, including predators, parasitoids, food competitors, scavengers, and mycetophages, were all known to be associated with the southern pine beetles. The attractant, including the volatiles of plant origin, appeared to function as a common chemical signal to several members of a multisegmented insect-tree complex (Dixon and Payne 1980), showing that particular phytochemicals may have a much more intricate role than one a prima facie would expect.

Economy of nature

After discussion of the complexity of nature, attention may be focused on the economy of nature, two principles which may, or may not, conflict with each other.

Modern biological thinking relies heavily on the economy principle. Ever since nature was viewed as a system based on both thermodynamics and modern economics, and new techniques became available for quantifying the flow and use of energy in an ecosystem, many relationships, including those between plants and insects, have been scrutinized in terms of 'cost-benefit' relationships. The notion that plants are protected against harmful insects and other organisms by producing a chemical umbrella, immediately raises the question about what costs are involved in the production and maintenance of the defence system. Amazingly little hard information on this topic seems available. The first complication encountered when studying resource allocation in plants is the fact that resources must be divided among plant structures and functions. This dichotomy is reflected in the approaches which either emphasize a biochemical–physiological analysis or the use of ecological parameters. The results of both methods are as yet difficult to integrate. Using biochemical information, it is possible to calculate in energy terms the direct costs required to generate some specific compounds and the cost of maintenance of the cellular machinery needed to construct them. The situation becomes more difficult when maintenance costs of sequestration and turnover have to be added or when the cost of whatever is the limiting factor is unknown (Gulmon and Mooney 1986). In an ecological context costs should be expressed in terms of fitness, such as growth rate, dry weight, or seed production.

An interesting example of variations in resource allocation was found in the neotropical tree *Cecropia peltata*. When seedlings were grown under strictly uniform conditions growth varied considerably between individuals and was negatively correlated with leaf tannin concentrations. After some time young trees with low amounts of tannins in their leaves had produced almost twice as many leaves as trees with leaf tannin concentrations which were about five times higher. The protective role generally attributed to high tannin concentrations, showed up in a field experiment. All high and low tannin trees were in a large forest light gap subjected to naturally occurring herbivores. After 10 days the losses of leaf surfaces eaten by insects were 80 per cent higher in low tannin plants than in high-tannin plants (Coley 1986).

The *Cecropia* tree experiment strikingly illustrates that, since plant energy or mineral resources are finite, resources channelled into chemical defence are often used at the expense of plant growth and reproduction. This concept is particularly relevant for crop production. Breeding programmes primarily aiming at high yields have in many cases produced cultivars with markedly reduced levels of secondary compounds. As a result crop plants generally contain lower levels of secondary metabolites than their wild relatives. The differences may even be 100-fold or more. Higher yield, however, is often paid for by reduced insect resistance levels, and several examples of positive correlations between secondary compound levels and insect resistance levels of different cultivars can be found in the literature (Fig. 1.6) (e.g. Kogan 1986).

Nature may economize in several ways to limit the costs of chemical defence in plants. Two examples may illustrate the idea of defence

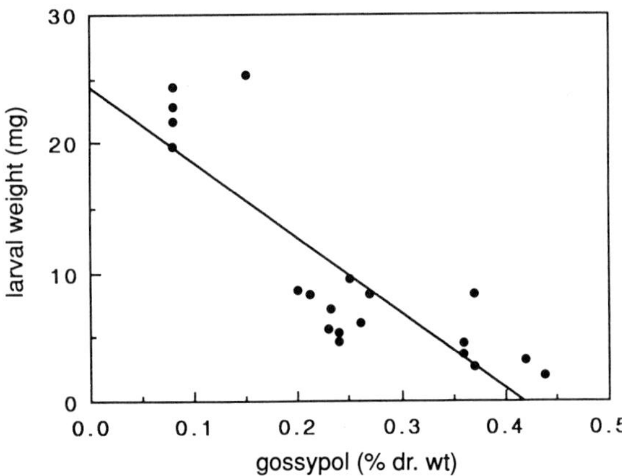

Fig. 1.6. Growth of first instar *Heliothis virescens* larvae in the field during 5 days on intact plants of 20 cotton cultivars with different gossypol concentrations (Reprinted with permission from Hedin *et al.* 1983. Copyright 1983 American Chemical Society.)

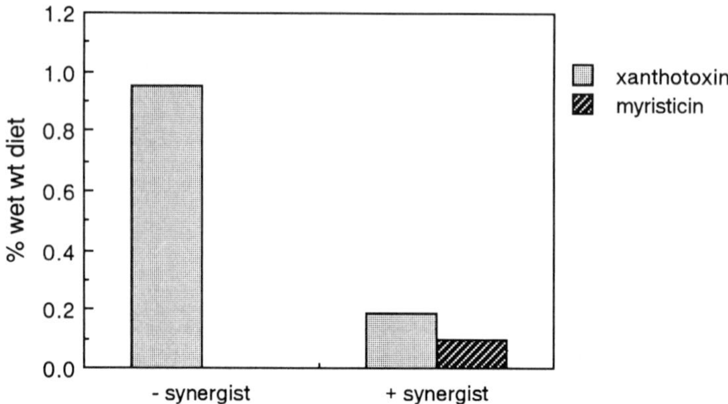

Fig. 1.7. Toxicity (LC_{50}) of xanthotoxin when added to the artificial diet of *Helicoverpa zea* larvae either alone or in combination with the synergist myristicin (data from Berenbaum and Neal 1985, with permission).

optimalization. The first concerns the production of synergistic compounds, which can significantly reduce the metabolic cost involved in producing toxic compounds without a concomitant loss in efficacy. Myristicin, for example, is an essential oil component co-occurring widely with furanocoumarins in the Umbelliferae. It increases the toxicity of xanthotoxin, a furanocoumarin, to larvae of *Helicoverpa zea* almost fivefold. The LC_{50} dose of xanthotoxin in the presence of 0.1 per cent myristicin is five times lower than when applied singly (Fig. 1.7). Thus, the production of every mg of myristicin can save the plant synthesizing 77 mg of furanocoumarin, a substantial saving at least in carbon allocation and possibly energy expenditure (Berenbaum and Neal 1985).

Another aspect which should be taken into account when considering cost–benefit analyses of allelochemicals production is the evolutionary reliability of a trait. It undoubtedly pays, even if the costs are high, to develop compounds which provide protection to as many herbivores and pathogens as possible, each of which may differ in its capacity to develop resistance.

An example of a physiologically very effective secondary compound is found in the bark of the Chinese tree *Melia toosendan*. The triterpenoid toosendanin acts as a feeding deterrent to, for example, larvae of the cabbage white butterfly at concentrations of less than 5 ppm. This compound strongly stimulates the deterrent cell in one of the maxillary taste hairs of the caterpillar in a dose-dependent way. This is what we might call the traditional way of signalling the presence of this

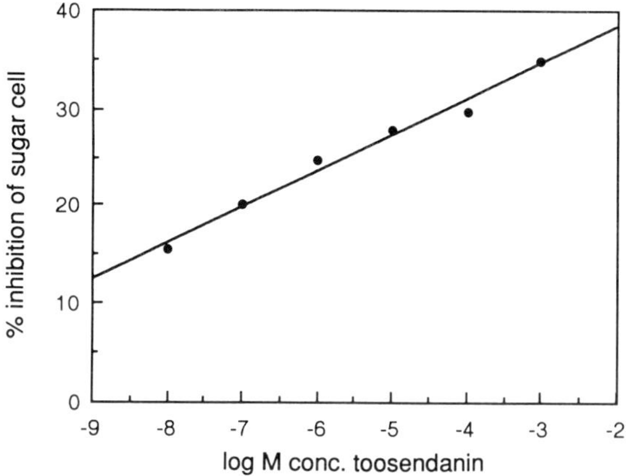

Fig. 1.8. Inhibition of sugar neuron responses in maxillary taste hair of *Pieris brassicae* larvae by toosendanin (Luo Lin-er and L. M. Schoonhoven, unpublished).

compound, although it remains a mystery how one cell type can respond to so many structurally quite different compounds. In addition the efficacy of toosendanin as a feeding deterrent is increased by the fact that this chemical inhibits the responsiveness of two cells which react to, respectively, sugars and glucosinolates, compounds which characteristically stimulate feeding in this insect (Fig. 1.8). Interestingly toosendanin inhibits the sugar cell more effectively at *higher* sugar concentrations, in contrast to the glucosinolate cell, which is more effectively inhibited at *lower* glucosinolate concentrations. Toosendanin thus inhibits feeding in this species by interacting with three taste cells, and because of their different response characteristics, it must be concluded that at the receptor membrane level three different molecular mechanisms are operative (Luo and Schoonhoven 1992; see also Chapter 2). Presumably it is difficult for an insect to develop resistance to a multi-tier system like the toosendanin mode-of-action on chemoreceptors. Indeed it has been reported that it is difficult to select for resistance in insects against terpenoid compounds derived from a related *Melia* tree species (Völlinger 1987).

Nature's economy may also be recognized in the plant's counterpart, i.e. the insect herbivore. Here it manifests itself in the efficacy of the insect's chemosensory system to detect even minor qualitative and quantitative differences in its food. Thus lepidopterous larvae, in spite of their very limited number of taste receptors, obtain highly detailed images of a plant's chemical make-up, allowing them not only to discriminate between different plant species, but also between different parts of the same plant (Schoonhoven 1987).

Conclusions

Biologists are nowadays aware of the fact that green plants, as primary producers, are the synthesizers and bankers of a biochemical wealth, presumably unsurpassed elsewhere on Earth or in the universe. Undoubtedly this chemical diversity influences many if not all ecological relationships. This notion has induced a surge of interest in chemical ecology, as evinced by the rapid growth of several journals devoted to this field. The role of insects in this branch of science is manifest from the fact that out of the nearly 200 research papers published in the *Journal of Chemical Ecology* in 1991, as much as 65 per cent dealt with insects.

Now that we have started to uncover the chemical richness of plants and the numerous responses of insects to it, one is tempted to think that the food web that links the various trophic levels of all communities is merely a reflection of an allelochemicals web. This is surely an oversimplification, as it does not take into account non-chemically mediated interactions between organisms. No doubt, however, plant secondary compounds are key regulatory factors of coevolutionary processes, and the relationships between plants, chemicals, and insects are seminal in understanding insect behaviour, ecology, and speciation. Since the plant–herbivore interface has been described as the major zone of interaction responsible for generating the diversity of terrestrial life forms (Ehrlich and Raven 1964), this understanding relates to basic issues in biology.

An important area for future research deals with the dynamic aspects of interspecific communications. Relationships based upon allelochemicals possess a built-in plasticity which is still difficult to fathom, but which becomes apparent when the time dimension is regarded. For example, concentrations and turnover rates of secondary compounds in plants depend on internal and external factors. Their biological half-lives vary greatly, from a few minutes to several months, and may even be variable in the same plant (Luckner 1990). Insects may learn and become imprinted on certain host plants, and their chemoreceptors may alter depending on earlier experiences or nutritional state. Clearly, interactions based on the production of chemical signals and on their perception by insects are anything but static or predictable.

What conclusions from the applied point of view can be drawn from our knowledge on the biology of phytochemicals and insects behaviour? First it is a reassuring thought that green plants, by producing attractants and repellents, feeding stimulants and feeding deterrents, toxins, and insect hormone mimics, have manipulated their enemies since the earliest evolutionary interactions. Recent attempts to develop insect control measures based on behaviour-modifying chemicals have met in some cases with considerable success, for example in control methods employing sex pheromones (Ridgway *et al.* 1990), or neem-based insecticides (Isman *et al.* 1991). It is perhaps a good omen that so far only few instances of behavioural resistance to behaviour-modifying substances have been reported, in contrast to the evidence of physiological resistances to pesticides in hundreds of insect species (Gould 1991). From the achievements made so far, and perhaps more so from the failures experienced, it is clear that a thorough knowledge of the biology of the system under study is a *conditio-sine-qua-non* for success. Moreover, once we realize that natural defence systems are complex and form part of an intricate ecological system, it cannot be expected that permanent control measures can be based upon simple applications of one particular chemical, however interesting it might be on first sight. On the other hand, if we can unravel the chemical principles underlying multitrophic interactions, and understand the subtle rules on which the economy of nature is founded, we will have strong tools, necessary to develop a sustainable agriculture, which does not further exhaust and damage the natural environment.

Plants, although suffering from herbivores, are, due to their phytochemicals, at the same time the masters of herbivores.

References

Abbott, H. C. de S. (1887). Comparative chemistry of higher and lower plants. *Am. Nat.*, **21**, 800–10.

Berenbaum, M. and Neal, J. J. (1985). Synergism between myristicin and xanthotoxin, a naturally cooccurring plant toxicant. *J. Chem. Ecol.*, **11**, 1349–58.

Blaney, W. M., Schoonhoven, L. M., and Simmonds, M. S. J. (1986). Sensitivity variations in insect chemoreceptors: a review. *Experientia*, **42**, 13–19.

Coley, P. D. (1986). Costs and benefits of defense by tannins in a neotropical tree. *Oecologia*, **70**, 238–41.

Darwin, C. (1859). *On the origin of species by means of natural selection.* Murray, London.

Dettner, K. (1989) Chemische Ökologie. *UWSF-Z. Umweltchem. Ökotox.*, **4**, 29–36.

Dixon, W. N. and Payne, T. L. (1980) Attraction of entomophagous and associate insects of the southern pine beetle to beetle- and host tree-produced volatiles. *J. Georgia Entomol. Soc.*, **15**, 378–89.

Dobson, H. E. M., Bergström, G., and Groth, I. (1990). Differences in fragrance chemistry between flower parts of *Rosa rugosa* Thunb. (Rosaceae). *Israel J. Bot.*, **39**, 143–56.

Ehrlich, P. R. and Raven, P. H. (1964). Butterflies and plants: a study in coevolution. *Evolution*, **18**, 586–608.

Gould, F. (1991). Arthropod behavior and the efficacy of plant protectants. *Annu. Rev. Entomol.*, **36**, 305–30.

Greene, E. (1989). A diet-induced developmental polymorphism in a caterpillar. *Science*, **343**, 643–6.

Gulmon, S. L. and Mooney, H. A. (1986). Costs of defense and their effects on plant productivity. In *On the economy of plant form and function* (ed. T. J. Givnish), pp. 681–98. Cambridge University Press.

Hedin, P. A., Jenkins, J. N., Collum, D. H., White, W. H., and Parrott, W. L. (1983). Multiple factors in cotton contributing to resistance to the tobacco budworm, *Heliothis virescens* F. In *Plant resistance to insects*, ACS Symposium 208 (ed. P. A. Hedin), pp. 347–65. American Chemical Society, Washington, DC.

Holden, C. (1989). Entomologists wane as insects wax. *Science*, **246**, 754–6.

Isman, M. B., Koul, O., Arnason, J. T., Stewart, J., and Salloum, G. S. (1991). Developing a neem-based insecticide for Canada. *Mem. Entomol. Soc. Can.*, **159**, 39–47.

Jermy, T. (1976). Insect–hostplant relationship: co-evolution or sequential evolution? *Symp. Biol. Hung.*, **16**, 109–13.

Kogan, M. (1986). Plant defense strategies and host-plant resistance. In *Ecological theory and integrated pest management practice* (ed. M. Kogan), pp. 83–134. Wiley, New York.

Lex, T. (1954). Duftmale an Blüten. *Z. Vergl. Physiol.*, **36**, 212–34.

Luckner, M. (1990). *Secondary metabolism in microorganisms, plants, and animals* (3rd edn). Springer, Berlin.

Luo Lin-er and Schoonhoven, L. M. (1992). Sensory responses to the triterpenoid antifeedant toosendanin. In *Proc. 8th Int. Symp. Insect–Plant Relationships* (ed. S. B. J. Menken, J. H. Visser, and P. Harrewijn), pp. 153–4. Kluwer Academic Publishers, Dordrecht.

Pham-Delegue, M. H., Etievant, P., Guichard, E., Marilleau, R., Douault, P., Chauffaille, J., and Masson, C. (1990). Chemicals involved in honeybee–sunflower relationship. *J. Chem. Ecol.*, **16**, 3053–65.

Raina, A. K., Stadelbacher, E. A., and Warthen, J. D. (1991). Role of host plants in the reproductive behavior of females of *Heliothis* and *Helicoverpa* species. In *Proc. Conf. Insect Chem. Ecol.*, Tábor 1990, pp. 261–76. Academia Prague and SPB Academic, The Hague.

Ridgway, R. L., Silverstein, R. M., and Inscoe, M. N. (1990). *Behavior modifying chemicals for insect management*. Marcel Dekker, New York.

Schoonhoven, L. M. (1987). What makes a caterpillar eat? The sensory code underlying feeding behavior. In *Perspectives in chemoreception and behavior* (ed. R. F. Chapman, E. A. Bernays, and J. G. Stoffolano), pp. 69–97. Springer, New York.

Schoonhoven, L. M. (1990). Insects and host plants: 100 years of 'botanical instinct'. *Symp. Biol. Hung.*, **39**, 3–14.

Schoonhoven, L. M. and Blom, F. (1988). Chemoreception and feeding behavior in a caterpillar: towards a model of brain functioning in insects. *Entomol. Exp. Appl.*, **49**, 123–9.

Smith, A. E. and Secoy, D. M. (1981). Plants used for agricultural pest control in western Europe before 1850. *Chem. Ind.*, 3 January 1981, 12–17.

Strong, D. R., Lawton, J. H., and Southwood, R. (1984). *Insects on plants.* Blackwell Scientific, Oxford.

Takabayashi, J., Dicke, M., and Posthumus, M. A. (1991). Variation in composition of predator-attracting allelochemicals emitted by herbivore-infested plants: relative influence of plant and herbivore. *Chemoecology*, **2**, 1–6.

Turlings, T. C. J., Tumlinson, J. H., Eller, F. J., and Lewis, W. J. (1991). Larval-damaged plants: source of volatile synomones that guide the parasitoid *Cotesia marginiventris* to the micro-habitat of its hosts. *Entomol. Exp. Appl.*, **58**, 75–82.

Verschaffelt, E. (1910). The cause determining the selection of food in some herbivorous insects. *Proc. Roy. Acad. Sci., Amst.*, **13**, 536–42.

Vet, L. E. M. and Dicke, M. (1992). Ecology of infochemical use by tritrophic enemies in a tritrophic context. *Annu. Rev. Entomol.*, **37**, 141–72.

Vet, L. E. M. and Groenewold, A. W. (1990). Semiochemicals and learning in parasitoids. *J. Chem. Ecol.*, **16**, 3119–35.

Visser, J. H. (1986). Host odor perception in phytophagous insects. *Annu. Rev. Entomol.*, **31**, 121–44.

Völlinger, M. (1987). The possible development of resistance against neem seed kernel extract and deltamethrin in *Plutella xylostella*. In *Proc. 3rd Int. Neem Conf.*, Nairobi 1986, pp. 543–54. TZ-Verlagsgesellschaft, Rossdorf.

Williams, H. J., Elzen, G. W., and Vinson, S. B. (1988). Parasitoid–host-plant interactions, emphasizing cotton (Gossypium). In *Novel aspects of insect–plant interactions* (ed. P. Barbosa and D. K. Letourneau), pp. 171–200. Wiley, New York.

Yan Fu-shun, Evans, K. A., Stevens, L. H., Beek, T. A. van, and Schoonhoven, L. M. (1990). Deterrents extracted from the leaves of *Ginkgo biloba*: effects on feeding and contact chemoreceptors. *Entomol. Exp. Appl.*, **54**, 57–64.

2. The chemistry of azadirachtin and other insecticidal constituents of Meliaceae

W. KRAUS, M. BOKEL, M. SCHWINGER, B. VOGLER,* R. SOELLNER, D. WENDISCH,[†] R. STEFFENS, and U. WACHENDORFF[‡]

**Department of Chemistry, University of Hohenheim, 70593 Stuttgart, Germany;* [†]*Central Research and Development, Bayer AG, 51273 Leverkusen, Germany;* [‡]*Agricultural Division, Bayer AG, 51273 Leverkusen, Germany*

Introduction

The insecticidal properties of extracts of Meliaceae plants have been known in Europe for a long time. As early as 1877 a 'Synopsis on plants' was published (Leunis 1877) in which under the heading *Melia azedarach* is mentioned 'the leaves kill insects'. This statement was ignored for decades, and investigation into the active principles only started in the middle of the twentieth century (see Kraus 1984, 1986; Taylor 1984; Kraus *et al.* 1987*a*). Here, some of our earlier and more recent work on constituents of the following Meliaceae species, which show insect antifeedant and/or insect growth regulation activity, is discussed:

(1) *Azadirachta indica* (neem tree): seed kernels, leaves, and bark, collected in India, Indonesia, and Togo;

(2) *Melia azedarach*: fruit, collected in Greece;

(3) *Melia dubia*: fruit, collected in Sri Lanka;

(4) *Melia toosendan*: fruit, collected in China, Canton area;

(5) *Khaya nyasica*: fruit, collected in South Africa.

Most of the compounds isolated so far from this plant material belong to the class of tetranortriterpenoids (Kraus 1984) and pentanortriterpenoids, respectively (Kraus and Cramer 1981*a*; Kraus 1984), related to the meliacane skeleton (**1**).

We are interested in these compounds not only because of the determination of new structures; our main interest is directed to structure–activity relationships, which may lead in the future to the development of a new generation of insecticides that cause fewer problems in the environment and to human health than the ones used nowadays.

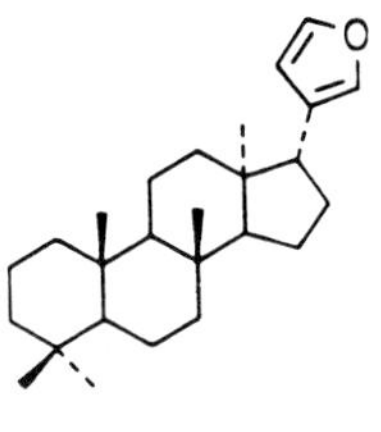

1

Biologically active tetra- and penta-nortriterpenoids

Azadiron group

A few simple examples of tetranortriterpenoids are represented by structures (**2**)–(**5**): These compounds have been known for a long time as constituents of neem seed extracts (see Kraus 1984, 1986; Taylor 1984;

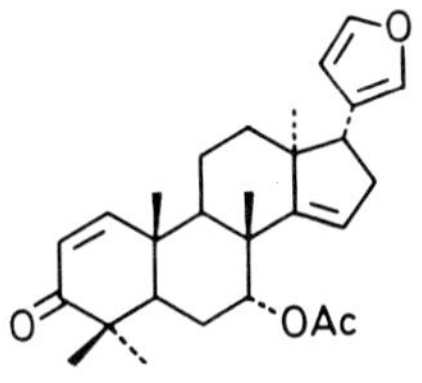

OAc OAc

2 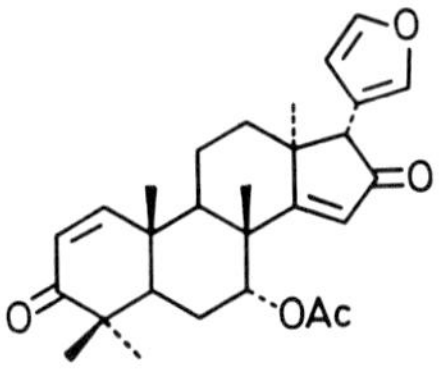 3

OAc OAc

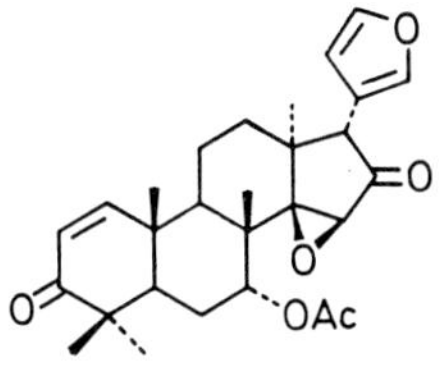

4 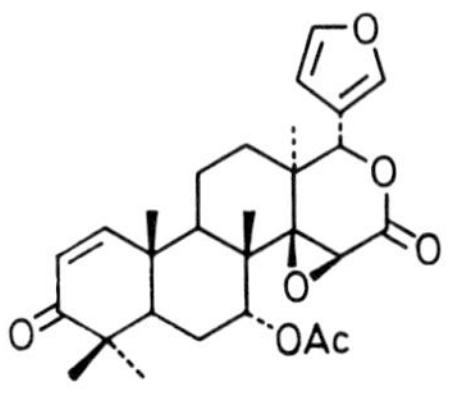 5

Table 2.1 Antifeedant activity of tetra- and penta-nortriterpenoids

	EC_{50} (ppm)	EC_{100} (ppm)
Azadiron (**2**)	5 500	
7-Deacetylazadiron	500	
Azadiradion (**3**)	320	> 5 000
17β-Hydroxyazadiradion	500	
14-Epoxyazadiradion (**4**)	1 300	
Gedunin (**5**)	930	
Khivorin (**6**)	50	100
Nimbinen (**20**)	140	
6-Deacetylnimbinen (**21**)	35	1 000
Nimbandiol (**18**)	100	1 000
6-Acetylnimbandiol (**19**)	90	1 000
Nimbin (**14**)	50	1 000
6-Deacetylnimbin (**15**)	35	1 000
6-Deacetylnimbinal (**16**)	10	500
Nimbolide (**22**)	70	2 500
28-Deoxonimbolide (**23**)	65	1 000
1,3-Diacetylvilasinin (**28a**)	13	250
1-Tigloyl-3-acetylvilasinin (**28c**)	10	500
1-Senecioylvilasinin (**28d**)	10	250
1-Tigloyl-3,12-diacetylvilasinin (**29g**)	$\gg$ 500	
Salannin (**24**)	14	250
3-Deacetylsalannin (**25**)	20	> 250
Salannol (**26**)	10	500
Salannolacetate (**27**)	9	100
Azadirachtin (**43**)	13	250

Kraus *et al.* 1987*a*). In our bioassay using the Mexican bean beetle, *Epilachna varivestis*, and, in certain cases, also *Spodoptera littoralis*, as the test insects we found that although the activities of these compounds were not very high (Table 2.1), there were distinct differences in activity (Kraus 1984; Schwinger *et al.* 1984; Schwinger 1985). Azadiron (**2**) was by far the least active (EC_{50} = 5.500 ppm), azadiradion (**3**), which biogenetically is the oxidation product of (**2**), is a much better antifeedant (EC_{50} = 320 ppm), similar to 7-deacetylazadiron and 17β-hydroxyazadiron (EC_{50} = 500 ppm, Table 2.1). The oxidation product of (**3**), epoxyazadiradion (**4**), is slightly less active (EC_{50} = 1.300 ppm), similar to gedunin (**5**), derived from (**4**) by oxidation (EC_{50} = 930 ppm). The relatively low activity of these compounds may be due to the presence of an enone system in Ring A, since saturated compounds like khivorin (**6**), isolated from *Khaya nyasica* (Taylor 1969), are considerably more

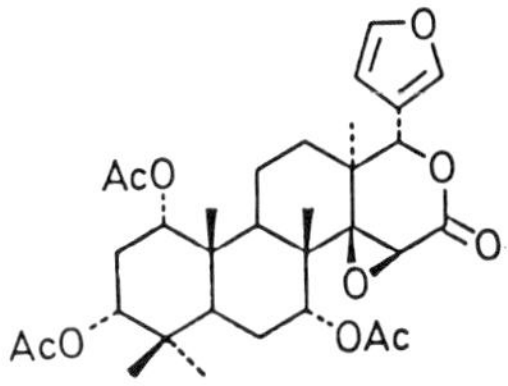

6

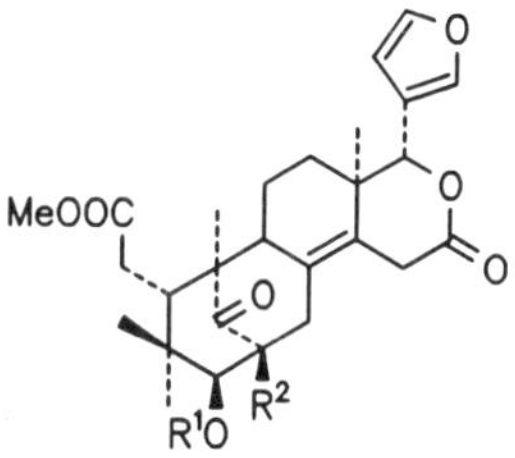

7 R^1 = Tig, R^2 = OAc

8 R^1 = i.-BuCO, R^2 = OAc

9 R^1 = Tig, R^2 = OH

10 R^1 = OH, R^2 = OH

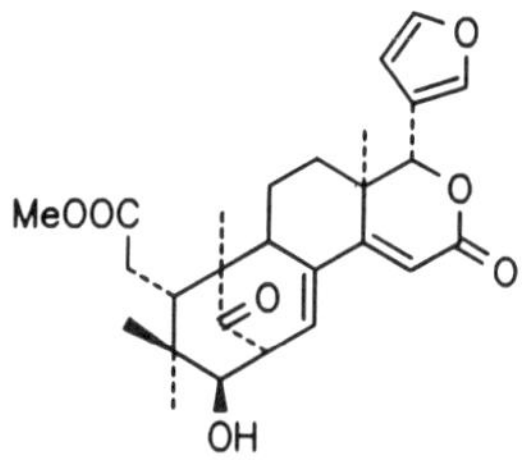

11

active (EC_{50} = 50 ppm; Keller 1988). (**6**) seems to be the only antifeedant found in *Khaya* extracts so far; a number of mexicanolide analogues (**7**)–(**11**) which have been isolated recently from *Khaya* seeds (Keller 1988) turned out to be entirely inactive.

Nimbin, nimbolide, and Salannin group

C-*seco* compounds related to structures (**12**; nimbin type) and (**13**; nimbolide or salannin type), have been isolated from *Azadirachta indica*. Compounds of type (**12**) like nimbin (**14**), 6-deacetylnimbin (**15**), 6-deacetylnimbinal (**16**), the nimbandiols (**18**, **19**), nimbinenes (**20**, **21**), and the nimbolides (**22**) and (**23**) show moderate antifeedant activities with EC_{50} values in the range of 50–100 ppm (Table 2.1). 28-Deoxonimbolide (**23**) was obtained from neem oil (Kraus *et al.* 1987*b*, 1988; Bokel *et al.* 1990); 6-deacetylnimbinal (**16**) and nimbinol (**17**) were isolated from neem leaf extracts (Kraus *et al.* 1987*b*, 1988; Bokel *et al.* 1990). (**16**)

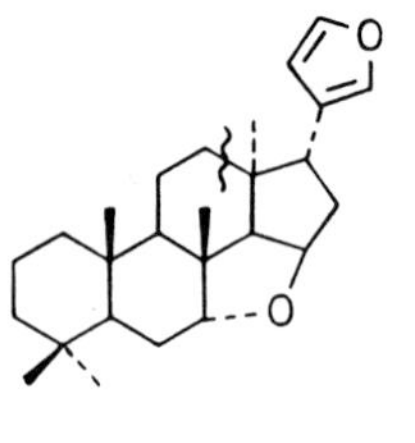

12

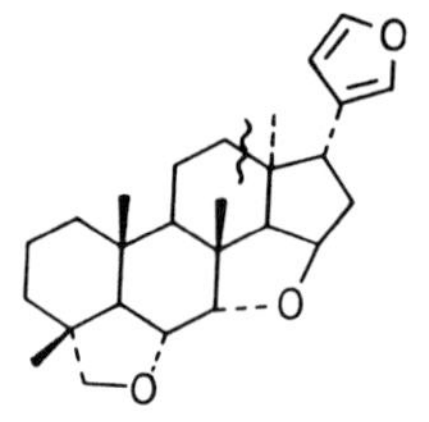

13

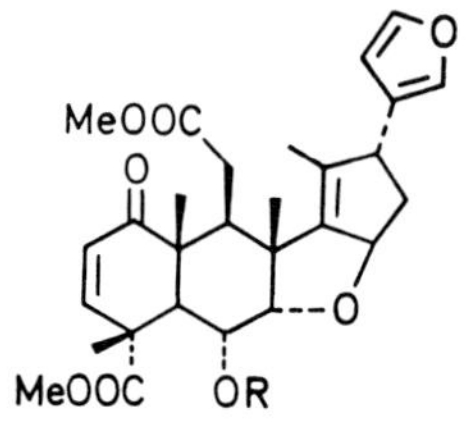

14 R = OAc

15 R = H

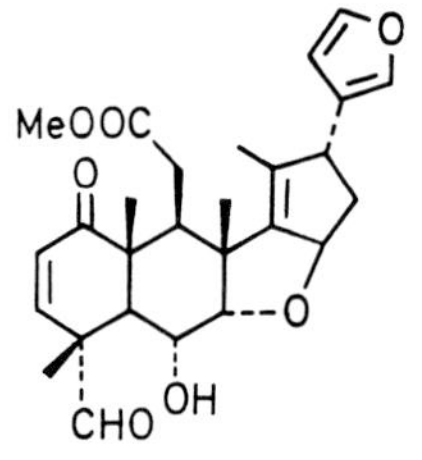

16

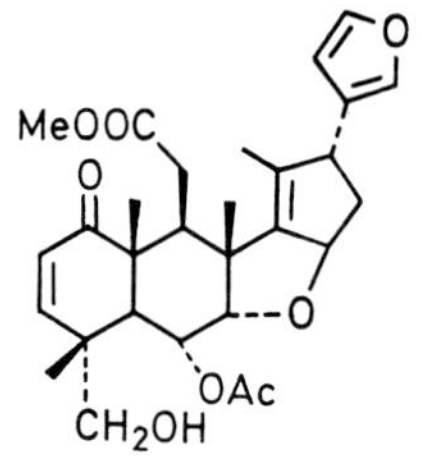

17

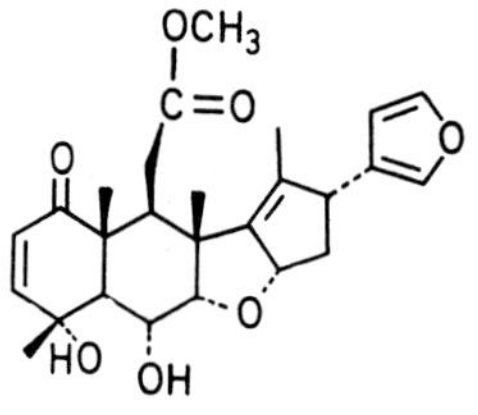

18

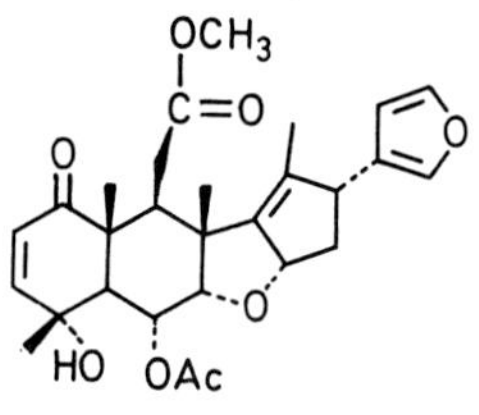

19

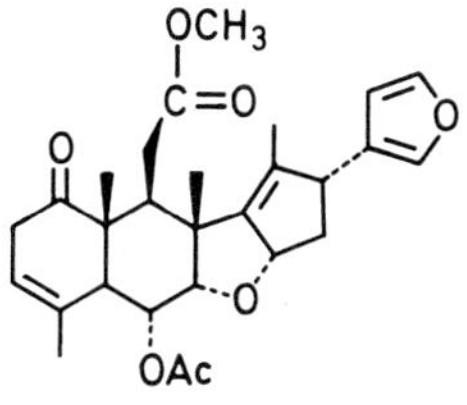

20 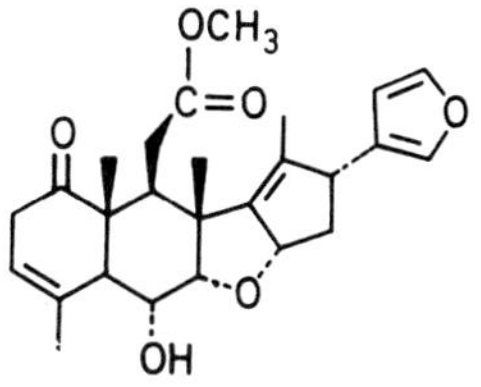 **21**

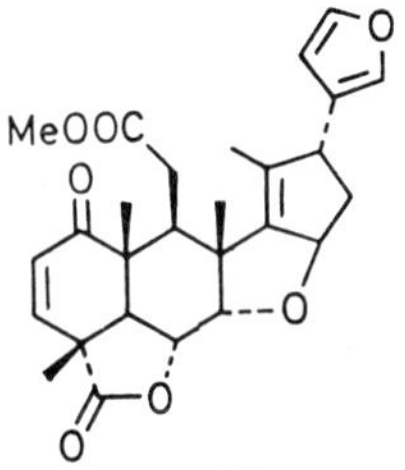

22 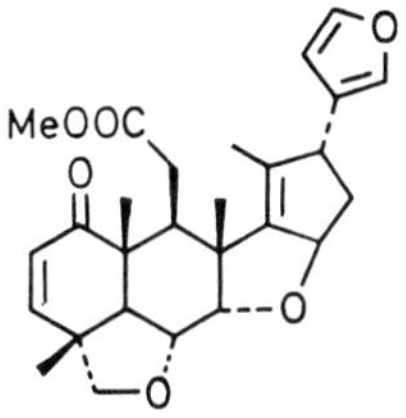 **23**

was converted into (**17**) via acetylation with acetic anhydride/pyridine followed by reduction with sodium borohydride (Kraus *et al.* 1987*b*, 1988; Bokel *et al.* 1990). (**16**) has been reported as a constituent of neem also by Rojatkar *et al.* (1989).

Much lower EC_{50} values (10–20 ppm) were observed with salannin (**24**) and its analogues (**25**)–(**27**) (Table 2.1) (Schwinger *et al.* 1984; Schwinger 1985; Kraus 1986; Kraus *et al.* 1987*a*,*b*), isolated from *Azadirachta indica* (Kraus and Cramer 1981*b*).

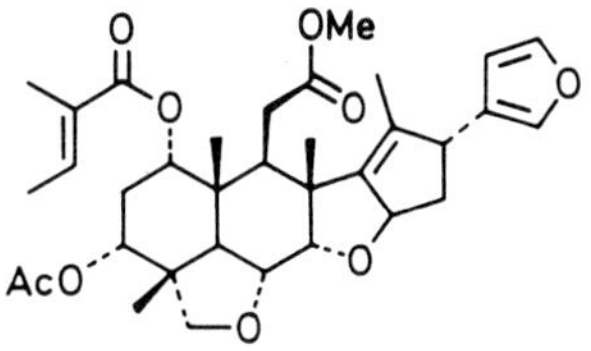

24 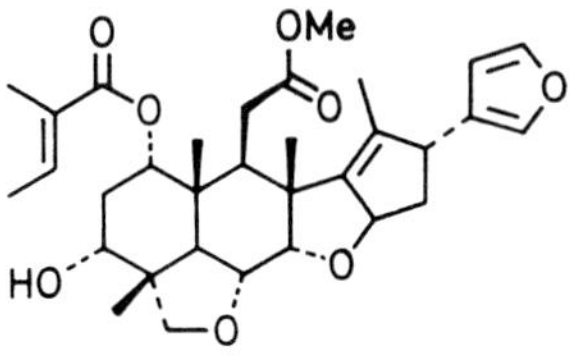 **25**

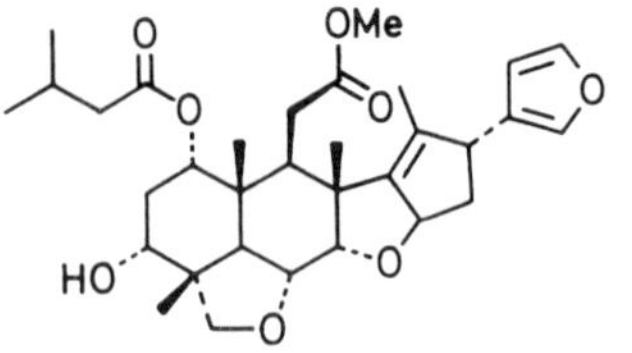

26

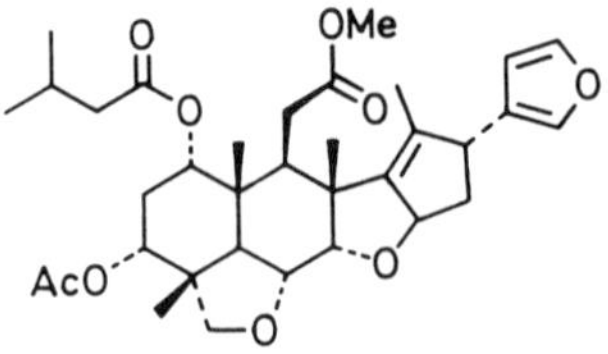

27

Vilasinin group

Vilasinins related to structure **(28)** are present in extracts of several Meliaceae species. Compounds **(28a)**–**(28d)** were isolated from *Azadirachta indica* (Kraus and Cramer 1981*b*; Kraus 1984), compound **(28e)** from *Melia dubia* (Purushothaman *et al.* 1984; Herr 1989). During the insect antifeedant tests these compounds turned out to be as active as azadirachtin (Table 2.1). Complete feeding inhibition was observed with concentrations as low as 250 ppm.

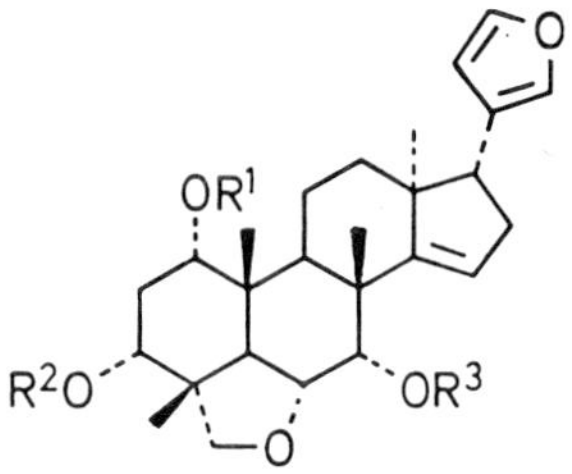

28

	R^1	R^2	R^3	
(a)	Ac	Ac	H	*A. indica*
(b)	Ac	H	Tig	*A. indica*
(c)	Tig	Ac	H	*A. indica*
(d)	Sen	H	H	*A. indica*
(e)	Tig	H	Tig	*M. dubia*

Quite different results were obtained with 12-substituted vilasinins **(29)**. These compounds were isolated from extracts from *Azadirachta indica*, *Melia azedarach*, *Melia dubia*, and *Melia toosendan* (Pöhnl 1985; Kraus

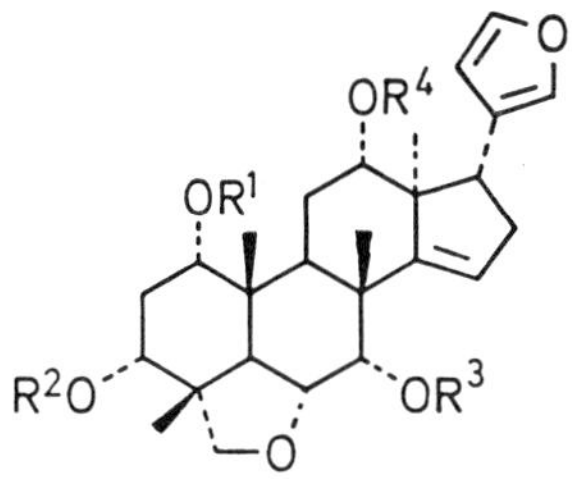

29

	R^1	R^2	R^3	R^4	
(a)	H	H	Tig	Ac	*A. indica*
(b)	Ac	H	H	Bz	*M. azedarach*
(c)	Ac	H	H	Cin	*M. azedarach*
(d)	Ac	Ac	H	Cin	*M. azedarach*
(e)	Ac	Ac	H	Tig	*M. dubia* *M. toosendan*
(f)	Tig	Ac	H	H (α)	*M. dubia*
(g)	Tig	Ac	H	H (ß)	*M. dubia*
(h)	Tig	Ac	H	Tig	*A. indica*

1986; Kraus *et al.* 1987*a*; Kaufmann-Horlacher 1990). Except **(29e)** all compounds were entirely biologically inactive. **(29e)** caused a feeding inhibition only as low as 40 per cent at a concentration of 100 ppm (Kaufmann-Horlacher 1990). Similar results were observed with the 12-oxovilasinin derivative **(30)** and the dihydroxyvilasinin derivative **(31)**, isolated from *Azadirachta indica* and *Melia dubia*, respectively (Kraus *et al.* 1987*b*; Herr 1989).

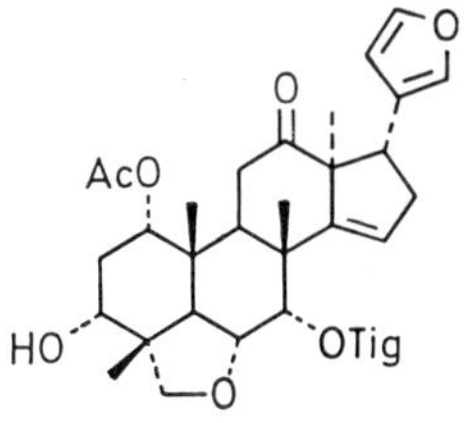

30 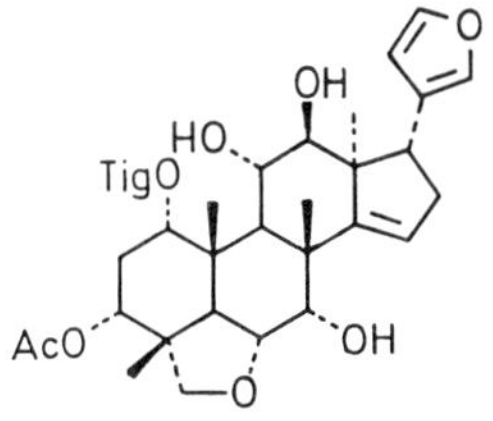 **31**

These findings apply also to structures **(32)**–**(34)**, isolated from *Azadirachta indica* and *Melia dubia*, respectively (Kraus 1987*b*; Herr 1989): **(32)** and **(33)** show insect antifeedant activity, **(34)** does not (Schwinger 1985). From these results it seems to be clear that the biological activities of the vilasinin derivatives depend strongly on the substitution pattern of ring C whereas the ester groups attached to the other rings do not play any important role. Substituents at C-12, however, decrease drastically the antifeedant activities of the compounds.

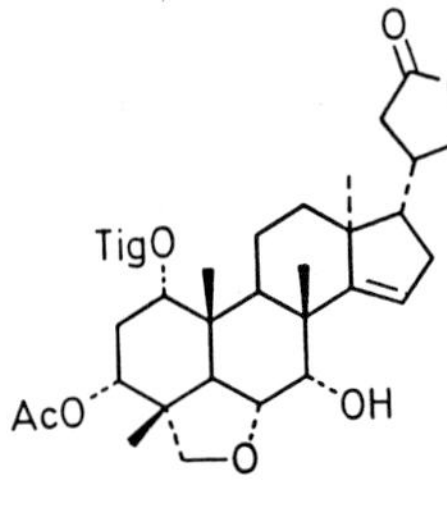

32

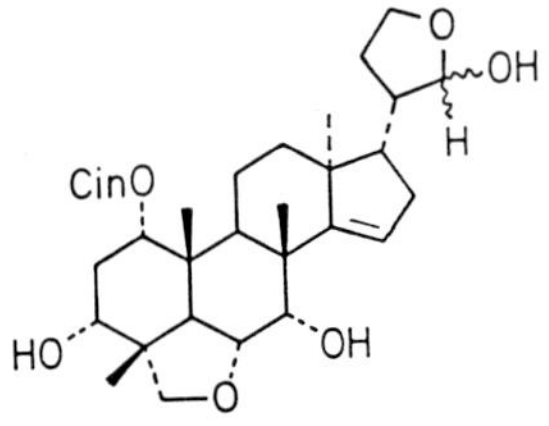

33

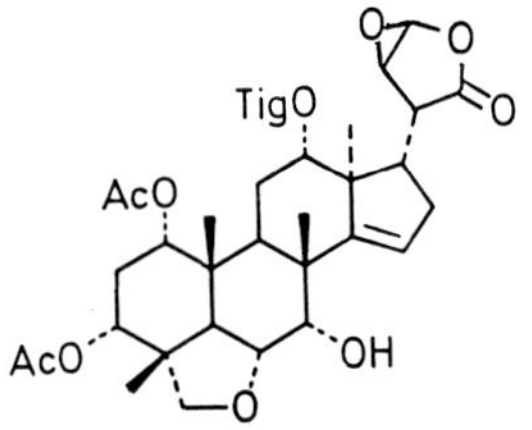

34

Nimbolinin group

Some time ago the nimbolinins **(35a)**, **(36a)**, and **(36b)** were isolated from *Melia azedarach* (Kraus and Bokel 1981; Kraus 1984; Pöhnl 1985); these showed some biological activity towards *Epilachna varivetis* and *Spodoptera littoralis* (Schwinger 1985). Recently, a number of nimbolinin derivatives, **(35b–e)**, **(36c)**, and **(37)**, were found in extracts of *Melia toosendan* fruit (Kaufmann-Horlacher 1990).

The first group, compounds **(35)**, are mixtures of epimers which cannot be separated from each other because of rapid equilibration. In compounds **(36)** hydroxyl groups are attached to C-1 instead of ester groups. Only one epimer is found in each of these structures, probably

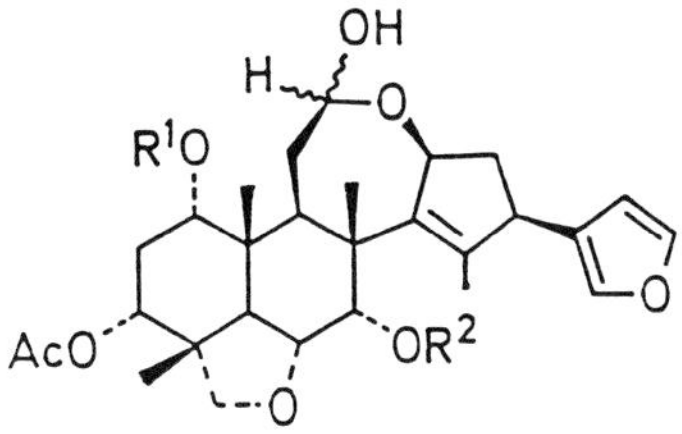

35

	R^1	R^2	
(a)	Bz	H	*M. azedarach*
(b)	Tig	H	*M. toosendan*
(c)	Tig	Tig	*M. toosendan*
(d)	Tig	Bz	*M. toosendan*
(e)	Ac	Tig	*M. toosendan*

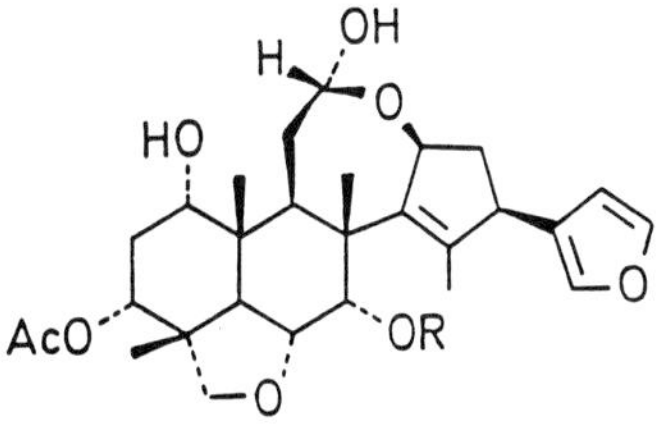

36

	R	
(a)	Tig	*M. azedarach*
(b)	Bz	*M. azedarach*
(c)	Cin	*M. toosendan*

because of strong hydrogen bonding from 1-OH to the oxygen functions attached to C-12. No hydrogen bonding is possible in compounds **(37)**; however, only one epimer was found in these cases. This result led us to assume that the acetals **(37)** may not be genuine natural products, but may be formed in thermodynamically controlled reactions during

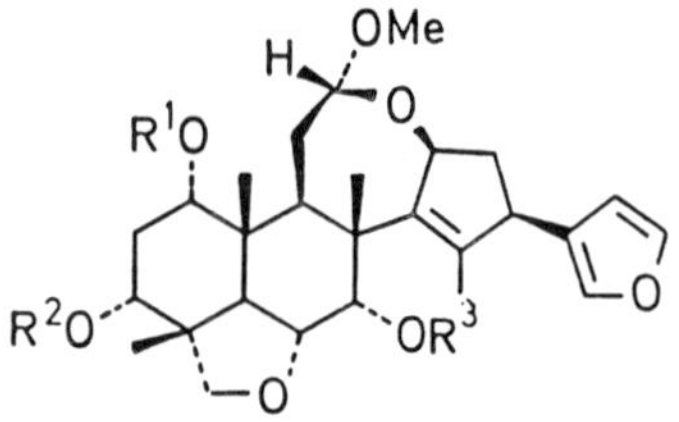

37

	R^1	R^2	R^3
(a)	H	Ac	Cin
(b)	Tig	Ac	H
(c)	Bz	H	H
(d)	Ac	Ac	Tig

extraction and work up with methanol as the solvent from the epimeric hemiacetals as the probably more stable products bearing the methoxy groups in α-configuration. All nimbolinin derivatives show antifeedant activity against *Epilachna varivestis* with EC values comparable to the vilasinins (**28**) (Kaufmann-Horlacher 1990).

Sendanin group

Meliatoxins (**38**), sendanins (**39**), and trichilins (**40**) are also present in *Melia dubia* extracts (Herr 1989). These compounds were tested in a screening programme against several types of insects. Insecticidal activity was found in most cases with EC_{100} values as low as 50–100 ppm.

Azadirachtin group

The most interesting structures up to now are related to the azadirachtin skeleton (**41**). Since the structures of azadirachtin and 22,23-dihydro-23β-methoxyazadirachtin were assigned as (**43**) and (**44**), respectively (Kraus *et al.* 1985*a*,*b*, 1987*c*; Broughton *et al.* 1986), and the structure of 3-tigloylazadirachtol was assigned as (**42**) (Klenk *et al.* 1986), a number of azadirachtols (**45**) and (**46**) and azadirachtin analogues (**47**)–(**52**) have been isolated. 11-Demethoxycarbonylazadirachtin (**52**) was also recently

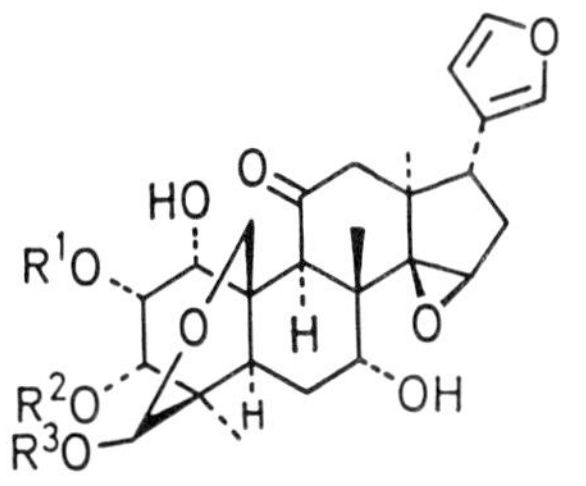

38

	R^1	R^2	R^3
(a)	Ac	H	^{i}Bu
(b)	Ac	H	2-Me-bu
(c)	H	Ac	^{i}Bu
(d)	H	Ac	2-Me-bu

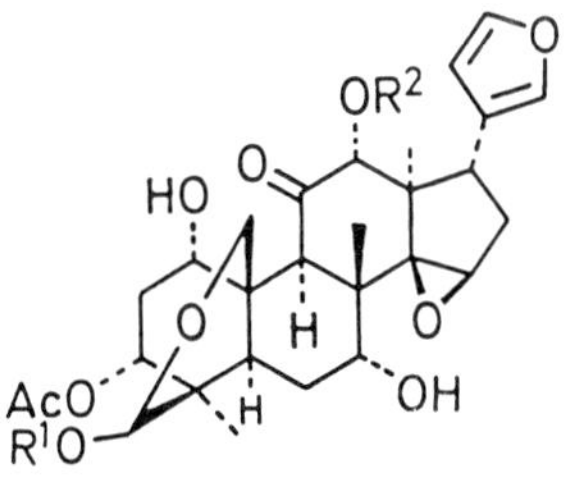

39

	R^1	R^2
(a)	H	H
(b)	^{i}Bu	H
(c)	H	Ac
(d)	Ac	Ac

reported as a constituent of neem by Nagasampagi *et al.* (1990) and other groups (Govindachari *et al.* 1992; Rojatkar and Nagasampagi 1993). 3-De-acetylazadirachtin (**48**) was also prepared from azadirachtin (**43**) by methanolysis (Rembold *et al.* 1987; Yamasaki and Klocke 1987).

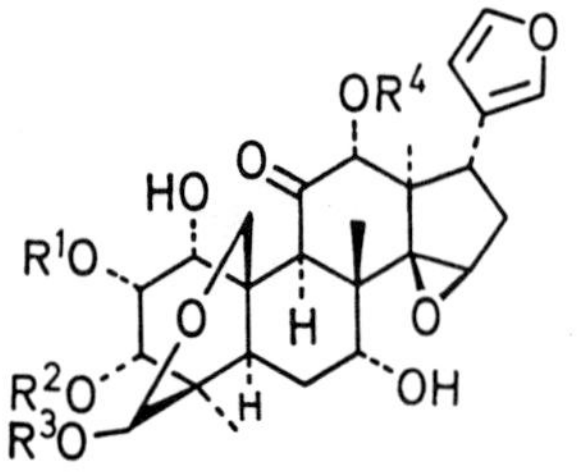

40

	R^1	R^2	R^3	R^4
(a)	H	H	H	Ac
(b)	H	Ac	^{i}Bu	H
(c)	H	Ac	2-Me-Bu	H
(d)	Ac	H	^{i}Bu	H
(e)	Ac	H	2-Me-Bu	H

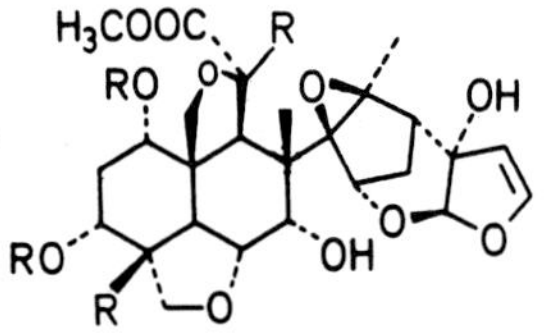

41

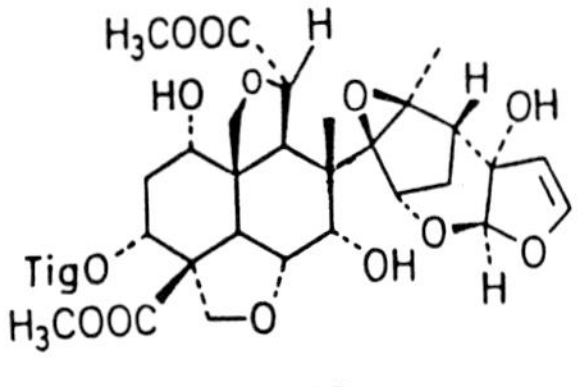

42

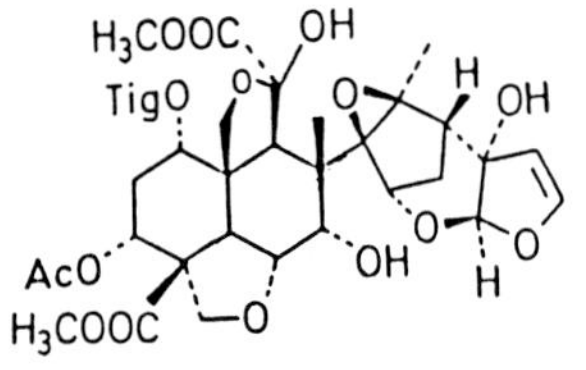

43

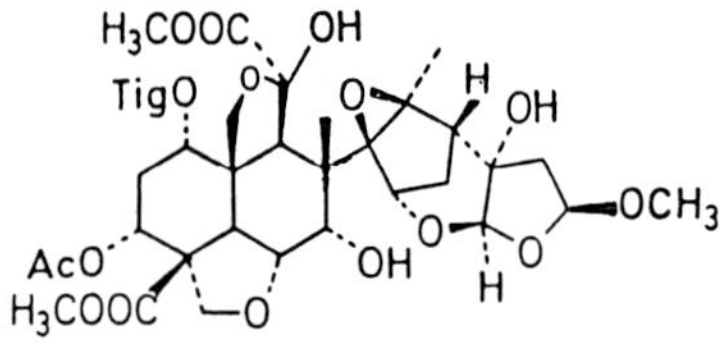

44

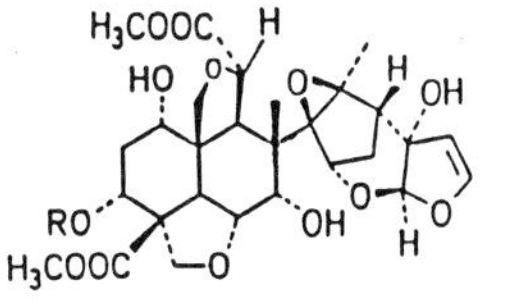

42 R = 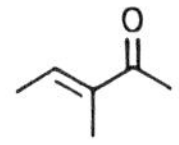
45 R = 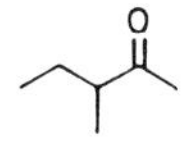
46 R =

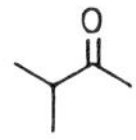

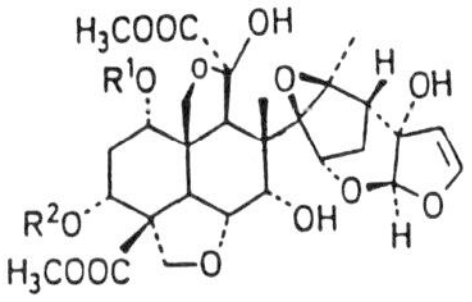

43 R^1 = R^2 = Ac
47 R^1 = R^2 =
48 R^1 = R^2 = H
49 R^1 = R^2 = Ac
50 R^1 = R^2 = Ac
51 R^1 = R^2 =

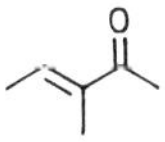

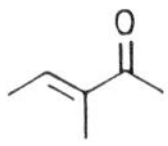

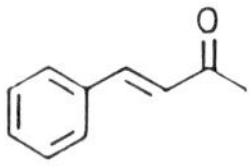

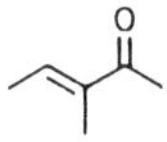

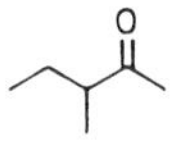

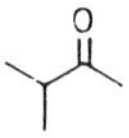

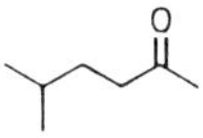

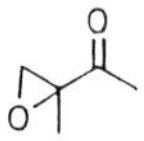

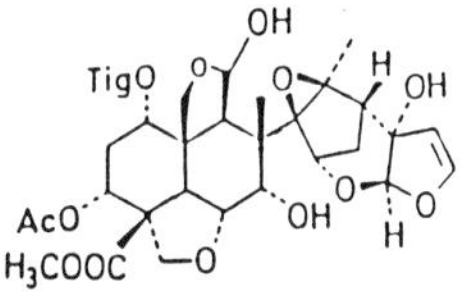

52

Structurally very closely related to azadirachtin **(43)** are the meliacarpins **(53)**–**(56)** isolated from *Melia azedarach* **(53, 55)** (Kraus *et al.* 1985*a*; Pöhnl 1985; Kraus 1986) and from *Melia toosendan* **(56)** (Kaufmann-Horlacher 1990). Most recently, a meliacarpin derivative, 1-tigloyl-3-acetylmeliacarpin **(54)**, was found for the first time in *Azadirachta indica* extracts (Nagasampagi *et al.* 1990; Thiele 1991; Govindachari *et al.* 1993). In addition, from these extracts 1-tigloyl-3-acetylmeliacarpinin **(57)** was

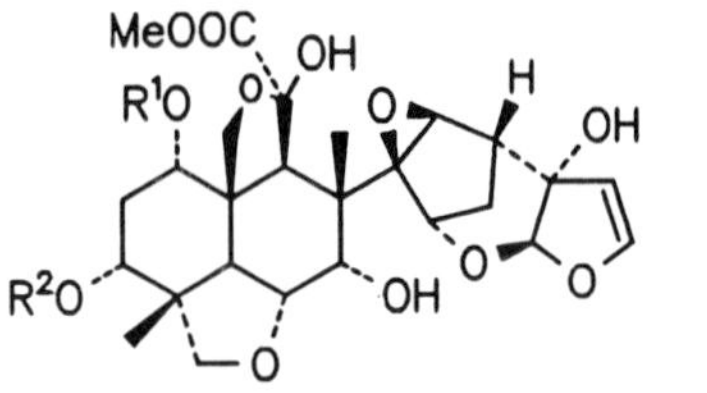

53 R^1 = Cin, R^2 = Fer

54 R^1 = Tig, R^2 = Ac

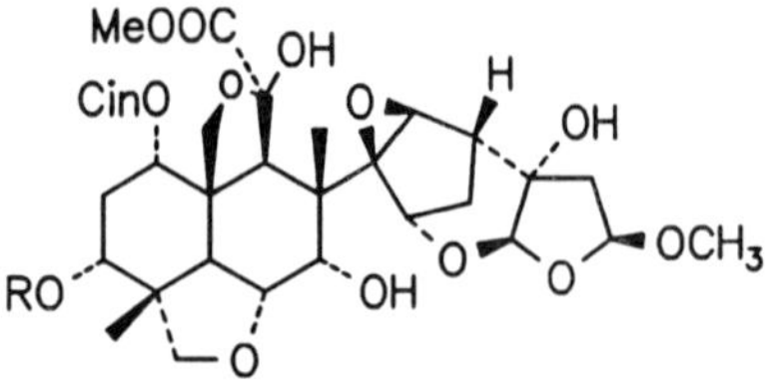

55 R = Fer

56 R = Cin

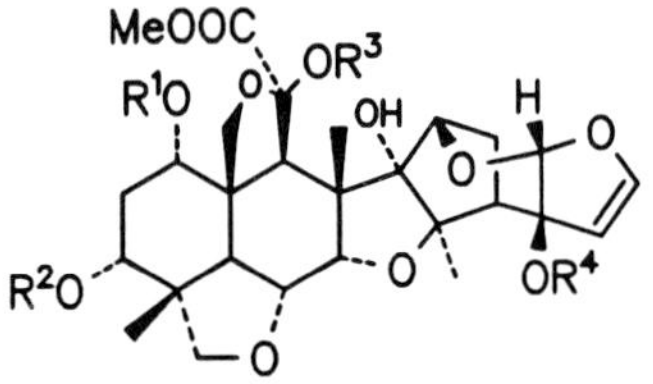

57 R^1 = Tig, R^2 = Ac, R^3 = H, R^4 = H

58 R^1 = Tig, R^2 = Ac, R^3 = Me, R^4 = H

59 R^1 = Tig, R^2 = H, R^3 = Me, R^4 = Ac

isolated (Kraus and Zhou-Halwart, unpublished; Zhou-Halwart 1993) which is related to 1-tigloyl-3-acetyl-11-methoxyazadirachtinin (**58**) from *Azadirachta indica* (Kraus *et al.* 1987*c*), and to 1-tigloyl-11-methoxy-20-acetyl-meliacarpinin (**59**), earlier found in *Melia azedarach* extracts (Kraus *et al.* 1985*a*; Pöhnl 1985; Kraus 1986). Interestingly, all meliacarpins are much less active insect antifeedants, but comparable to azadirachtin (**43**) with respect to the insect growth regulation properties.

Biogenetic pathways to azadirachtins and meliacarpins

There are some hypotheses in the literature about the biogenesis of these compounds (Taylor 1984). If we assume, according to Taylor (1984), aldehyde (**60**) as an intermediate, then meliacarpins, azadirachtins, and azadirachtols as well may be formed through a reaction sequence as outlined in Fig. 2.1. However, intermediate (**63**) could be formed from (**61**) via (**62**) or (**64**), as outlined in Fig. 2.2.

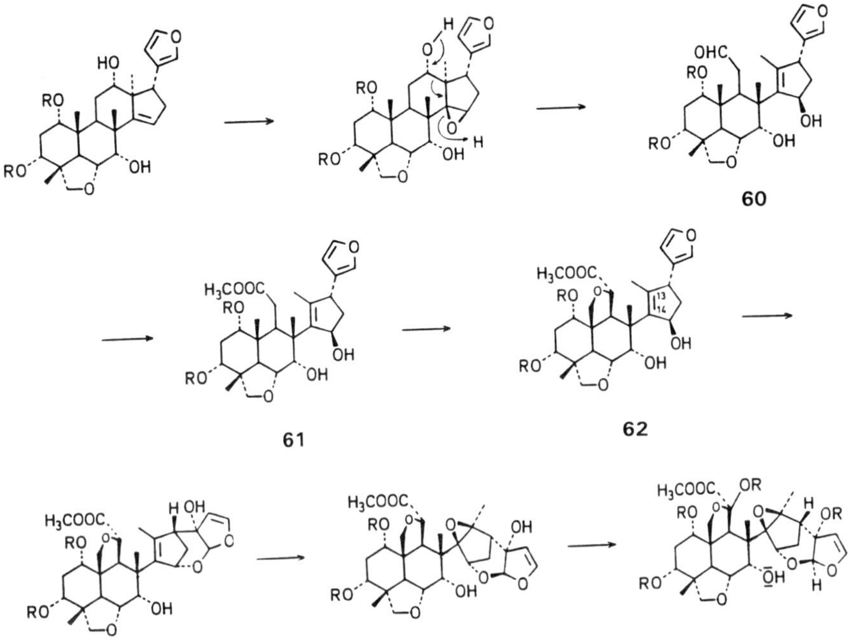

60

61

62

63

Fig. 2.1. Possible biogenetic pathway leading to azadirachtin.

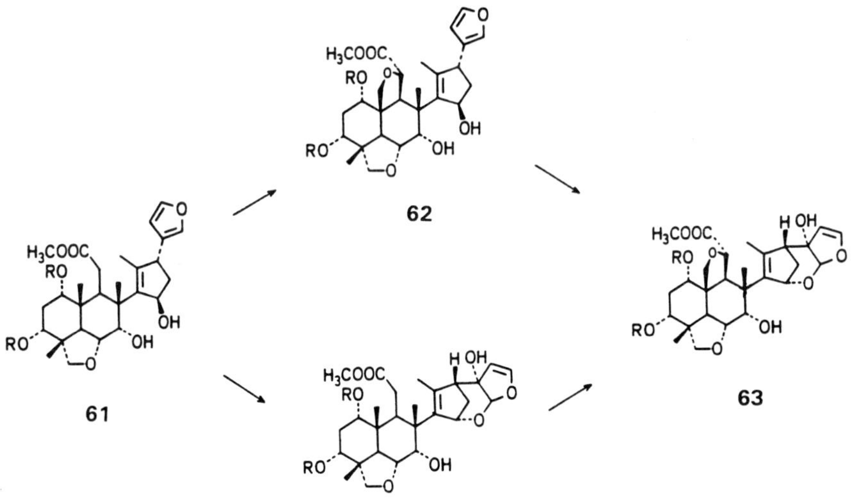

62

61

63

64

Fig. 2.2. Two plausible pathways leading to intermediate 63.

Recently ketoester **(65)** has been isolated from extracts of *Azadirachta indica* (Kraus *et al.* 1989). In this compound C-19 and C-29 are not yet oxidized but the 'eastern part' of the molecule is already finished. On this basis we suggest that intermediate **(63)** is formed from **(61)** via **(64)** as outlined in Fig. 2.3, which is now our proposal for the biosynthesis of azadirachtin and its analogues.

Azadirachtinins **(58)** and meliacarpinins **(57)** and **(59)** are most probably formed from azadirachtin and meliacarpin, respectively, in intramolecular nucleophilic substitutions (Fig. 2.4). This type of conversion was also found to occur during treatment of azadirachtin analogues with acid (Ley *et al.* 1989).

Very recently aldehyde **(66)** was found in *Azadirachta indica* extracts (Kraus and Zhou-Halwart, unpublished; Zhou-Halwart 1993). This can be considered as, possibly, a very early intermediate in the biogenetic pathway leading to the azadirachtin group via epoxy rearrangement of the dioxatriquinane moiety attached to C-8 into the bridged acetal present in meliacarpins and azadirachtins.

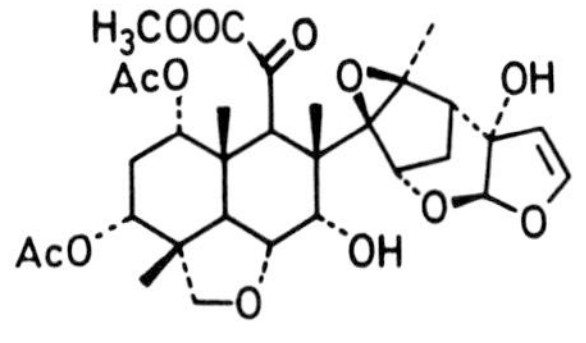

65

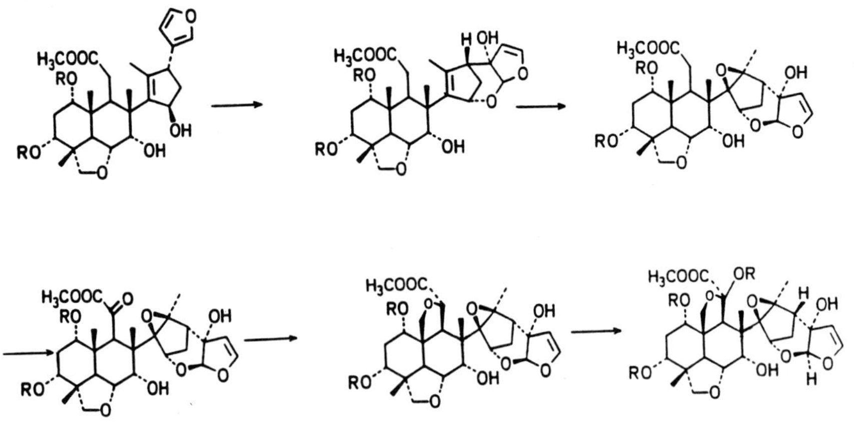

Fig. 2.3. Authors' proposal for final steps in azadirachtin biogenesis.

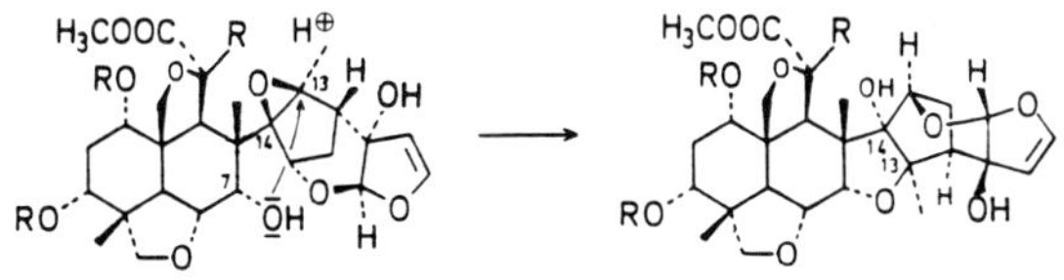

Fig. 2.4. Possible mechanism leading from azadirachtin and meliacarpin to azadirachtinins and meliacarpinins.

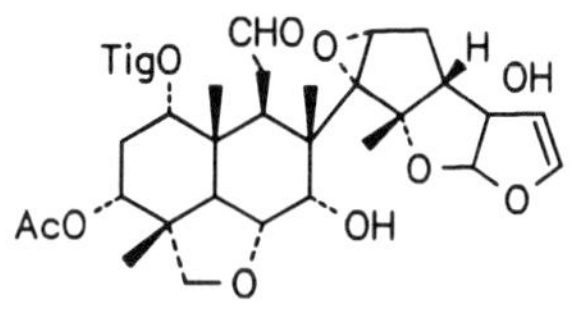

66

Structure–activity relationships

The test methods which we have used during our investigations have been published elsewhere (Schwinger *et al.* 1984; Kraus 1986; Kraus *et al.* 1987*b*). These tests show clearly that there are some relationships between structure and antifeedant activity as long as only one insect species and one skeleton type are considered, e.g. *Epilachna varivestis* and vilasinins. Application of these compounds to other insects like *Spodoptera littoralis* may give different results (Schwinger 1985). This applies also to the other types of compounds and other insect species.

No antifeedant activity, but a remarkable insecticidal activity towards a number of insect species was observed with the compounds belonging to the sendanin group.

Insect growth regulation (IGR) activity is found only with compounds belonging to the azadirachtin group. Out of all the compounds investigated so far azadirachtin (**43**) is the only one showing, depending on concentration, antifeedant as well as IGR activity. All the azadirachtin analogues including azadirachtols and meliacarpins cause very low or no feeding inhibition but are toxic to insects or impact metamorphosis. Oral application of the compounds to L_4 larvae of *Epilachna varivestis* resulted in severe deformations of pupae and/or adult beetles which were

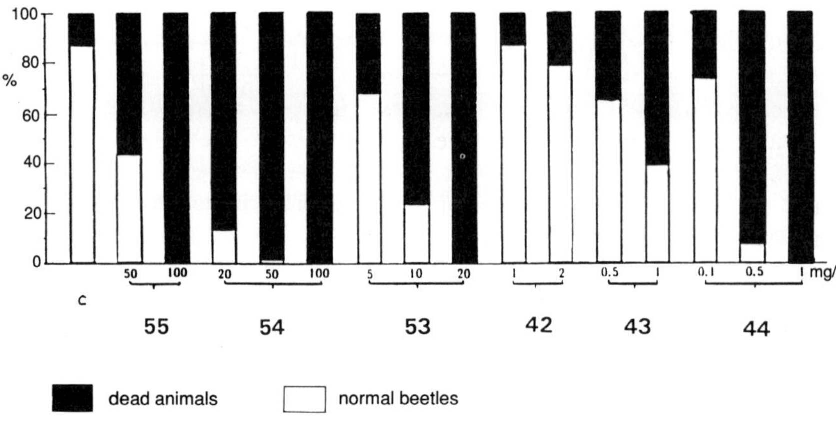

Fig. 2.5. IGR activities of meliacarpins and azadirachtins. [(**42**) 3-tigloylazadirachtol; (**43**) azadirachtin; (**44**) 22,23-dihydro-23β-methoxyazadirachtin; (**53**) 1-cinnamoyl-3-feruloylmeliacarpin; (**54**) 1-tigloyl-3-acetylmeliacarpin; (**55**) 1-cinnamoyl-3-feruloyl-22,23-dihydro-23β-methoxymeliacarpin.

not able to survive. After application of 22,23-dihydro-23β-methoxyazadirachtin (**44**), however, the larvae remained and finally died in the larval stage. The structure–IGR activity relationship in the azadirachtin and meliacarpin family is summarized in Fig. 2.5. From this chart it can be seen that small changes in structure cause remarkable changes in biological activity. The meliacarpins (**53**)–(**55**) are much less active than 3-tigloylazadirachtol (**42**) and the azadirachtins (**43**) and (**44**). The most active principle investigated in this connection, is 22,23-dihydro-23β-methoxyazadirachtin (**44**) (EC_{100} = 1 ppm), as far as IGR activity is concerned.

Outlook

The compounds, mentioned in this review, are too complicated in structure for cheap large scale synthesis and too difficult to isolate for direct usage in plant protection. Furthermore the biological activities are considerably lower than the activity of the synthetic pesticides presently on the market. However, if more results on new compounds are obtained, these may provide a basis for the construction of simpler leading structures that can be synthesized more easily in order to develop a new generation of pesticides.

Acknowledgements

We acknowledge the engaged collaboration of S. Baumann, P. Beck, A. Bruhn, R. Cramer, B. Ehhammer, H. Gutzeit, B. Herr, I. Kaufmann-Horlacher, R. Keller, U. Keller, I. Klaiber, M. Klingele, P. Kümel, G. Nagl, S. Reeb, G. Schwinger, S. Thiele, and Y. Zhou-Halwart. Plant material was kindly provided by C. M. Ketkar, Khadi & Village Industries Commission, Poona, India, and GTZ, Sce. Protection des Vegetaux Projet Allemand, Lomè, Togo (*Azadirachta indica*), Dr A. H. M. Jayasuriya, National Herbarium, Peradeniya, Sri Lanka (*Melia dubia*), and Prof. Chiu Shin Foon, South China Agricultural University, Guangzhou, China (*Melia toosendan*). Part of this work was financially supported by Deutsche Forschungsgemeinschaft and Fonds der Chemischen Industrie.

References

Bokel, M., Cramer, R., Gutzeit, H., Reeb, S., and Kraus, W. (1990). Tetranortriterpenoids related to nimbin and nimbolide from *Azadirachta indica* A. Juss (Meliaceae). *Tetrahedron*, **40**, 775–82.

Broughton, H. B., Ley, S. V., Slawin, A. M. Z., Williams, D. J., and Morgan, E. D. (1986). X-ray crystallographic structure determination of detigloyldihydroazadirachtin and reassignment of the structure of the limonoid insect antifeedant azadirachtin. *J. Chem. Soc., Chem. Commun.*, 46–7.

Govindachari, T. R., Sandhya, G., and Ganesh Raj, S. P. (1992). Azadirachtins H and I: Two new tetranortriterpenoids from *Azadirachta indica. J. Nat. Prod.*, **55**, 596–601.

Herr, B. (1989). Isolierung und Strukturaufklärung neuer Limonoide aus *Melia dubia* Cav. (Meliaceae). Dissertation, Universität Tübingen.

Kaufmann-Horlacher, I. (1990). Neue Tetranortriterpenoide aus *Melia toosendan* Sieb. et Zucc. (Meliaceae). Dissertation, Universität Hohenheim.

Keller, R. (1988). Biologisch aktive Naturstoffe aus den Samen von *Khaya nyasica* (Meliaceae) und aus Kartoffelwurzeln (Solanaceae)—Isolierung und Strukturaufklärung. Dissertation, Universität Hohenheim.

Klenk, A., Bokel, M., and Kraus, W. (1986). 3-Tigloylazadirachtol, an insect growth regulating constituent of *Azadirachta indica. J. Chem. Soc., Chem. Commun.*, 523.

Kraus, W. (1984). Biologically active compounds from Meliaceae. *Stud Org. Chem.*, **17**, 331–45.

Kraus, W. (1986). Constituents of neem and related species. A revised structure of azadirachtin. *Stud. Org. Chem.*, **26**, 237–56.

Kraus, W. and Bokel, M. (1981). Neue Tetranortriterpenoide aus *Melia azedarach* Linn. (Meliaceae). *Chem. Ber.*, **114**, 267.

Kraus, W. and Cramer, R. (1981*a*). Pentanortriterpenoide aus *Azadirachta indica* A. Juss (Meliaceae). *Chem. Ber.*, **114**, 2375.

Kraus, W. and Cramer, R. (1981*b*). Neue Tetranortriterpenoide mit insektenfraßhemmender Wirkung aus Neem-Öl. *Liebigs Ann. Chem.*, 181.

Kraus, W., Bokel, M., Cramer, R., Klenk, A., and Pöhnl, H. (1985*a*). Constituents of neem and related species. A revised structure of azadirachtin. In *3rd Int. Conference on Chemistry and Biotechnology of Biologically Active Natural Products, Sofia 1985, Abstracts of papers*, Vol. 4, pp. 446–50. Bulgarian Academy of Sciences, Sofia.

Kraus, W., Bokel, M., Klenk, A., and Pöhnl, H. (1985*b*). The structure of azadirachtin and 22,23-dihydro-23β-methoxyazadirachtin. *Tetrahedron Lett.*, **26**, 6435.

Kraus, W., Baumann, S., Bokel, M., Keller, U., Klenk, A., Klingele, M., *et al.* (1987*a*). In *Natural pesticides from the neem tree and other tropical plants (Proc. 3rd Int. Neem Conf., Nairobi 1986)* (ed. H. Schmutterer and K. R. S. Ascher), pp. 111–25. GTZ, Eschborn.

Kraus, W., Baumann, S., Bokel, M., Cramer, R., Grimminger, W., Hendlmeier, R., *et al.* (1987*b*). Insect feeding and development controlling constituents of Meliaceae. In *Proc. 1st Princess Chulabhorn Science Congress, Int. Congress on Natural Products, Bangkok 1987*, Vol. II, pp. 554–63. Chulabhorn Foundation, Bangkok.

Kraus, W., Bokel, M., Bruhn, A., Cramer, R., Klaiber, I., Klenk, A., *et al.* (1987*c*). Structure determination by NMR of azadirachtin and related compounds from *Azadirachta indica* A. Juss (Meliaceae). *Tetrahedron*, **43**, 2817.

Kraus, W., Bokel, M., Cramer, R., Gutzeit, H., Pöhnl, H., and Zhou, Y. (1988). Tetranortriterpenoids related to nimbin and vilasinin from *Azadirachta indica* and *Melia azedarach*. In *16th Int. Symposium on the Chemistry of Natural Products, Kyoto 1988, Abstracts of papers*, p. 488.

Kraus, W., Gutzeit, H., and Bokel, M. (1989). 1,3-Diacetyl-11,19-deoxa-11-oxo-meliacarpin, a possible precursor of azadirachtin, from *Azadirachta indica* A. Juss (Meliaceae). *Tetrahedron Lett.*, **30**, 1797.

Leunis, J. (1877). *Synopsis der Pflanzenkunde.* Hahn'sche Buchhandlung, Hannover.

Ley, S. V., Anderson, J. C., Blaney, W. M., Jones, P. S., Lidert, Z., Morgan, E. D., *et al.* (1989). Insect antifeedants from *Azadirachta indica*: Chemical modification and structure–activity relationships of azadirachtin and some related limonoids. *Tetrahedron*, **45**, 5175.

Nagasampagi, B. A., Rojatkar, S. R., and Kulkarni, M. M. (1990). Three new antifeedant tetranortriterpenoids from *Azadirachta indica*. In *17th IUPAC International Symposium on the Chemistry of Natural Products, New Delhi 1990, Abstracts of papers*, p. 242.

Pöhnl, H. (1985). Neue Verbindungen der Azadirachtin-Reihe und andere biologisch aktive Tetranortriterpenoide aus *Melia azedarach* L.—Isolierung und Strukturaufklärung. Dissertation, Universität Tübingen.

Purushothaman, K. K., Duraiswamy, K., and Connolly, J. D. (1984). Tetranortriterpenoids from *Melia dubia*. *Phytochemistry*, **23**, 135–7.

Rembold, H., Forster, H., and Czoppelt, Ch. (1987). Structure and biological activity of azadirachtins A and B. In *Natural products from the neem tree and other tropical plants* (ed. H. Schmutterer and K. R. S. Ascher), pp. 149–60. GTZ, Eschborn.

Rojatkar, S. R. and Nagasampagi, B. A. (1993). 1-Tigloyl-3-acetyl-11-hydroxy-4β-methylmeliacarpin from *Azadirachta indica. Phytochemistry*, **32**, 213–14.

Rojatkar, S. R., Bhat, V. S., Kulkarni, M. M., Joshi, V. S., and Nagasampagi, B. A. (1989). Tetranortriterpenoids from *Azadirachta indica. Phytochemistry*, **28**, 203–5.

Schwinger, M. (1985). Über die fraßabschreckende Wirkung von Meliaceen-Inhaltsstoffen auf *Epilachna varivestis* (Muls.) und andere Insekten: Methoden, Versuchstechniken, Ergebnisse. Dissertation, Universität Hohenheim.

Schwinger, M., Ehhammer, B., and Kraus, W. (1984). Methodology of the *Epilachna varivestis* bioassay of antifeedants demonstrated with some compounds from *Azadirachta indica* and *Media azedarach*. In *Natural Pesticides from the neem tree (*Azadirachta indica *A. Juss) and other tropical plants, (Proc. 2nd Int. Neem Conf., Rauischholzhausen 1983)* (ed. H. Schmutterer and K. R. S. Ascher), pp. 181–98. GTZ, Eschborn.

Taylor, D. A. H. (1969). Limonoids from *Khaya nyasica* Stapf. ex Bak. *J. Chem. Soc. (C)*, 2439–42.

Taylor, D. A. H. (1984). The chemistry of the limonoids from meliaceae. In *Progress in the chemistry of organic natural products* (ed. L. Zechmeister), **45**, 1–102. Springer New York.

Thiele, S. (1991). Isolierung und Strukturaufklärung neuer Tetranortriterpenoide aus *Azadirachta indica* A. Juss (neem tree, Meliaceae) und Untersuchung der nematiziden Wirkung von Neem-Extrakten. Dissertation, Universität Hohenheim.

Yamasaki, R. B. and Klocke, J. A. (1987). Structure–bioactivity relationships of azadirachtin, a potential insect control agent. *J. Agric. Food Chem.*, **35**, 467–71.

Zhou-Halwart, Y. (1993). Isolierung und Strukturaufklärung neuer Tetranortriterpenoide aus *Azadirachta indica* A. Juss. (Meliaceae, Neem-Baum) mit insektenfraßhemmender Wirkung. Dissertation, Universität Hohenheim.

3. Progress in developing insecticides from natural compounds

BHUPINDER P. S. KHAMBAY and
NORVAL O'CONNOR

Department of Insecticides & Fungicides, AFRC Institute of Arable Crops Research, Rothamsted Experimental Station, Harpenden, Herts, AL5 2JQ, UK

Introduction

Approximately one third of world food production is lost through damage caused by pests. The most efficient method of controlling insect pests relies on the use of insecticides (compounds which kill insects) many of which act on the insect nervous system. Although current commercial insecticides have limited persistence in the field, the public perception of insecticides as highly undesirable and environmentally unfriendly chemicals stems from the persistence of DDT, ironically, an insecticide which has probably given more benefit to mankind than any other chemical invention. A great deal of research is currently being directed towards finding alternatives to insecticides. However, the present limitations in efficiency and effectiveness of such approaches ensures that for the foreseeable future, insecticides will remain the most effective means of insect control in the field. Therefore the search for newer insecticides with even greater selectivity towards pests and activity against resistant insects will continue.

Since an exhaustive treatment of the subject is not possible in the allotted space, this paper will be limited to a review of selected classes of natural compounds. An attempt will be made to examine the applicability of the structure–activity relationship (SAR) approach to assess their potential as lead compounds. For convenience, the sources of natural insect-active compounds have been grouped under three broad categories: microorganisms, animals, and plants

The main classes of insecticides, their origin, and importance are listed in Table 3.1. Clearly plants have been a major source of lead compounds for SAR programmes leading to commercial insecticides.

Microorganisms

Major discovery programmes in industry are based on screening microbial metabolites which are considered to be an abundant source of compounds

Table 3.1 Current commercial insecticides

Class	Insecticide market (per cent)	Origin
Organophosphates	36	Synthesis
Pyrethroids	25	Plant
Carbamates	21	Plant
Organochlorines	8	Synthesis
Others	10	Various

Ref: Smith (1992).

with novel structures. These can be divided into two general categories; peptidic and non-peptidic. Two well known examples of the latter are avermectin, produced by *Streptomyces avermitilis* (Fig. 3.1) (Fisher 1989) and the related but simpler milbemycin, first isolated from *Streptomyces* B-41–146 (Mishima *et al.* 1975). These compounds exhibit anthelmintic and insecticidal activities. The major difference between the two series is the absence of the disaccharide side chain on C-13 and the saturation of the C22–C23 double bond in the milbemycins. SAR studies have been limited by the complexity of total synthesis (Danishefsky *et al.* 1989). Modifications leading to retention or an increase in insecticidal

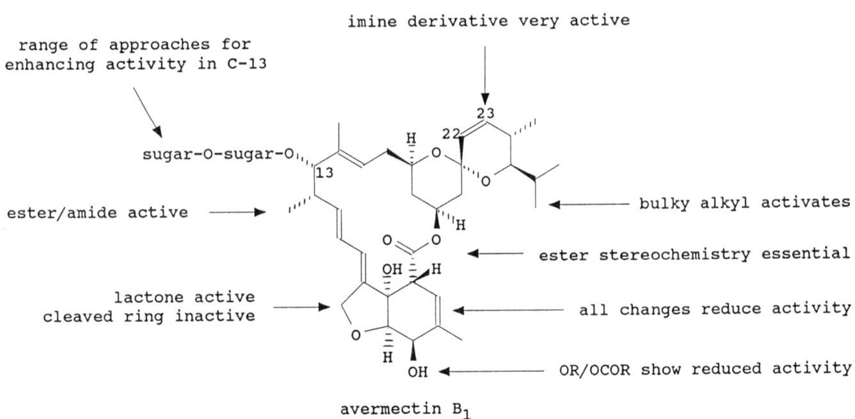

Fig. 3.1. Avermectins/milbemycins.

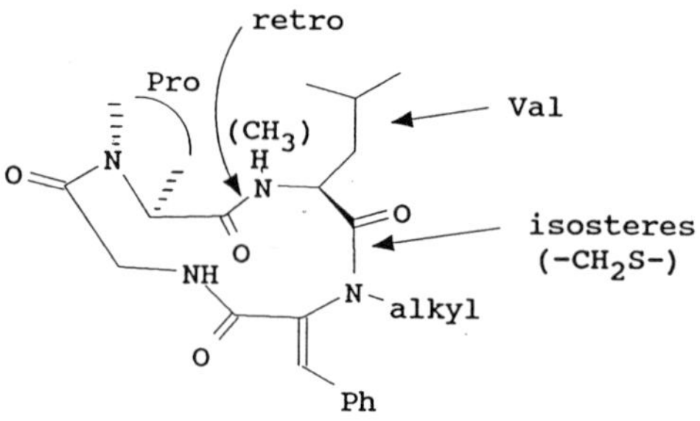

Fig. 3.2. Tentoxin.

activity have generally been limited to the upper part of the molecule with minor exceptions (Blizzard *et al.* 1990). For example, introduction of an amino group in the 4" position of the α-L-oleandrosyl-α-L-oleandrosyl unit leads to an altered spectrum of insecticidal activity (Fisher 1989). Saturation of C22–C23 (to give the ivermectins) generally increases activity, whereas saturation of C10–C11 increases the flexibility of the molecule without loss of activity. Removal of the disaccharide in both the ivermectin and avermectin series to give the 13-deoxy compounds, leads to a substantial increase in activity. In contrast, the aglycone (13-OH) analogues show a marked decrease in activity in the avermectin series but an increase in the ivermectins. In conclusion, general simplification of the avermectin structure has led to an increase in activity. However, the use of such compounds has been limited by high mammalian toxicity.

Bacillus thuringiensis (BT) toxins are probably the best known examples of peptidic secondary metabolites from microorganisms, but since they have nothing to offer currently to the discussion on SAR studies, they will not be considered further.

Another example of a peptidic compound is tentoxin, a cyclic tetrapeptide produced by *Alternaria alternata*. Although tentoxin (Fig. 3.2) is a herbicide and not an insecticide, the chemistry and approach used is particularly interesting as it relates to new work on neuropeptides at Rothamsted. Edwards *et al.* (1988) have demonstrated the extent to which peptidic features in tentoxin can be replaced without loss of herbicidal activity leading to an increase in flexibility and metabolic stability.

Animals

Parallel work on assessment of insect neuropeptides as potential lead structures for novel insecticides is currently in progress at Rothamsted.

Table 3.2 Examples of insect neuropeptides

Proctolin—	hindgut contraction in cockroach Arg-Tyr-Leu-Pro-Thr
FMRF-amide—	cardioexcitatory effect in molluscs Phe-Met-Arg-PheNH_2
AKH 1—	lipid mobilization in locusts pGlu-Leu-Asn-Phe-Thr-Pro-Asn-Trp-Gly-ThrNH_2
Others—	diuretic, red pigment concentrating, egg-laying hormones

A few examples of neuropeptides are listed in Table 3.2 to illustrate the range of activities. Neuropeptides are generally extremely active molecules, typically at picomolar concentration. However, when applied to insects (orally or topically) they show negligible effects, primarily because they are highly polar molecules (and therefore penetrate poorly), and are easily metabolized in the haemolymph. The work is directed at addressing both these issues simultaneously, by replacing peptidic features with less hydrophilic groups in the molecule.

One area of research is based on the diuretic neuropeptides (Fig. 3.3) found in a species of house cricket, *Acheta domesticus*. Coast *et al.* (1990) established that a shorter pentapeptide sequence containing the Trp–GlyNH_2 fragment accounted for most of the observed activity. The authors' collaborative work with Goldsworthy *et al.* (1992) has since established that the amide group of the C-terminus can be replaced by a methyl ester function and that the free amino group of the N-terminus is also not critical for activity. For example, the Phe group can be replaced by the $PhCH_2CH_2CO$ group with retention of activity. Other substitutions have established that, in general, increasing the hydrophobicity at the N-terminus leads to increased activity. In addition, the OH of the tyrosine residue can be replaced by H or OMe with retention of activity, also increasing lipophilicity.

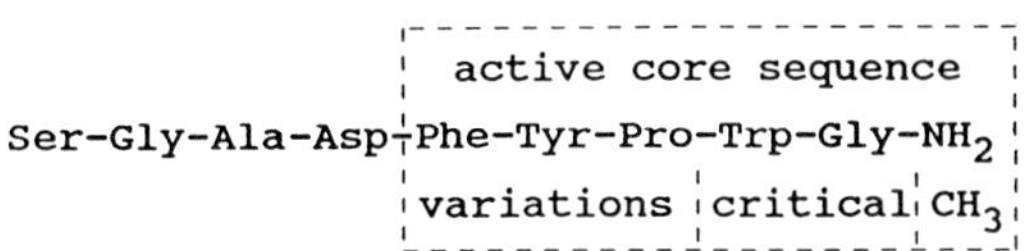

Fig. 3.3. SAR work on achetakinins.

i) bulky anchoring group with moderate hydrophobicity

ii) 2,6-halogens: I>Br>Cl>F>H

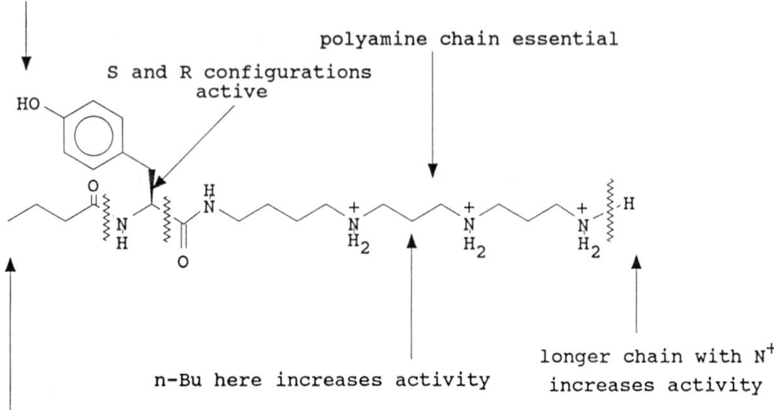

i) hydrophobicity and/or aromaticity increases activity

ii) C_{10} increases activity (x16); further changes reduce activity

iii) hydrophilic groups remove activity

Fig. 3.4. Philanthotoxin-433.

In conclusion, activity is retained even when several peptidic features are modified. Such approaches to structural modifications of peptidic compounds are common in pharmaceutical research, but have not yet been reported for insecticide research. The project described here is essentially fundamental in nature but, given the progress made so far, there is clear indication that neuropeptides may serve as novel lead structures for an SAR programme.

Another example of a highly active but polar compound from animals is philanthotoxin-433 (PhTX433) (Fig. 3.4), a spider venom. Although slightly more stable towards metabolism than neuropeptides, it is too polar to penetrate the cuticle. SAR studies on every part of the molecule have been carried out (Chiles *et al.* 1992). For example, the cationic polyamine chain has been shown to be critical for activity, and activity can be further improved by increasing the length of the polyamine. At the opposite end of the molecule, hydrophobicity and/or aromaticity increases activity.

In summary, although extensive SAR studies have led to an over 30-fold increase in activity, none of the compounds tested shows an increase in overall lipophilicity of the molecule. This remains a substantial barrier to the development of externally active compounds from venoms.

Plants

Only a small percentage (Grainge and Ahmed 1988) of plants have been screened for insecticidal activity. In addition, many such studies are not complete (e.g. see section below on *Annona* species) and often the bioassay procedures used have been inadequate or inappropriate. Consequently, plants remain a major untapped resource for potential leads to novel insecticides.

The key requirement for identification of insecticidal compounds is the availability of an appropriate bioassay system. In establishing such a bioassay, five major aspects require careful consideration:

(1) *in-vitro* or *in-vivo* screens;

(2) species for primary and secondary screens;

(3) throughput of samples;

(4) exposure methods (e.g. topical application, residual);

(5) decision on threshold activity (i.e. level at which the observed activity can be considered to be significant).

Table 3.3 is a comparison of the standard bioassays used at Rothamsted and in industry. All the considerations listed are inter-related and ultimately affect the ability of the systems to identify reliably low levels of activity against relevant insects which could be due to small amounts

Table 3.3 Bioassay systems

Rothamsted	Industry
In vivo (contact activity)	*In vivo* + *in vitro*
4 species: HF, MB, PX, TU Stable, homogeneous strains (including resistant and susceptible)	Broader spectrum (typically 10 species) Market orientated
Low/medium throughput	High throughput
Topical applications Requires small quantities Accuracy (± 10 per cent)	Residual and spray techniques (less sensitive)

of highly active compounds. The screen at Rothamsted is based on four species:

Musca domestica — housefly (HF)
Phaedon cochleariae — mustard beetle (MB)
Plutella xylostella — diamondback moth (PX)
Tetranychus urticae — mite (TU)

Strains of the three species other than TU have been established over many years and are stable and homogeneous. Although three of the species are not major agricultural pests, their usefulness in identifying broad spectrum insecticidal activity has been well proven. For example, the SAR work on pyrethroids leading to the identification of the photostable pyrethroids (Elliott *et al.* 1973) was based on bioassay against HF and MB only. In contrast, the standard primary screen in industry is based on a larger number of species usually of economic importance. High throughput of samples is also an essential feature, which influences the method of exposure and generally leads to lower sensitivity of the bioassay. Consequently, the probability of missing a highly active compound present in low concentration is increased relative to that for a more time consuming and sensitive bioassay procedure.

A typical procedure at Rothamsted is based on a sequential extraction of finely ground plant material with petroleum ether (b.p. 60–80 °C), ethyl acetate, and aqueous ethanol. Each extract is then tested by topical application (20 μg per insect). When the overall activity is low, selection of extracts for further examination is based on variation in activity between the three extracts. Typically when the variation is greater than 30 per cent, the most active fraction is subjected to short column chromatography. Only when a significant increase in activity has been realized is there further purification to the pure compound(s), often using a simpler and a quicker bench top bioassay (e.g. mosquito larvae). In a simple check, the bioassay described above easily identified as highly insecticidal even the stems and leaves of a locally grown pyrethrum plant, *Tanacetum* (= *Chrysanthemum*) *cinerariifolium* (Compositae = Asteraceae) (100 per cent kill of MB at 20 μg per insect), although the natural pyrethrins occur primarily in the flowers.

Two examples are shown in Fig. 3.5 to illustrate this approach. In both cases, the crude petroleum ether extracts showed activity at 20 μg per insect. For the *Hexalobus monopetalus* (Annonaceae) extract, further purification by column and thin layer chromatography showed insecticidal activity was due to the major component (**I**) with an LD_{50} of *c.* 9 μg per insect against MB. In contrast, the activity in *Vismia*

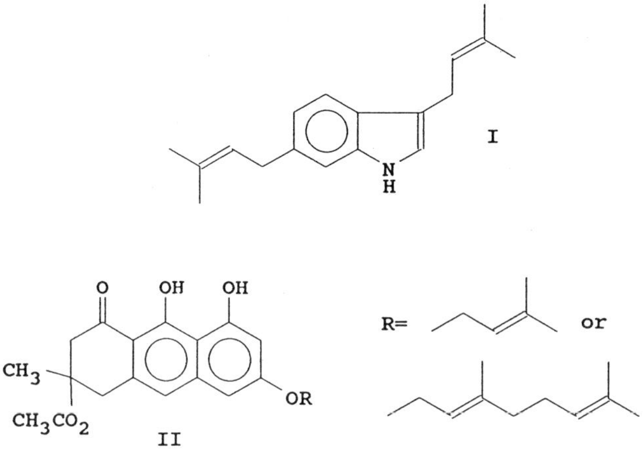

Fig. 3.5. Examples of isolated compounds.

guineensis (Hypericaceae) was attributed to a minor component **(II)** which upon purification had a higher level of activity (4 μg per insect against MB). Other examples have shown that even more active compounds with LD_{50} values of less than 1 μg per insect (MB) can be isolated from crude extracts showing activity only at doses greater than 10 μg per insect.

The next important decision lies in the selection of lead compounds for further SAR studies. Ideally, a lead compound should require optimization in one respect only. For example, pyrethrins had low mammalian and high insecticidal activity but poor photostability. However, such ideal examples are few and most require optimization of more than one property. Further, the minimum level of insecticidal activity expected of a lead structure has changed over the years, as evidenced by industrial attitudes. Given the very high costs of development, industry prefers to screen secondary metabolites from microorganisms in the hope of finding compounds such as avermectins and milbemycins, with commercially interesting levels of activity. However, to date no further compounds with such high levels of insecticidal activity have been reported, and interest in screening higher plants is reviving.

The rest of this chapter is devoted to examining new and established examples of insecticidal compounds from plants.

Pyrethroids

The hexane extract from flowers of *Tanacetum cinerariifolium* contains six insecticidal compounds. Elliott *et al.* (1975) chose pyrethrin I (Fig. 3.6)

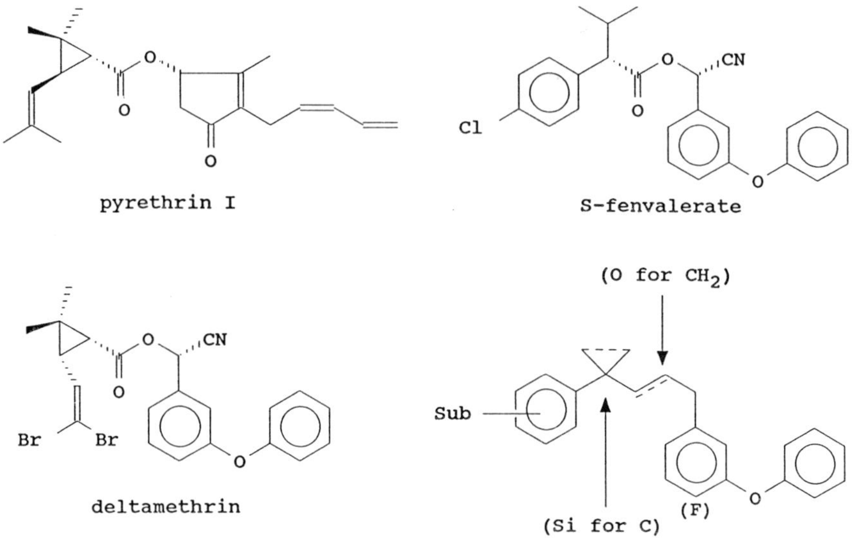

pyrethrin I

S-fenvalerate

deltamethrin

(O for CH_2)

Sub

(Si for C)

(F)

Fig. 3.6. Pyrethroids.

as the lead structure for optimization of insectidal activity. SAR studies based on systematic variations within the molecular structure eventually led to the discovery of photostable pyrethroids such as deltamethrin. In common with pyrethrin I, deltamethrin contains the ester function and three chiral centres in which the absolute stereochemistry is critical for activity. Almost simultaneously, Ohno *et al.* (1976) discovered fenvalerate. Although it was less active than deltamethrin, it had a simpler structure devoid of the cyclopropane ring and incorporating only two chiral centres. The only common feature with pyrethrin I was the ester function. The next major breakthrough was the discovery of non-ester pyrethroids based on the fenvalerate structure. These compounds are achiral, and have the ester linkage replaced by an alkane, alkene (*E*), or ether linkage. Work at Rothamsted (Elliott 1985) established the relationship between the dimethyl group and the central linkage. Optimum combination of the isopropyl group was with the ester linkage, whilst the *gem*-dimethyl group was most effective with a non-ester linkage. Further SAR work also established that in compounds with an alkene linkage, the 1,1-disubstituted cyclopropane group was particularly effective compared with the *gem*-dimethyl group. Bushell (1989) further demonstrated that the *gem*-dimethyl group can be replaced by a single

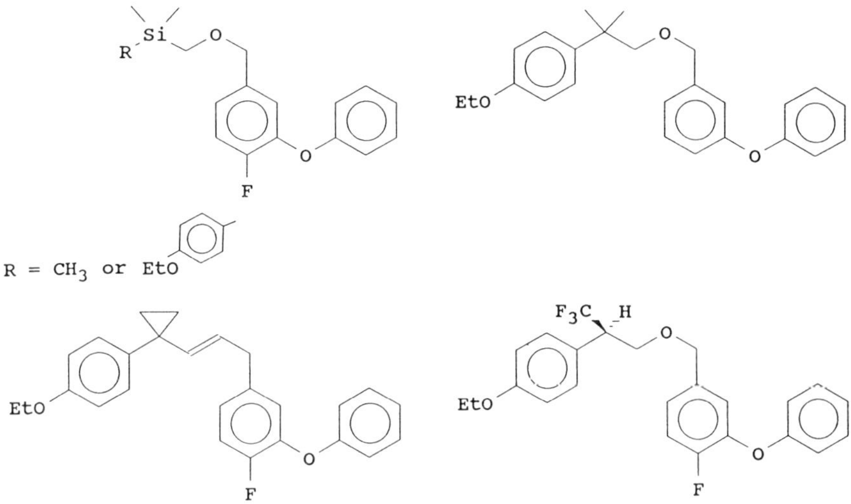

R = CH_3 or EtO

Fig. 3.7. Examples of non-ester pyrethroids.

CF_3 group to give active compounds. The non-ester pyrethroids are generally more selective towards rice pests and in general show very low fish toxicity. Isosteric compounds (Fig. 3.7) containing Si are exceptionally good in this respect. Introduction of a fluorine atom in the 4-position of the alcoholic component in both ester and non-ester pyrethroids increases selectivity. For example, activity towards mustard beetles increases two- to eight-fold; in contrast there is negligible effect towards houseflies.

In conclusion, insecticidal activity of the synthetic pyrethroids can now be increased over 1000 fold relative to the lead compound, pyrethrin I (for reviews see Elliott 1989; Naumann 1990). Photostability has also improved by several orders allowing their use in the field. In addition toxicity towards mammals has been reduced compared to the natural compound, and the spectrum of activity has increased to include several orders of insects (e.g. mites). SAR work at Rothamsted (unpublished) has also demonstrated that the level of resistance towards houseflies is influenced by the nature of the alcoholic component, thus pointing towards an effective means of prolonging the field use of pyrethroids.

N-Alkylamides

Insecticidal activity of a number of natural isobutylamides from Compositae and Rutaceae (Jacobson 1971) has been known for a long time.

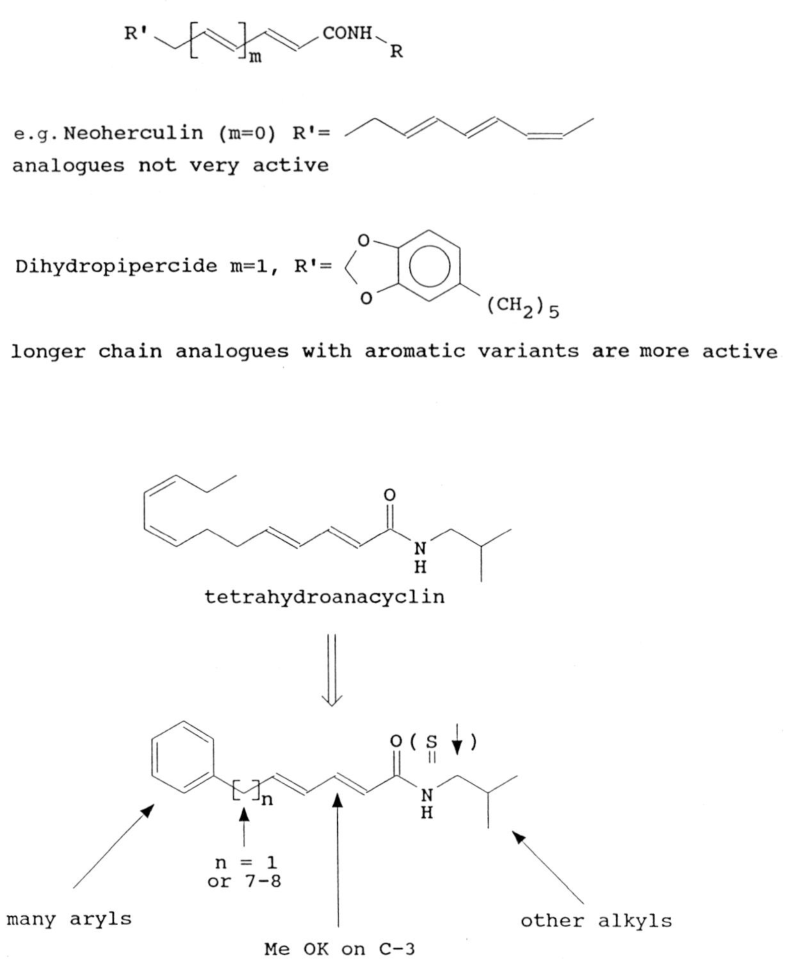

R' CONH R
m

e.g. Neoherculin (m=0) R'=
analogues not very active

Dihydropipercide m=1, R'=
O
O
$(CH_2)_5$

longer chain analogues with aromatic variants are more active

O
N
H
tetrahydroanacyclin

O (S)
N
H
n
n = 1
or 7-8
many aryls
other alkyls
Me OK on C-3

Fig. 3.8. *N*–Alkylamides.

Elliott *et al.* (1987) reviewed the literature (Fig. 3.8) and concluded that, in the absence of sufficient reliable bioassay data, there were no obvious natural alkylamides suitable as leads for SAR studies. Blade's (1989) earlier work on a monoenamide structure (as in neoherculin) did not lead to significantly active compounds. Miyakado *et al.* (1989) used the long chain dienamide, dihydropipercide, as a lead structure and reported compounds with improved activity. Elliott on the other hand used a small-chain synthetic dieneamide as a lead compound, conceptually derived from tetrahydroanacyclin by cyclizing the terminal *Z*,*Z*-diene chain into a phenyl ring, as had been done successfully in the pyrethroid work (Fig. 3.8). Elliott *et al.* (1987, 1989) published a seven-paper series of SAR studies in which they identified the diene-amide separated by at least a single methylene group (Fig. 3.8, $n = 1$) from the aromatic group as the key combination of features necessary for activity. Many variations of the amine group and substitution on the aromatic ring were found to be successful. Methyl substitution on the diene chain was restricted to position 3 only. Increasing the chain length to $n = 7-8$ showed a second peak in activity towards MB but significantly diminished activity towards HF. A further special feature of *N*-alkylamides is that they are more toxic to a pyrethroid resistant (*kdr*) strain of HF than to the susceptible strain (Elliott *et al.* 1986), and may thereby reveal potential against economically important resistant insect strains.

The SAR work to date has shown that most regions of the molecule are replaceable. However, although the overall activity has been improved, it remains moderate. Furthermore, *N*-alkylamides are highly selective compounds between insect species, an aspect sometimes influenced by even minor changes in structure.

Annona species

This is another species long known for several types of biological activity (e.g. cytotoxicity, antitumor, antimalarial, antimicrobial, antifeedant, pesticidal). The activity of *Annona squamosa* (Annonaceae) was for a long time associated only with anonaine (Fig. 3.9), a benzylisoquinoline alkaloid. However, recently another class of polar compounds with insecticidal activity, the acetogenins, have been isolated from these plants (Lieb *et al.* 1990; Mikolajczak *et al.* 1990; Rupprecht *et al.* 1991; Ohsawa *et al.* 1991). The acetogenins occur in several *Annona* species and can be

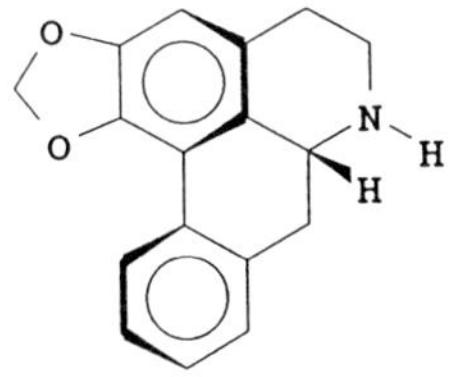

Fig. 3.9. Anonaine.

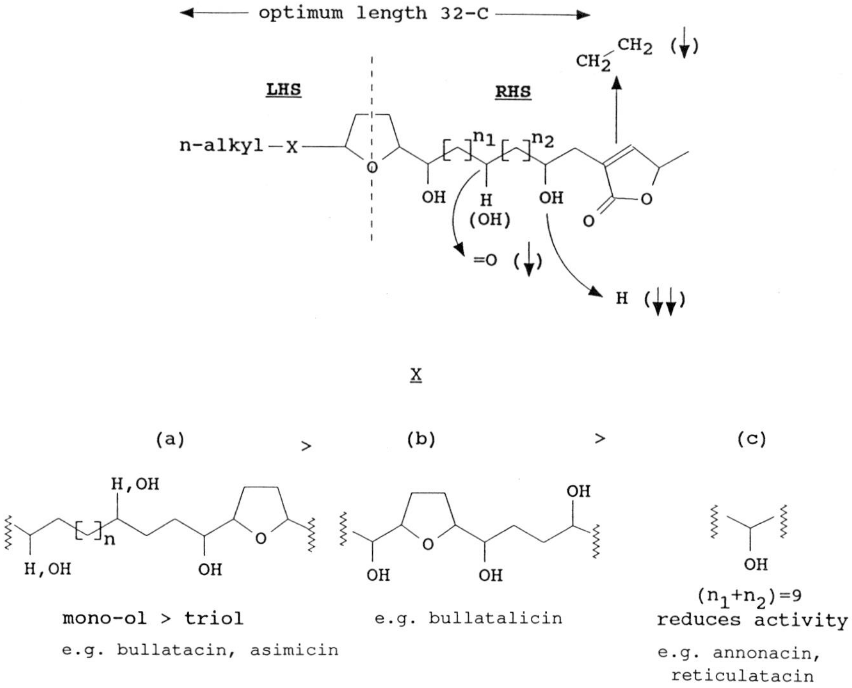

optimum length 32-C

LHS RHS

n-alkyl—X

OH H (OH) OH

=O (↓)

H (↓↓)

CH_2 CH_2 (↓)

X

(a) > (b) > (c)

H,OH H,OH OH O

mono-ol > triol

e.g. bullatacin, asimicin

OH O OH OH

e.g. bullatalicin

OH

$(n_1+n_2)=9$ reduces activity

e.g. annonacin, reticulatacin

Fig. 3.10. Annonaceous acetogenins.

classified into three broad categories (Fig. 3.10) based on the environment of the central tetrahydrofuran ring:

(a) adjacent bis-tetrahydrofuran rings e.g. bullatacin, asimicin;

(b) non-adjacent bis-tetrahydrofuran rings e.g. bullatalicin;

(c) single tetrahydrofuran ring e.g. annonacin, reticulatacin.

The activity diminishes in the order of compounds listed under (a) > (b) > (c).

Only a limited amount of synthesis has been reported to develop SARs. However, these compounds are believed to be generally biocidal, therefore SAR work will have to be directed towards optimizing at least two aspects. This is a clear example of a well known insecticidal plant where past investigations have been inadequate.

Lignans

Lignans, highly oxygenated phenylpropanoid dimers, are compounds not generally associated with insecticidal activity (Harmatha and Nawrot 1988). However, Yamauchi and Taniguchi (1991) recently identified haedoxan A (Fig. 3.11) from *Phryma leptostachya* as a highly active lignan. They deduced that the 1,4-benzodioxane framework, the methylenedioxy group, and the absolute stereochemistry were essential for activity.

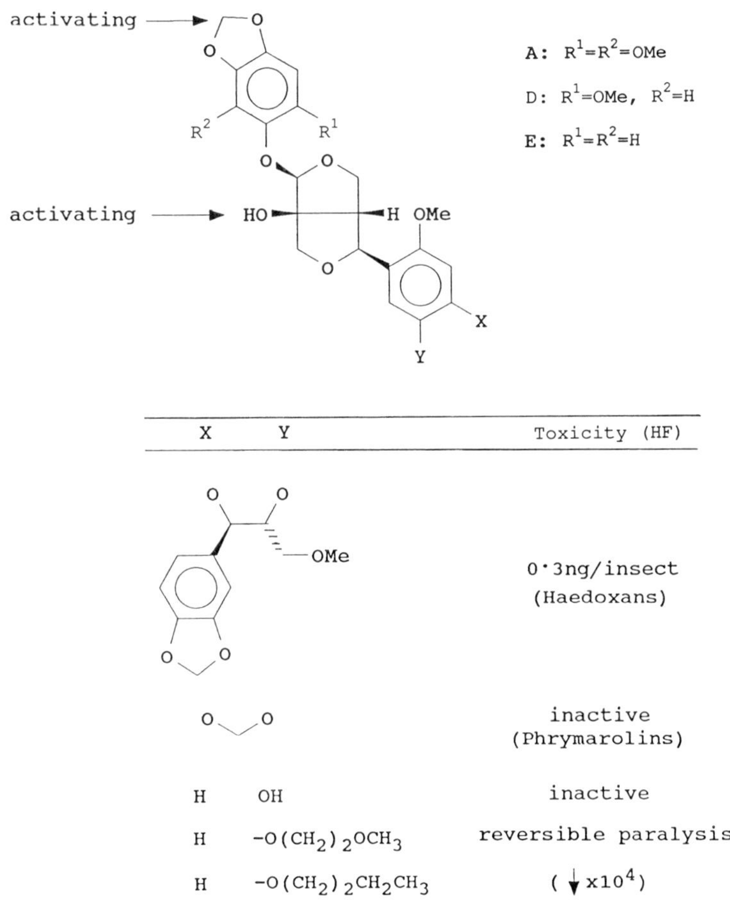

Fig. 3.11. Phryma lignans.

Although high insecticidal activity has been found in lignans more SAR investigations are required to improve upon the natural compound and to reduce its mammalian toxicity.

Phototoxins

Polyacetylenes and thiophenes (e.g. α-terthienyl, Fig. 3.12) are two examples of compounds which have been isolated from Asteraceae and shown to be highly active (LD_{50} 40 ppb against HF) but only when exposed to UV radiation (Wat *et al.* 1981; Christensen and Lam 1990; Fields *et al.* 1991) or in the presence of singlet oxygen.

Phototoxins act by a novel mode of action and therefore have great potential as a new family of insecticides. However, two major disadvantages are that they are general biocides and are rapidly photodegraded.

Naphthoquinones

Plumbagin has been known since 1828 (Fig. 3.13) and is present in many species of Plumbaginaceae (e.g. *Plumbago europea*) known for medicinal properties. Plumbagin has been claimed to have several types of activity, including insecticidal and antifeedant, and has been shown to be more active than nicotine sulphate (Gujar 1990).

Although a limited number of variations (mostly natural) are available for SAR studies the simple structure of plumbagin indicates much potential for further synthesis.

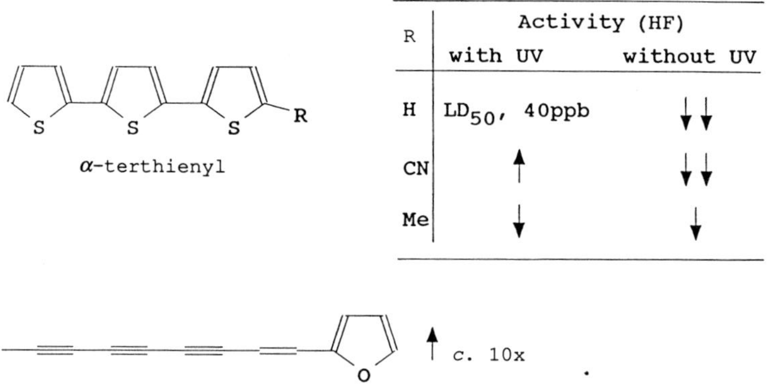

R	Activity (HF)	
	with UV	without UV
H	LD_{50}, 40ppb	↓↓
CN	↑	↓↓
Me	↓	↓

Fig. 3.12. Asteraceae phototoxins.

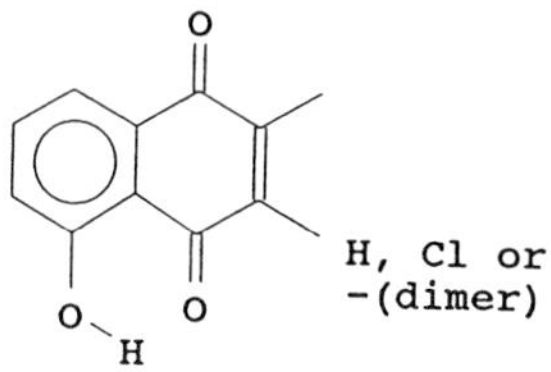

Fig. 3.13. Plumbagin insecticides.

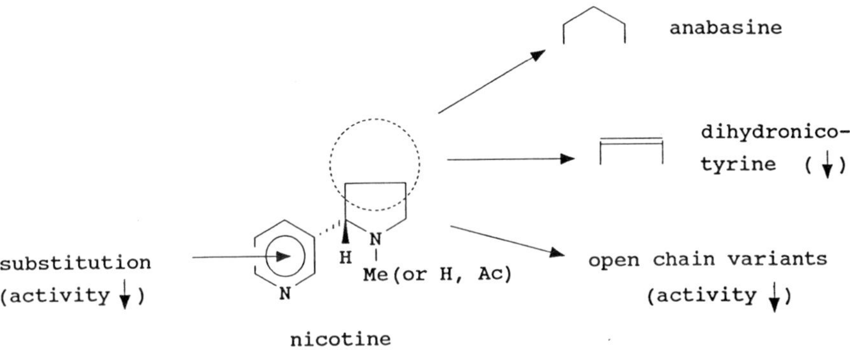

Fig. 3.14. Nicotinoids.

Nicotinoids

Use of nicotine (Fig. 3.14) as an insecticide has also been known for at least 200 years. It is believed to act by blocking acetylcholine receptors (Corbett *et al.* 1984). Its use in agriculture has declined since the advent of synthetic pyrethroids mainly due to its high mammalian toxicity and low persistence. Although some variations retain insecticidal activity (Schmeltz 1971) none is significantly more active than nicotine. Variations in the pyridine ring were shown to have a greater detrimental effect on activity than those in the pyrrolidine ring.

Coumarins

Members of the genus *Mammea* are amongst the most insecticidal plants. However, little information is reported on other types of activity (e.g. fish, mammals). Insecticidal activity of the leaves and seeds of the immature fruit of *Mammea americana* (Guttifercaeae = Clusiaceae) has been attributed to a coumarin (Fig. 3.15). The acetyl group at the 1′-position has been shown to be critical for activity.

Although much work on synthetic analogues has been done by Crombie (1989), no data on their activities is available.

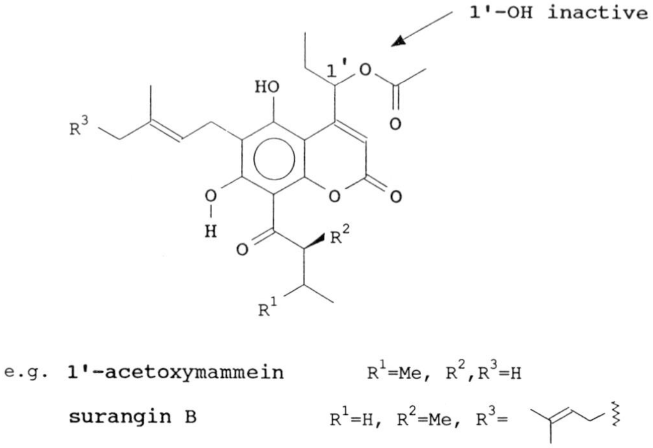

Fig. 3.15. Mammea coumarins.

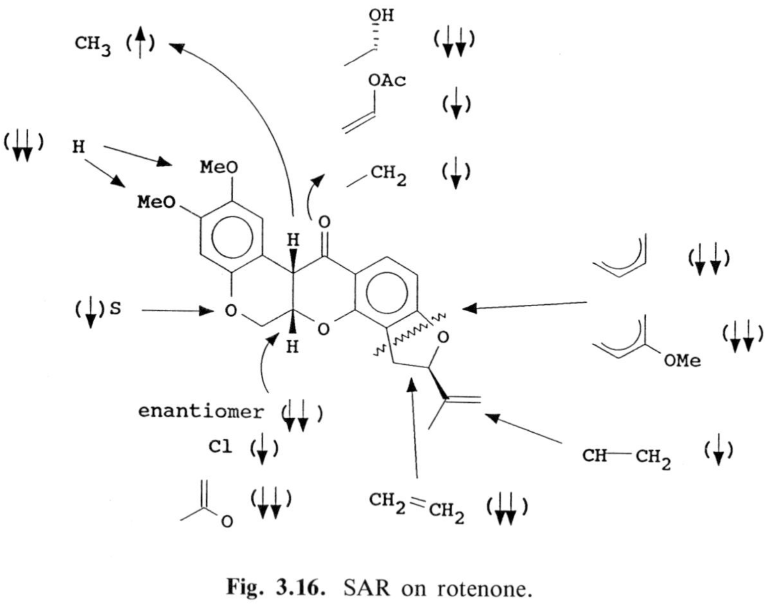

Fig. 3.16. SAR on rotenone.

Rotenoids

Ground roots from plants of the genera *Derris* and *Lonchocarpus* (Leguminosae) have been used as commercial insecticides for almost 150 years. The active principle is rotenone (Fig. 3.16) which is effective against a variety of insect species but is also well recognized as a fish poison (Fukami and Nakajima 1971). SAR studies on synthetic analogues have identified the key features necessary for activity (Crombie *et al.* 1992). However only one analogue, in which a $-CH_3$ group is introduced at position 12a, has been found to be more active than rotenone (J.L. Josephs, private communication). Use of rotenone as an insecticide will continue to be limited by its relatively high mammalian toxicity.

Veratrum alkaloids

Sabadilla preparations from *Schoenocaulon officinale* (Liliaceae) (Crosby 1971) have also been known for their insecticidal and mammalian toxicity for a long time. Recent SAR studies (Ujváry *et al.* 1991; I. Ujváry, private communication) have been based on variation of the 3-acyl group of veracevine (Fig. 3.17). The 3,5-dimethoxybenzoyl derivative was the most active compound and was almost twice as insecticidal as the natural 3,4-dimethoxy derivative, but without an increase in mammalian toxicity.

The complexity of the structure has limited SAR studies to simple modifications of the natural compound.

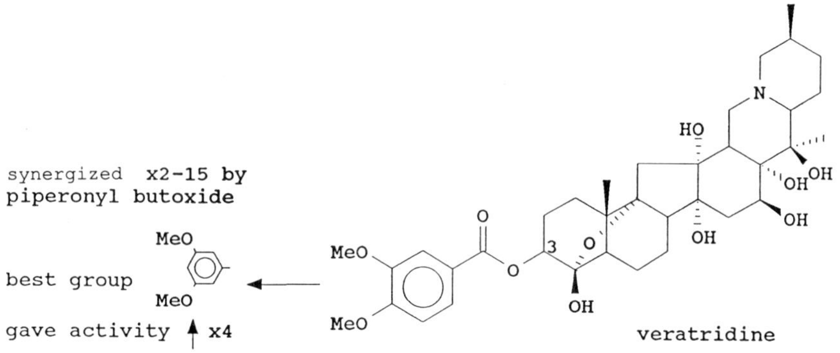

Fig. 3.17. Veratrum alkaloids.

Conclusion

There are two major considerations in choosing a lead structure for optimization studies. First, the level of activity and secondly, the number of features in the compound that require optimization. Ideally this should be only one (e.g. photostability) as in the pyrethrins.

Plants remain a vastly under-utilized resource for the discovery of novel insecticidal compounds. However, the discovery process demands a well thought-out and systematic approach in which the bioassay system plays a key role. Past studies should be assessed carefully before deciding not to undertake work in an area.

References

Blade, R. J. (1989). Some aspects of synthesis and structure-activity in insecticidal lipid amides. In *Recent advances in the chemistry of insect control II* (ed. L. Crombie), pp. 151–69. Royal Society of Chemistry, London.

Blizzard, T. A., Mrozik, H., Preiser, F. A., and Fisher, M. H. (1990). Synthesis, reactivity, and bioactivity of avermectin B_1-3,4-oxide. *Tetrahedron Lett.*, **31,** 4965–8.

Bushell, M. J. (1989). Synthesis of some fluorinated non-ester pyrethroids. In *Recent advances in the chemistry of insect control II* (ed. L. Crombie), pp. 125–41. Royal Society of Chemistry, London.

Chiles, G., Choi, S.-K., Eldefrawi, A., Eldefrawi, M., Fushiya, S., Goodnow, Jr, R., *et al.* (1992) Philanthotoxin-433 and analogs, noncompetitive antagonists of glutamate and nicotinic acetylcholine receptors. In *Neurotox 91: Molecular basis of drug and pesticide action* (ed. I. R. Duce). Elsevier, London.

Christensen, L. P. and Lam, J. (1990). Acetylenes and related compounds in Cynareae. *Phytochemistry*, **29,** 2753–85.

Coast, G. M., Holman, G. M., and Nachman, R. J. (1990). The diuretic activity of a series of cephalomyotropic neuropeptides, the achetakinins, in isolated Malpighian tubules of the house cricket (*Acheta domesticus*). *J. Insect Physiol.*, **36,** 481–8.

Corbett, J. R., Wright, K., and Baillie, A. C. (1984). Insects acting elsewhere in the nervous system. In *The biochemical mode of action of pesticides* (2nd edn), pp. 141–78. Academic Press, London.

Crombie, L. (1989). Natural models for the design of insect control compounds: the mammeins. In *Recent advances in the chemistry of insect control II* (ed. L. Crombie), pp. 23–51. Royal Society of Chemistry, London.

Crombie, L., Josephs, J. L., Cayley, J., Larkin, J., and Weston, J. B. (1992). The rotenoid core structure: modifications to define the requirements of the toxophore. *Bio-org. Med. Chem. Lett.*, **2,** 13–6.

Crosby, D. G. (1971). Minor insecticides of plant origin. In *Naturally occurring insecticides* (ed. M. Jacobson and D. G. Crosby), pp. 177–239. Dekker, New York.

Danishefsky, S. D., Armistead, D. M., Wincott, F. E., Selnick, H. G., and Hungate, R. (1989). The total synthesis of avermectin A_{1a}. *J. Am. Chem. Soc.*, **111**, 2967–80.

Edwards, J. V., Dailey, Jr, O. D., Bland, J. M., and Cutler, H. G. (1988). Approaches to structure–function relationships for naturally occurring cyclic peptides. A study of tentoxin. In *Biologically active natural products: Potential use in agriculture*, American Chemical Society Symposium Series, No. 380 (ed. H. G. Cutler), pp. 35–56. American Chemical Society, Washington, DC.

Elliott, M. (1985). Lipophilic insect control agents. In *Recent advances in the chemistry of insect control* (ed. N. F. Janes), pp. 73–102. Royal Society of Chemistry, London.

Elliott, M. (1989). The pyrethroids: early discovery, recent advances and the future. *Pestic. Sci.*, **27**, 337–51.

Elliott, M., Farnham, A. W., Janes, N. F., Needham, P. H., Pulman, D. A., and Stevenson, J. H. (1973). A photostable pyrethroid. *Nature*, **246**, 169–70.

Elliott, M., Janes, N. F., and Graham-Bryce, I. J. (1975). Retrospect on the discovery of a new insecticide. In *Proc. 8th Brit. Insect. Fung. Conf.*, pp. 373–9. BCPC Publications, Croydon, UK.

Elliott, M., Farnham, A. W., Janes, N. F., Johnson, D. M., Pulman, D. A., and Sawicki, R. M. (1986). Insecticidal amides with selective potency against a resistant (super-*kdr*) strain of houseflies (*Musca domestica* L.). *Agric. Biol. Chem.*, **50**, 1347–9.

Elliott, M., Farnham, A. W., Janes, N. F., Johnson, D. M., and Pulman, D. A. (1987). Synthesis and insecticidal activity of lipophilic amides. Parts 1–6. *Pestic. Sci.*, **18**, 191–201; 203–9; 211–21; 223–8; 229–38; 239–44.

Elliott, M., Farnham, A. W., Janes, N. F., Johnson, D. M., and Pulman, D. A. (1989). Synthesis and insecticidal activity of lipophilic amides. Part 7. Alternative aromatic groups for phenyl in 6-phenylhexa-2,4-dienamides. *Pestic. Sci.*, **26**, 199–208.

Fields, P. G., Arnason, J. T., Philogène, B. J. R., Aucoin, R. R., Morand, P., and Soucy-Breau, C. (1991). Phototoxins as insecticides and natural plant defences. *Memoirs of the Entomological Society of Canada*, **159**, 29–38.

Fisher, M. H. (1989). Novel avermectin insecticides and miticides. In *Recent advances in the chemistry of insect control II* (ed. L. Crombie), pp. 52–68. Royal Society of Chemistry, London.

Fukami, H. and Nakajima, M. (1971). Rotenone and the rotenoids. In *Naturally occurring insecticides* (ed. M. Jacobson and D. G. Crosby), pp. 71–97. Dekker, New York.

Goldsworthy, G., Coast, G., Wheeler, C., Cusinato, O., Kay, I., and Khambay, B. (1992). Structure and functional activity of neuropeptides. In *16th Sym-*

posium of the Royal entomological society (ed. J. M. Crampton and P. Eggleston), pp. 205–225. Academic Press, London.

Grainge, M. and Ahmed, S. (1988). *Handbook of plants with pest control properties*. Wiley, New York.

Gujar, G. T. (1990). Plumbagin, a naturally occurring naphthoquinone. Its pharmacological and pesticidal activity. *Fitoterapia*, **61**, 387–94.

Harmatha, J. and Nawrot, J. (1988). The role of lignans in plant–insect interaction. In *Endocrinological frontiers in physiological insect ecology* (ed. F. Sehnal, A. Zabza, and D. L. Denlinger), pp. 81–6. Wrocław Technical University Press.

Jacobson, M. (1971). The unsaturated isobutylamides. In *Naturally occurring insecticides* (ed. M. Jacobson and D. G. Crosby), pp. 137–76. Dekker, New York.

Lieb, F., Nonfon, M., Wachendorff-Neumann, U., and Wendisch, D. (1990). Annonacins and annonastatin from *Annoa squamosa*. *Planta Medica*, **56**, 317–19.

Mikolajczak, K. J., Madrigal, R. V., Rupprecht, J. K., Hui, Y.-H., Lui, Y.-M., Smith, D. L., and McLaughlin, J. L. (1990). Sylvaticin: a new cytotoxic and insecticidal acetogenin from *Rollinia sylvatica* (Annonaceae). *Experientia*, **46**, 324–7.

Mishima, H., Karabayashi, M., and Tamura, C. (1975). Structures of milbemycin β1, β2, and β3. *Tetrahedron Lett.*, 711–14.

Miyakado, M., Nakayama, I., and Ohno, N. (1989). Insecticidal unsaturated isobutylamides: from natural products to agrochemical leads. In *Insecticides of plant origin*, American Chemical Society Symposium, No. 387 (ed. J. T. Arnason, B. J. R. Philogène, and P. Morand), pp. 173–87. American Chemical Society, Washington, DC.

Naumann, K. (1990). *Chemistry of plant protection. Volume 4. Synthetic pyrethroid insecticides: structures and properties*, and *Volume 5. Synthetic pyrethroid insecticides: chemistry and patents*. Springer-Verlag, Berlin.

Ohno, N., Fujimoto, K., Okuno, Y., Mizutani, T., Hirano, M., Haya, N., *et al.* (1976). 2-Arylalkanoates, a new group of synthetic pyrethroid esters not containing cyclopropanecarboxylates. *Pestic. Sci.*, **7**, 241–6.

Ohsawa, K., Atsuzawa, S., Mitsui, T., and Yamamoto, I. (1991). Isolation and insecticidal activity of three acetogenins from seeds of pond apple, *Annona glabra* L. *J. Pestic. Sci.*, **16**, 93–6.

Rupprecht, J. K., Hui, Y.-H., and McLaughlin, J. L. (1990). Annonaceous acetogenins: a review. *J. Nat. Prod.*, **53**, 237–78.

Schmeltz, I. (1971). Nicotine and other tobacco alkaloids. In *Naturally occurring insecticides* (ed. M. Jacobson and D. G. Crosby), pp. 99–136. Dekker, New York.

Smith, C. (1992). *Pyrethroid pesticides—the market*. PJB Publications Ltd, Richmond, UK.

Ujváry, I., Eya, B. K., Grendell, R. L., Toia, R. F., and Casida, J. E. (1991). Insecticidal activity of various 3-acyl and other derivatives of veracevine relative to the veratrum alkaloids veratridine and cevadine. *J. Agric. Food Chem.*, **39**, 1875–81.

Wat, C.-K., Prasad, S. K., Graham, E. A., Partington, S., Arnason, T., Towers, G. H. N., and Lam, J. (1981). Photosensitization of invertebrates by natural polyacetylenes. *Biochemical Systematics and Ecology*, **9**, 59–62.

Yamauchi, S. and Taniguchi, E. (1991). Synthesis and insecticidal activity of lignan analogs. *Agric. Biol. Chem.*, **55**, 3075–84.

4. Exploiting behaviourally active phytochemicals in crop protection

JOHN A. PICKETT, LESTER J. WADHAMS, and CHRISTINE M. WOODCOCK

Department of Biological & Ecological Chemistry, AFRC Institute of Arable Crops Research, Rothamsted Experimental Station, Harpenden, Herts, AL5 2JQ, UK

Introduction

Rapid and efficient destruction of invertebrate pests can be achieved by using synthetic pesticides that selectively disrupt the nervous system. Alternatives to these agents are urgently sought, partly because of real problems such as pesticide resistance but also because of misconceptions concerning the toxicity and synthetic nature of the compounds involved. Although some early organochlorine compounds persisted to an unacceptable extent in the environment and accumulated in predators at the top of the food chain, these have now been replaced by the much less persistent organophosphorus, carbamate, and pyrethroid insecticides. Certainly, with the pyrethroids, the likelihood of environmental pollution or human hazard is extremely slight when correctly used in agriculture. None the less, non-target invertebrates can be at risk, and, much more importantly, there are serious problems of control failure through the selection of populations of pests resistant to these agents.

The pyrethroid insecticides were developed from natural products, particularly pyrethrin I, an insecticidal component of the plant *Tanacetum* (= *Chrysanthemum*) *cinerariifolium* (Compositae = Asteraceae) (Elliott *et al.* 1978). Many other plant-derived biocides are currently under investigation, although none show as much promise as pyrethrin I for the development of new insecticides (Pickett 1992*a*). It would be naive to imagine that natural insecticides, or derived analogues, would be benign towards ourselves or other non-target organisms merely because they are natural products and Ames (Ames and Gold 1990; Ames *et al.* 1990) has provided convincing statistics against such a view. Although the pyrethroids are highly selective between insect pests and human beings, they are at present under attack simply because they are nerve poisons. Therefore, in the long term, it may be necessary to move away from compounds using toxic modes of action, whether natural or not, to allay such fears which, sadly, seem to stem from a lack of understanding of risk assessment (Pickett 1992*b*).

If an inventory is made of the wide range of chemical substances that influence the development of invertebrate pests or plant diseases, many will have a purely regulatory and non-toxic mode of action and, on this basis alone, warrant investigation. Nonetheless, to be useful in pest control, they must provide efficient methods for controlling the unwanted organisms. As such, those agents that affect regulatory processes within the organism provide difficult targets. This is in part because they are often highly labile, hydrophilic compounds, e.g. the neuropeptides in insects (Grimmelikhuijzen *et al.* 1987), that do not readily penetrate the cuticle. In fact, the most promising of these leads are those where death would ensue through disrupting the normal action of the regulator. Regulatory signals which act externally on the organism, i.e. semiochemicals such as pheromones and host attractants, avoid this problem. By interacting with the peripheral nervous system, these materials can work extremely rapidly, but, because they act only as signals, there are no direct toxic effects, certainly not at the levels required for behavioural activity. Considerable effort has been directed at pheromones as potential pest control agents, but pheromones represent only one component of the chemical ecology of pests. It is now accepted that for reliable control, other aspects of chemical ecology must be understood and appropriate semiochemicals employed together with pheromones in pest management regimes (Pickett 1991). Furthermore, it is now apparent that merely by preventing initial colonization of a crop, or by interfering with normal mate location, a lasting effect on the pest population will not necessarily result. It is therefore necessary, having prevented immediate damage by deploying semiochemicals, to reduce the population but, in line with the general objectives, by a mechanism that could not be considered hazardous. This second aspect of such an integrated strategy can best be provided by biological control agents. However, to avoid criticisms implicit in using toxic modes of action, the most appropriate biological agents will decrease populations by pathogenicity rather than by antibiosis.

Plants contain a wide range of semiochemicals, some of which may also act as biocides at high concentrations. The semiochemicals generally, including those found in plants, fall into four main categories:

(1) pheromones, which signal between organisms of the same species, for example methyl jasmonate (Farmer and Ryan 1990) acting as a wound signal between individual plants;

(2) allomones, which favour the emitting organism, and would include chemicals that interfere with feeding behaviour (antifeedants) and

chemicals that attract beneficial insects such as hymenopteran pollinators of flowers;

(3) kairomones that favour the receiving organism, including those chemicals used by herbivores to find plant hosts;

(4) a group of semiochemicals receiving increased attention, the synomones, which can benefit both emitting and receiving organisms.

This latter category includes those chemicals released by plants during herbivore attack that stimulate parasitism of the herbivore, thus benefiting both the parasite and the plant (Dicke *et al.* 1990; Vinson 1992; see also Chapter 6). Semiochemicals such as pheromones used by pest and disease species may also be the same as or similar to phytochemicals. In this review, behaviour-controlling phytochemicals will primarily be considered, but insect metabolites will also be discussed where they are known to occur as phytochemicals, even if produced in different ecological contexts.

Strategies for use

Behaviour-controlling phytochemicals, and semiochemicals generally, will in the future find use within push–pull, or stimulo–deterrent diversionary, strategies (SDDS) (Pickett *et al.* 1991). In such strategies, the harvestable crop is protected by means of repellent semiochemicals such as plant-derived antifeedants, by employing repellent cultivars or those with a reduced attractancy, and by using chemicals from non-host-plants which interfere with the location of the host-plant. Aggregation away from the crop is then encouraged by attractants such as sex pheromones, oviposition pheromones, and host attractants, and also by trap crop plants producing large quantities of attractants. On the trap crop is placed either a highly selective but slow-acting regulant insecticide or, more desirably, a pathogenic biological control agent. Thus, the crop to be harvested is protected and the population of pests is reduced by the biological agent on the trap crop, where a slow mode of action can be tolerated. Conditions on the trap crop may be modified to benefit the biological agent.

As well as regimes employing both a harvestable crop and a trap crop, it is possible to develop SDDS within individual plants by, for example, protecting the growing tip or seed-bearing part of the plant, whilst causing aggregation of pests within the lower leaf canopy where the biological control agent is placed. This has been demonstrated under simulated field conditions using mustard plants and the Coleopteran *Phaedon cochleariae* (Chrysomelidae). By selectively spraying the top of the plant with the antifeedant ajugarin I, a clerodane from *Ajuga remota* (Labiatae = Lamiaceae), the beetles were driven down to feed on the lower leaves. Here, they were destroyed by a slow-acting but highly selective growth regulant insecticide before they could return to the upper part of the plant as it grew away from the antifeedant (Griffiths *et al.* 1991).

Building on this success, the push–pull or stimulo–deterrent effect has now been demonstrated in a field experiment on the coleopterous pest *Sitona lineatus* (Curculionidae) (L.E. Smart, personal communication). Here, the crop was field beans (*Vicia faba*, Leguminosae = Fabaceae), the repellent was a neem (*Azadirachta indica*, Meliaceae) extract, and the attractant was the aggregation pheromone 4-methyl-3,5-heptanedione. The combined effect caused a highly significant deterrent to colonizing the crop, with aggregation where the attractant was placed. The next step will be to introduce a biological control agent on to the plots where aggregation has been enhanced. For these studies, the antifeedants were selected from a wide range of compounds known to have such effects (e.g. Griffiths *et al.* 1988). The aggregation pheromone for *S. lineatus* had previously been identified as part of a programme to devise a monitoring trap for this pest (Blight *et al.* 1991). However, for full exploitation of SDDS, it is necessary to define much more extensively the components of pest chemical ecology and for this, special analytical methods have been developed.

Identification of behaviourally active phytochemicals

Since precolonization of the crop is the stage that should be targeted for most economic intervention, volatile compounds are the most important semiochemicals to be considered. The main method for obtaining samples of these materials is adsorption on to a porous material, for example Porapak Q, and then elution for gas chromatographic (GC) analysis. Particular care must be adopted in removing contamination from the adsorbent and from the eluting solvents (Blight 1990). These methods can be augmented by vacuum distillation (Pickett and Griffiths 1980; Pickett and Stephenson 1980) and also by extraction using liquid carbon dioxide, preferably at less than the critical temperature to avoid any thermolysis of the plant material (Pickett *et al.* 1987). Due consideration must also be given to whether enzymic activity arising through tissue damage should be eliminated or encouraged.

Once volatiles are isolated from the plant, it is essential that they are stored sealed in glass ampoules under nitrogen at −20 °C. High resolution capillary columns, for example a 50 m quartz capillary column with a bonded silicone phase, give excellent resolution of such mixtures

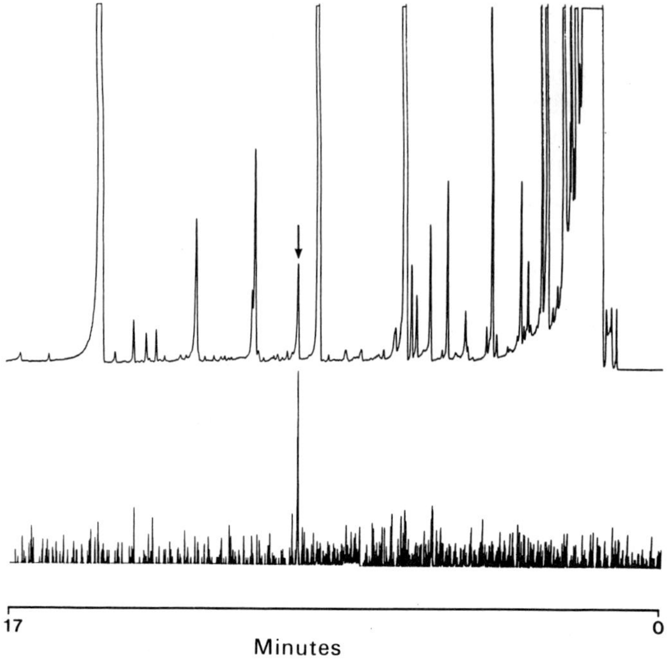

Fig. 4.1. Coupled GC-single cell recording on antenna of raspberry beetle (*Byturus tomentosus*, Byturidae): (*top*) gas chromatogram of volatiles from flowers of hawthorn (*Crataegus monogyna*, Rosaceae); (*bottom*) corresponding impulse frequency response of olfactory cell.

with a flame ionization detector. Extremely complicated chromatograms are usually obtained, with hundreds of peaks varying tremendously in height. Plants from the same cultivar grown under different agronomic conditions show considerable variation in composition, which is aggravated by the presence of pests or disease (Blight 1990). Minor components are often involved in important interactions with pest organisms and direct biological monitoring of the GC effluent is very effective in locating activity in complex extracts. This has been brought to a high level of sophistication with invertebrate pests, particularly insects, using electrophysiological preparations from the olfactory organs. Originally, recordings were made from the entire antenna, but through pioneering work by Wadhams (1990), it has been shown that recordings from single olfactory sensilla, or even single nerve cells within the sensilla, can provide much valuable information. Thus, using sophisticated 'plumbing', so that the GC effluent reaches the flame ionization detector simultaneously with it reaching the antennal preparation, the column effluent can be monitored for electrophysiological activity (Wadhams 1990). A typical trace is shown in Fig. 4.1.

For the coleopterous pest *Ceutorhynchus assimilis* (Curculionidae) of oilseed rape, *Brassica napus* (Cruciferae = Brassicaceae), single-cell responses have been detected for a wide range of compounds, including terpenoids, fatty acid derivatives, and a number of aromatic compounds, as well as catabolites of the glucosinolates known to be characteristic of crucifers (Blight *et al.* 1992). The isothiocyanates are important catabolic products resulting from the action of thioglucosidase (myrosinase) enzymes brought into contact with the glucosinolates on damage to the plant. The most commonly occurring cell responds generally to these compounds, while other cells detect particular homologues, for example those with alkenyl groups or the 2-phenylethyl isothiocyanate (Blight *et al.* 1989). These differing cell types may convey information on the growth stage of the plant to the prospective pest (Milford *et al.* 1989). During these studies, it was also found that the best known glucosinolate catabolite, allyl isothiocyanate, produced by myrosinase action upon sinigrin (allylglucosinolate), was unlikely to be an important semiochemical for this particular insect. Good field catches were obtained using water traps baited with isothiocyanates, including those having the greatest electrophysiological activity, i.e. the 3-butenyl and 4-pentenyl isothiocyanates (Blight *et al.* 1992).

For *S. lineatus*, electrophysiological studies have provided phytochemical synergists for the aggregation pheromone already mentioned and the compounds so far demonstrated to have activity in field trials include the green leaf volatiles linalol, (*Z*)-3-hexenyl acetate, and (*Z*)-3-hexenol (Blight *et al.* 1991). Although these are common plant components, the cells which respond to them are extremely specific (e.g. Fig. 4.2). A number of other active peaks have recently been detected in a range of plant extracts (C.M. Woodcock, unpublished) and these are currently being identified.

Although invertebrates, and insects in particular, have been used most frequently in electrophysiological investigations, it has recently been demonstrated that similar approaches can be adopted for terrestrial molluscs (Garraway *et al.* 1992) and certain nematodes (Jones *et al.* 1991). For contact-acting phytochemicals, new methods are being developed, including liquid chromatography (LC)-coupled gustatory preparations (L.J. Wadhams, unpublished).

Once an electrophysiologically active GC peak has been noted, GC-coupled mass spectrometry provides a tentative structure (Pickett 1990), although some further microchemistry or use of GC-coupled Fourrier transform infra-red spectroscopy may be necessary. Authentic compound

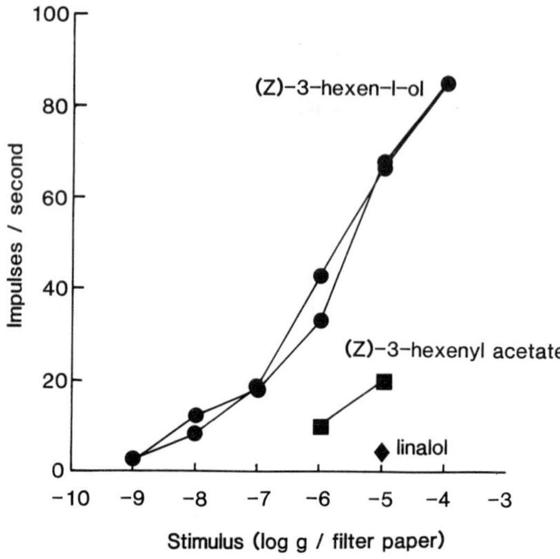

Fig. 4.2. Dose-response of a *Sitona lineatus* olfactory cell to three green leaf volatiles.

is then obtained by synthesis or isolation from another known source, often a plant, and the structure confirmed by high-field nuclear magnetic resonance spectroscopy for comparison by GC and peak enhancement studies. Further electrophysiological and behavioural bioassays confirm the compound as behaviourally active and enable a final account to be made of the full activity of the natural product.

Kairomone inhibition

Methods for identification of attractant phytochemicals can be equally effective for other groups of behaviourally active phytochemicals. Thus, compounds that inhibit the attractancy of host-plant kairomones are sought by GC-coupled electrophysiological studies using volatiles from non-host-plants. The taxonomically diverse insects *Ceutorhynchus assimilis* (Curculionidae) and the black bean aphid, *Aphis fabae* (Homoptera, Aphididae), have antennal cells responding similarly to isothiocyanates (see Fig. 4.3), which are host-plant components for *C. assimilis* but not for *A. fabae*. Indeed, these compounds are from plants upon which *A. fabae* cannot normally survive. This phenomenon has been found to be common to many other herbivore host/non-host situations. Therefore,

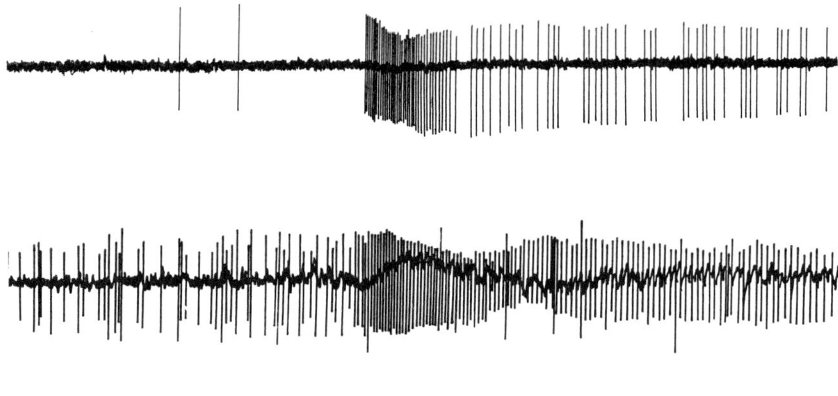

Fig. 4.3. Response of olfactory cells from insects in different orders to 4-pentenyl isothiocyanate: (*top*) at 10^{-7} g for *Ceutorhynchus assimilis* (host-plant compound); (*bottom*) at 10^{-6} g for *Aphis fabae* (non-host-plant compound). Horizontal bar = 1s stimulus.

it must be concluded that the hehavioural response to these agents is determined within the central nervous system and accordingly, studies on this aspect are proceeding (Pham-Delègue *et al.* 1992). Where there is a highly developed receptor system for non-host-plant compounds, these offer prospects for inhibition of kairomonal attraction. Indeed, it has recently been demonstrated that isothiocyanates can prevent the attraction of *A. fabae* to volatiles from its host-plant, *Vicia faba* (Nottingham *et al.* 1991). It has also been possible to interfere with the attraction of hop aphids (*Phorodon humuli*) to volatiles from their secondary host-plant, the hop *Humulus lupulus* (Cannabaceae) during the spring migration into the crop (C.A.M. Campbell, personal communication).

Phytochemicals as pheromones

As well as some phytochemicals acting as plant pheromones, plants can contain chemicals which are identical to insect pheromones, either by coincidence or evolutionary design. For example, many plants produce small amounts of (*E*)-β-farnesene, a compound which is also employed by aphids as an alarm pheromone signifying attack by a predator or parasitoid. Aphid colonization is possible because most plants also produce β-caryophyllene, which acts as an alarm pheromone inhibitor

(Dawson *et al.* 1984), but a rapid increase in the level of (*E*)-β-farnesene, as released by an individual under attack, can still be detected by other aphids. However, the long foliar trichomes of the solanaceous plant *Solanum berthaultii* release (*E*)-β-farnesene in such a way that an alarm response is elicited in aphids (Gibson and Pickett 1983) and here there is a clear benefit to the plant.

The Far Eastern vine *Actinidia polygama* (Actinidiaceae) produces various iridoids, or cyclopentanoid monoterpenoids, which cause aggregation of lacewings (Sakan *et al.* 1970). This phenomenon has now been explained by the recent identification of aphid sex pheromones as cyclopentanoids (Dawson *et al.* 1987, 1990*b*): the lacewings are presumably using these chemical signals as kairomones to find their prey and therefore aggregate upon the plant. At the same time, it could be that the plant is using iridoid production to attract predators against herbivore infestation. In addition, it has now been established that parasitoids of aphids, particularly in the genus *Praon*, are using aphid sex pheromones as kairomones (Hardie *et al.* 1991).

Production of semiochemicals as phytochemicals

Recently, it has been shown (Olagbemiro *et al.* unpublished) that (*Z*)-5-hexadecenoic acid can be obtained from plants in the Chenopodiaceae and converted into the mosquito oviposition pheromone, (5*R*,6*S*)-6-acetoxy-5-hexadecanolide, which is of potential value in controlling disease vectors (Dawson *et al.* 1989*a*). This route provides an alternative to expensive acetylene, Wittig, or Baeyer–Villiger chemistry in the production of the pheromone (Dawson *et al.* 1990*a*). Studies on oilseed rape are elucidating the biosynthetic pathways to those glucosinolates that on catabolism give the attractants to which such pests as *C. assimilis* respond (Dawson *et al.* 1989*b*). Attempts are being made to use molecular biology to eliminate this biosynthesis within oilseed rape plants, to provide less attractive crops for use within SDDS (Pickett 1989). This work benefits by the fact that some cultivars of the crucifer, *Arabidopsis thaliana*, already under detailed study in molecular biology programmes because of its relatively small genome, do not appear to produce the alkenyl-glucosinolates (Hogge *et al.* 1988).

Various plants producing iridoids, particularly the labiates (Lamiaceae) *Nepeta mussinii* (= *racemosa*) and *N. cataria*, are valuable sources of aphid sex pheromone components or precursors, allowing production of moderate amounts of chemicals for field trials (Dawson *et al.* 1990*b*). The biosynthetic pathways can also be studied in such plants more conveniently than in aphids. The proposed open chain precursors (Jensen

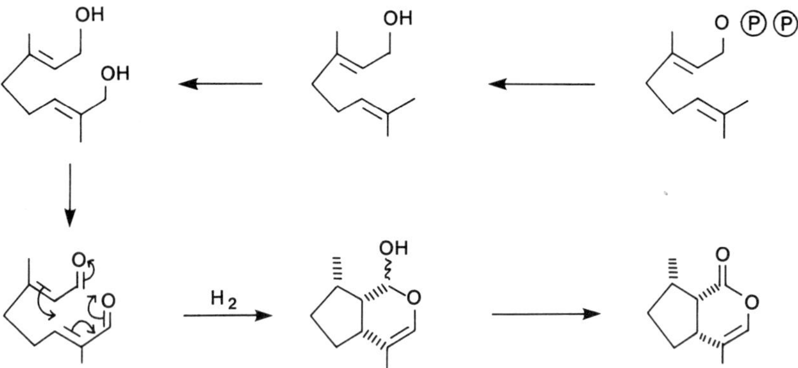

Fig. 4.4. Possible biosynthetic pathway from geranyl pyrophosphate to (4a*R*, 7*S*, 7a*S*)-nepetalactone. Note that the course of the latter part of this pathway is not fully elucidated but for brevity is depicted as incorporating a concerted reductive cyclization.

et al. 1989) of these cyclopentanoids have been synthesized with radio-labelling (Fig. 4.4). This has already allowed sequencing of a partial clone of the mixed function oxidase protein from the haem binding site towards the carboxyl terminus that is responsible for the first oxidative conversion of geraniol to 10-hydroxygeraniol (Hallahan *et al.* unpublished). Also, the oxidoreductase for the next stage has been purified to homogeneity (Hallahan *et al.* unpublished). Similar studies on the enzymes causing the reductive cyclization of the open chain compounds into the cyclopentanoid system could allow isolation of genes which, if cloned into a yeast or *E. coli*, would allow production of these materials, or their precursors, in fermentors.

The sex pheromones of lepidopterous insect pests have received considerable attention. Various innovative formulations have been devised to overcome problems of field use pertaining to their volatility and instability in light and air (Jutsum and Gordon 1989). An alternative approach would be to release these agents from plants. At present, it is impossible to consider transferring an entire biosynthetic pathway from a moth to a plant. However, certain plants already have a pathway that could give rise to the lepidopterous sex pheromone compounds, but which lacks the necessary desaturation step (Dawson *et al.* 1989*b*). Therefore, studies on the enzymology of lepidopterous sex pheromone desaturases are in progress (Rodriguez *et al.* 1992), with the long-term objective of transferring the associated genes to augment plant biosynthetic pathways.

Table 4.1 Exploitation of semiochemicals

1.	Identification: e.g. coupled GC-electrophysiology
2.	Integrated strategies: e.g. SDDS
3.	Production: biotechnology
4.	Molecular genetics: modify plant secondary metabolism

Pheromone components could then be produced within the essential oil of the plant and released into the field to interfere with normal mate location. In the long term, the approach of genetically supplementing crop plant secondary metabolism may prove generally valuable in improving pest and disease resistance (Pickett and Woodcock 1992). As a precedent, genetic augmentation of secondary metabolism in *Nicotiana tabacum* has been shown to result in biosynthesis of a phytoalexin known as a valuable protectant in other plants (Hain *et al.* 1990).

Conclusions

Although some of the objectives described require substantial effort and innovation to be achieved, the tremendous demand for use of biocontrol agents in pest control must be met with scientific endeavour. Attempts to use behaviourally active phytochemicals must be made, bearing in mind the four main issues exemplified and discussed here and which are summarized in Table 4.1.

References

Ames, B. N. and Gold, L. S. (1990). Misconceptions on pollution and the causes of cancer. *Angew. Chem., Int. Ed. Engl.*, **29**, 1197–208.

Ames, B. N., Profet, M., and Gold, L. S. (1990). Dietary pesticides (99.99% all natural). *Proc Natl. Acad. Sci. USA*, **87**, 7777–81.

Blight, M. M. (1990). Techniques for isolation and characterization of volatile semiochemicals of phytophagous insects. In *Chromatography and isolation of insect hormones and pheromones* (ed. A. R. McCaffery and I. D. Wilson), pp. 281–8. Plenum, New York/London.

Blight, M. M., Pickett, J. A., Wadhams, L. J., and Woodcock, C. M. (1989). Antennal responses of *Ceutorhynchus assimilis* and *Psylliodes chrysocephala* to volatiles from oilseed rape. *Asp. Appl. Biol.*, **23**, 329–34.

Blight, M. M., Dawson, G. W., Pickett, J. A., and Wadhams, L. J. (1991). The identification and biological activity of the aggregation pheromone of *Sitona lineatus*. *Asp. Appl. Biol.*, **27**, 137–42.

Blight, M. M., Hick, A. J., Pickett, J. A., Smart, L. E., Wadhams, L. J., and Woodcock, C. M. (1992). Volatile plant metabolites involved in host plant recognition by the cabbage seed weevil, *Ceutorhynchus assimilis* Payk. In *Proc. 8th Int. Symp. on Insect Plant Relationships, Wageningen, 1992*, pp. 105–6. Kluwer, Dordrecht.

Dawson, G. W., Griffiths, D. C., Pickett, J. A., Smith, M. C., and Woodcock, C. M. (1984). Natural inhibition of the aphid alarm pheromone. *Entomol. Exp. Appl.*, **36**, 197–9.

Dawson, G. W., Griffiths, D. C., Janes, N. F., Mudd, A., Pickett, J. A., Wadhams, L. J., and Woodcock, C. M. (1987). Identification of an aphid sex pheromone. *Nature*, **325**, 614–16.

Dawson, G. W., Laurence, B. R., Pickett, J. A., Pile, M. M., and Wadhams, L. J. (1989*a*). A note on the mosquito oviposition pheromone. *Pestic. Sci.*, **27**, 277–80.

Dawson, G. W., Hallahan, D. L., Mudd, A., Patel, M. M., Pickett, J. A., Wadhams, L. J., and Wallsgrove, R. M. (1989*b*). Secondary plant metabolites as targets for genetic modification of crop plants for pest resistance. *Pestic. Sci.*, **27**, 191–201.

Dawson, G. W., Mudd, A., Pickett, J. A., Pile, M. M., and Wadhams, L. J. (1990*a*). Convenient synthesis of mosquito oviposition pheromone and a highly fluorinated analog retaining biological activity. *J. Chem. Ecol.*, **16**, 1779–89.

Dawson, G. W., Griffiths, D. C., Merritt, L. A., Mudd, A., Pickett, J. A., Wadhams, L. J., and Woodcock, C. M. (1990*b*). Aphid semiochemicals—a review, and recent advances on the sex pheromone. *J. Chem. Ecol.*, **16**, 3019–30.

Dicke, M., Van Beek, T. A., Posthumus, M. A., Ben Dom, N., Van Bokhoven, H., and De Groot, A. E. (1990). Isolation and identification of volatile kairomone that affects acarine predator–prey interactions. Involvement of host plant in its production. *J. Chem. Ecol.*, **16**, 381–96.

Elliott, M., Janes, N. F., and Potter, C. (1978). The future of pyrethroids in insect control. *Ann. Rev. Entomol.*, **23**, 443–69.

Farmer, E. E. and Ryan, C. A. (1990). Interplant communication: airborne methyl jasmonate induces synthesis of proteinase inhibitors in plant leaves. *Proc. Natl. Acad. Sci. USA*, **87**, 7713–16.

Garraway, R., Leake, L. D., Ford, M. G., and Pickett, J. A. (1992). The development of a chemoreceptive neurophysiological assay for the field slug, *Deroceras reticulatum* (Müll). *Pestic. Sci.*, **34**, 89–98.

Gibson, R. W. and Pickett, J. A. (1983). Wild potato repels aphids by release of aphid alarm pheromone. *Nature*, **302**, 608–9.

Griffiths, D. C., Hassanali, A., Merritt, L. A., Mudd, A., Pickett, J. A., Shah, S. J., *et al.* (1988). Highly active antifeedants against coleopteran pests. In *Proceedings of the Brighton Crop Protection Conference—Pests and Diseases*, pp. 1041–6. BCPC Publications, Thornton Heath.

Griffiths, D. C., Maniar, S. P., Merritt, L. A., Mudd, A., Pickett, J. A., Pye, B. J., *et al.* (1991). Laboratory evaluation of pest management strategies combining antifeedants with insect growth regulator insecticides. *Crop Protection*, **10**, 145–51.

Grimmelikhuijzen, C. J. P., Graff, D., Groeger, A., and McFarlane, I. D. (1987). *Neuropeptides in invertebrates.* NATO ASI Series A **141**, 105–32.

Hain, R., Bieseler, B., Kindl, H., Schröder, G., and Stöcker, R. (1990). Expression of a stilbene synthase gene in *Nicotiana tabacum* results in synthesis of the phytoalexin resveratrol. *Plant Mol. Biol.*, **15**, 325–35.

Hardie, J., Nottingham, S. F., Powell, W., and Wadhams, L. J. (1991). Synthetic aphid sex pheromone lures female parasitoids. *Entomol. Exp. Appl.*, **61**, 97–9.

Hogge, L. R., Reed, D. W., Underhill, E. W., and Haughn, G. W. (1988). HPLC separation of glucosinolates from leaves and seeds of *Arabidopsis thaliana* and their identification using thermospray liquid chromatography/mass spectrometry. *J. Chromatogr. Sci.*, **26**, 551–6.

Jensen, S. R., Kirk, O., and Nielsen, B. J. (1989). Biosynthesis of the iridoid glucoside cornin in *Verbena officinalis. Phytochemistry*, **28**, 97–105.

Jones, J. T., Perry, R. N., and Johnston, M. R. L. (1991). Electrophysiological recordings of electrical activity and responses to stimulants from *Globodera rostochiensis* and *Syngamus trachea. Revue de Nématologie*, **14**, 467–73.

Jutsum, A. R. and Gordon, R. F. S. (eds) (1989). *Insect pheromones in plant protection.* Wiley, Chichester.

Milford, G. F. J., Fieldsend, J. K., Porter, A. J. R., Rawlinson, C. J., Evans, E. J., and Bilsborrow, P. E. (1989). Changes in glucosinolate concentrations during the vegetative growth of single- and double-low cultivars of winter oilseed rape. *Asp. Appl. Biol.*, **23**, 83–90.

Nottingham, S. F., Hardie, J., Dawson, G. W., Hick, A. J., Pickett, J. A., Wadhams, L. J., and Woodcock, C. M. (1991). Behavioral and electrophysiological responses of aphids to host and nonhost plant volatiles. *J. Chem. Ecol.*, **17**, 1231–42.

Pham-Delègue, M. H., Blight, M. M., Le Métayer, M., Marion-Poll, F., Picard, A. L., Pickett, J. A., *et al.* (1992). Plant chemicals involved in honeybee–rapeseed relationships: behavioural, electrophysiological and chemical studies. In *Proc. 8th Int. Symp. on Insect Plant Relationships, Wageningen, 1992*, pp. 129–30. Kluwer, Dordrecht.

Pickett, J. A. (1989). *Towards zero pesticide residues: the biomanagement of pests and diseases of oilseed rape.* AFRC Institute of Arable Crops Research Report for 1989, pp. 79–82. IACR, Harpenden.

Pickett, J. A. (1990). Gas chromatography-mass spectrometry in insect pheromone identification: three extreme case histories. In *Chromatography and isolation of insect hormones and pheromones* (ed. A. R. McCaffery and I. D. Wilson), pp. 299–309. Plenum, New York/London.

Pickett, J. A. (1991). Pheromones: will their promise in insect pest control ever be achieved? *Bull. Entomol. Res.*, **81**, 229–32.

Pickett, J. A. (1992*a*). Potential of novel chemical approaches for overcoming insecticide resistance. In *Proceedings SCI Symposium. Resistance '91: Achievements and Developments in Combating Pesticide Resistance, July 1991,*

Harpenden, UK (ed. J. Denholm, A. L. Devonshire, and D. W. Hollomon), pp. 354–65. Elsevier Applied Science, London and New York.

Pickett, J. A. (1992*b*). Even safer insecticides? *Chem. Ind.*, **1**, 25–6.

Pickett, J. A. and Griffiths, D. C. (1980). Composition of aphid alarm pheromones. *J. Chem. Ecol.*, **6**, 349–60.

Pickett, J. A. and Stephenson, J. W. (1980). Plant volatiles and components influencing behaviour of the field slug, *Deroceras reticulatum* (Müll). *J. Chem. Ecol.*, **6**, 435–44.

Pickett, J. A. and Woodcock, C. M. (1992). Semiochemicals: molecular determinants of activity and biosynthesis. In *Royal Ent. Soc. Symposium Volume —Insect Molecular Science*, pp. 141–51. Academic Press, New York, San Francisco, London.

Pickett, J. A., Dawson, G. W., Griffiths, D. C., Hassanali, A., Merritt, L. A., Mudd, A., *et al.* (1987). Development of plant-derived antifeedants for crop protection. In *Pesticide science and biotechnology* (ed. R. Greenhalgh and T. R. Roberts), pp. 125–8. Blackwell Scientific Publications, Oxford.

Pickett, J. A., Wadhams, L. J., and Woodcock, C. M. (1991). New approaches to the development of semiochemicals for insect control. *Proceedings of the Conference on Insect Chemical Ecology, Tábor, 1990*, pp. 333–45. Academia Prague and SPB Acad. Publ., The Hague.

Rodriguez, F., Hallahan, D. L., Pickett, J. A., and Camps, F. (1992). Characterization of the Δ^{11}-palmitoyl-CoA-desaturase from *Spodoptera littoralis* (Lepidoptera: Noctuidae). *Insect Biochem. Molec. Biol.*, **22**, 143–8.

Sakan, T., Isoe, S., and Hyeon, S. B. (1970). Chemistry of attractants for Chrysopidae from *Actinidia polygama*. In *Control of insect behaviour by natural products* (ed. D. L. Wood, R. M. Silverstein, and M. Nakajima), pp. 237–47. Academic Press, New York.

Vinson, S. B. (1992). Chemical signals used by parasitoids. In *Proc. 4th European Workshop on Insect Parasitoids, Perugia, 3–5 April 1991* (ed. F. Bin), pp. 15–42. REDIA.

Wadhams, L. J. (1990). The use of coupled gas chromatography: electrophysiological techniques in the identification of insect pheromones. In *Chromatography and isolation of insect hormones and pheromones* (ed. A. R. McCaffery and I. D. Wilson), pp. 289–98. Plenum, New York/London.

5. Metabolic signals in the rhizosphere: catabolism of calystegins and betaines by *Rhizobium meliloti*

ARLETTE GOLDMANN, BRIGITTE MESSAGE, LOIC LECOEUR, MARIANNE DELARUE, MONIQUE MAILLE, and DAVID TEPFER

Laboratoire de Biologie de la Rhizosphère, Institut National de la Recherche Agronomique, F-78026, Versailles Cedex, France

Introduction

Certain small secondary metabolites of plants are implicated in the establishment and maintenance of specific plant–bacterium relationships. Some of these substances induce the transcription of bacterial genes implicated in the plant–bacterium interaction, e.g. virulence genes in the phytopathogen *Agrobacterium tumefaciens* (Stachel *et al.* 1985) and nodulation genes in the nitrogen-fixing symbiont *Rhizobium* (Firmin *et al.* 1986; Peters *et al.* 1986). In the former case, the inducing molecules are monocyclic phenols, such as acetosyringone, released from wounded tissues; in the latter, they are complex phenolic flavones or flavonoids from legume root exudates. These substances also serve as chemoattractants.

Other plant secondary metabolites are implicated in specific nutritional relationships between these bacteria and their hosts. In general, the production of secondary metabolites by plants and their selective catabolism by microorganisms are thought to influence the dynamics of the microbial populations that depend on the plant as a source of energy (Nutman 1965). We have called 'nutritional mediators' such metabolites that, upon release into the soil by certain species of plants, would provide an exclusive energy source to soil microorganisms that contain the genes necessary for their uptake and catabolism (Tepfer *et al.* 1988*b*).

Studies of the specific catabolism of plant secondary metabolites by soil bacteria originally focused on the opines, the substances synthesized in plant tissues genetically transformed by *Agrobacterium* (Tempé and Goldmann 1982). Opines are catabolized by *Agrobacterium* (Tempé *et al.* 1984) and *Pseudomonas* (Rossignol and Dion 1985). Members of the genus *Rhizobium* catabolize compounds produced by or in their legume hosts, homoserine from pea root exudates (Armitage *et al.* 1988) and rhizopine from alfalfa root nodules (Murphy *et al.* 1987). A general search for nutritional mediators focused on two types of plant secondary metabolites catabolized by *R. meliloti*: a family of previously undescribed compounds, called 'calystegins' (since they were first observed in a species belonging to the Convolvulaceae, *Calystegia sepium*), and several betaines produced by one of the legume hosts of *R. meliloti*, *Medicago sativa*. The ability of these molecules to selectively stimulate the growth of particular bacteria suggests that they could play a role in rhizosphere ecology through nutritional selection (Tepfer *et al.* 1988*b*).

The calystegins

Discovery of calystegins

The search for nutritional mediators began by screening a variety of plants for substances produced by subterranean organs in sufficient quantities and with sufficient specificity to make them probable candidates. Crude extracts of roots and leaves were analysed using high voltage paper electrophoresis and silver staining. The calystegins caught the authors' attention because they appeared abundantly in the roots of only three species (out of the 105 species, representing 23 families, that were screened; see Tepfer *et al.* 1988*b*): these are two species of morning glory (*Calystegia sepium* and *Convolvulus arvensis*) and *Atropa belladonna*. In order to have a constant source of calystegins, *C. sepium* roots transformed by *Agrobacterium rhizogenes* were cultured. These cultures produce kilogram quantities of roots in the space of a few weeks from an inoculum consisting of a few grams. Calystegins can be obtained from transformed roots or from rhizomes from pot-grown plants by methods previously used to prepare opines. Opines are thus coextracted and serve as internal standards in catabolic tests.

Chemical structures of calystegins

High pressure liquid chromatography was used to further purify calystegins, and mass spectrometry and nuclear magnetic resonance spectroscopy (NMR) were used to determine the structures. The two spots (calystegins A and B) that were detected on electrophoretograms were separated by column chromatography and analysed using HPLC. (A) is composed of 4 peaks, designated A1–4 and the lower spot (B) 2 peaks, B1–2. A3, B1, and B2 are predominant.

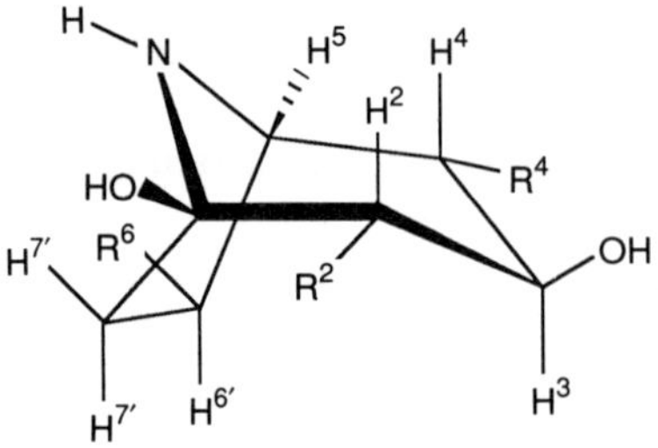

Calystegins	R^2	R^4	R^6	*M*	Formula
B_1	OH	H	OH	175	$C_7H_{13}O_4N$
B_2	OH	OH	H	175	$C_7H_{13}O_4N$
A_3	OH	H	H	159	$C_7H_{13}O_3N$

Fig. 5.1. Structures of calystegins.

Calystegins constitute a family of novel polyhydroxy-nortropanes (Goldmann *et al.* 1990); they are bicyclic, nitrogen-containing, hydroxylated, small molecules (Fig. 5.1). Although similar to tropane, they are characterized by a nortropane ring and an OH-group on the C-4. Other tropane derivatives, such as atropine and cocaine, have dramatic pharmacological properties and may be involved in the plant's interactions with its environment, perhaps as allelopathic substances that discourage foraging. A similarity between calystegins and the tropane alkaloids is thus perhaps not surprising, if calystegins are indeed involved in mediating the plant's relations with other organisms, i.e. soil bacteria. It remains to be seen whether calystegins also have pharmacological actions. Calystegins can be labelled with putrescine, the precursor to tropane, as expected from the structure. The structures of A3, B1, and B2 (see Fig. 5.1) are consistent with the presence of calystegins in *Atropa belladonna*, a species that also produces tropane alkaloids, although calystegins were not detected in other species that produce tropane alkaloids, such as *Hyoscyamus muticus*.

Catabolism of calystegins by *Rhizobium meliloti*

The authors elected to first search for catabolic genes encoding calystegin catabolism (*cac*) in genetically defined soil bacteria previously shown to associate with plants (*Rhizobium*, *Bradyrhizobium*, *Agrobacterium*, *Azospirillum*, and *Pseudomonas*). Forty-four bacterial species were tested for the ability to degrade calystegins. Only one of these, *Rhizobium meliloti* strain 41, was Cac^+. Thus, there was reason to suspect that calystegin catabolism would be limited to specific soil bacteria. It was determined that this catabolism in *R. meliloti* 41 is conferred by a large, previously cryptic plasmid, pRme41a, which is not essential for nodulation and symbiotic nitrogen fixation (Tepfer *et al.* 1988*b*). The organization of the *cac* genes was investigated, using insertional and deletional mutagenesis. They are numerous and spread out over a 30 to 40 kb region of pRme41a (Boivin *et al.* 1990*b*). Bacteria containing *cac* genes can use calystegins as a sole source of carbon and nitrogen (Tepfer *et al.* 1988*b*). Calystegins are thus selective agents in the laboratory. Their selective activity was confirmed by measuring the growth of Cac^+ and Cac^- bacteria in the rhizospheres of whole *C. sepium* plants growing *in vitro* (N. Amarger *et al.*, unpublished results).

Ecological significance of calystegin catabolism

The *cac* genes do not seem to be directly involved in the control of nitrogen-fixing symbiosis, since calystegins were not detected in any of the *R. meliloti* leguminous hosts tested. Furthermore, the symbiotic properties of the strain cured of pRme41a are not altered in *in-vitro* nodulation tests. Nevertheless, the *cac* genes could be involved in the survival of *Rhizobium* (or any bacterium) during its saprophytic (as opposed to symbiotic) life in the soil. If this is true, one would expect to find Cac^+ bacteria in the rhizospheres of plants that synthesize calystegins (Cas^+). The authors have searched for such bacteria both in the rhizospheres of plants that produce and do not produce calystegins. Cac^+ bacteria were indeed found in the rhizospheres of Cas^+ plants and not in the rhizospheres of other species (Tepfer *et al.* 1988*a*). The wild Cac^+ bacteria are a mixture of Gram positives and negatives (including *Pseudomonas* species) and their *cac* genes are only distantly or not at all related to those found in *R. meliloti* strain 41. Therefore, the Cac^+ phenotype is not simply the product of recent sharing of *cac* genes among similar soil bacteria. *Cac* genes probably originated long ago and diverged in diverse bacteria after interspecific transfer or, alternatively, they may have arisen independently in widely different soil bacteria. Specificity in plant–microbe interactions is likely to be the product of coevolution over long periods of interaction. It would be of interest to determine whether *cac* genes are plasmid borne in these wild Cac^+, as they are in *R. meliloti* 41.

Since calystegins constitute a new class of bicyclic polyhydroxy alkaloids, it was logical to determine whether they have the glucosidase-inhibiting activities known for such molecules—see Elbein and Molyneux (1987). They are indeed potent glycosidase antagonists (Molyneux *et al.*, manuscript submitted). This finding suggests new interpretations of their ecological importance.

The betaines

Although the calystegins have the properties expected of nutritional mediators, they are not produced by the legume hosts of *R. meliloti.* It seemed logical that *Rhizobium* would have evolved catabolic functions to benefit from secondary metabolites naturally present in the legume hosts. In an effort to identify possible nutritional mediators in the Leguminoseae, attention was drawn to a class of compounds named 'betaines', which are fully *N*-methylated derivatives of amino acids or related compounds (Fig. 5.2). They are not specific to legumes, but are often present in large quantities in their seeds (Tramontano *et al.* 1986). For example, trigonelline (nicotinic acid *N*-methylbetaine) is the methylated derivative of nicotinic acid, and (although found in a wide range of plant species) trigonelline is particularly abundant in seeds and roots of legumes and has been detected in their root exudates (Ikegami *et al.* 1982). This compound has also been reported to act as a plant hormone regulating the cell cycle in the roots of legumes (Evans and Tramontano 1984). Some betaines, such as glycine betaine and stachydrine (proline betaine or dimethylproline), are metabolites of general interest because they are osmoprotectants essential to the survival of diverse organisms under osmotic stress (Wyn Jones and Storey 1981; Csonka 1989). The authors set out to examine the possibility that betaines from *Medicago sativa* are nutritional mediators.

pSym-encoded catabolism of betaines by *Rhizobium meliloti*

Medicago sativa is one of the plant hosts of *R. meliloti* and synthesizes several betaines, which have been detected in the seeds. Among these are choline, trigonelline, stachydrine, and carnitine (trimethyl-hydroxybutyrobetaine) (Fig. 5.2). *R. meliloti* is known to use choline, glycine betaine (the oxidative product of choline), and stachydrine as sources of energy (Bernard *et al.* 1986; Gloux and Le Rudulier 1989). It was confirmed that strain RCR2011 uses all the betaines described in alfalfa as carbon and nitrogen sources. Efforts to localize genes involved in betaine catabolism

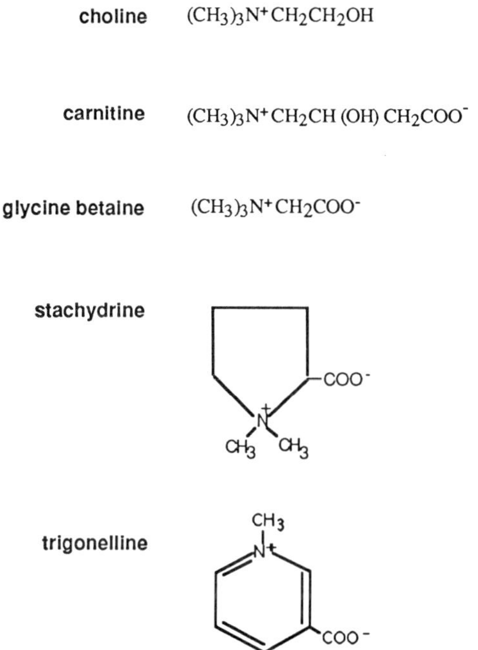

Fig. 5.2. Structures of the betaines. A characteristic trait is the quaternary amine shown as N^+.

were made using *R. meliloti* strain RCR2011 because of the availability of derivatives carrying deletions in and subclones from the pSym, the extra-chromosomal plasmid that carries most of the genes essential for the symbiosis. Strains carrying deletions in pSym were tested for their capacity to catabolize choline, glycine betaine, trigonelline, stachydrine and carnitine. Strain GMI766, a spontaneous mutant of RCR2011, carrying a deletion of more than 100 kb to the right of symbiotic genes *nifKDH*, could not catabolize the betaines, suggesting that to the right of *nifKDH* the symbiotic plasmid encodes functions essential to the use of these compounds (Goldmann *et al.* 1991). To locate the betaine catabolic genes on the pSym, three recombinant plasmids carrying pSym inserts, pGMI467, pGMI469, and pGMI471, were introduced into the strain GMI766 (Fig. 5.3). One of these, pGMI471, restored the catabolism of trigonelline, indicating that trigonelline catabolism genes are located on the pSym in a region between the *nifKDH* genes and the *nok/fixVI'* cluster. Plasmid pGMI467 carried genes important in the catabolism of stachydrine and carnitine, but located to the right of the trigonelline

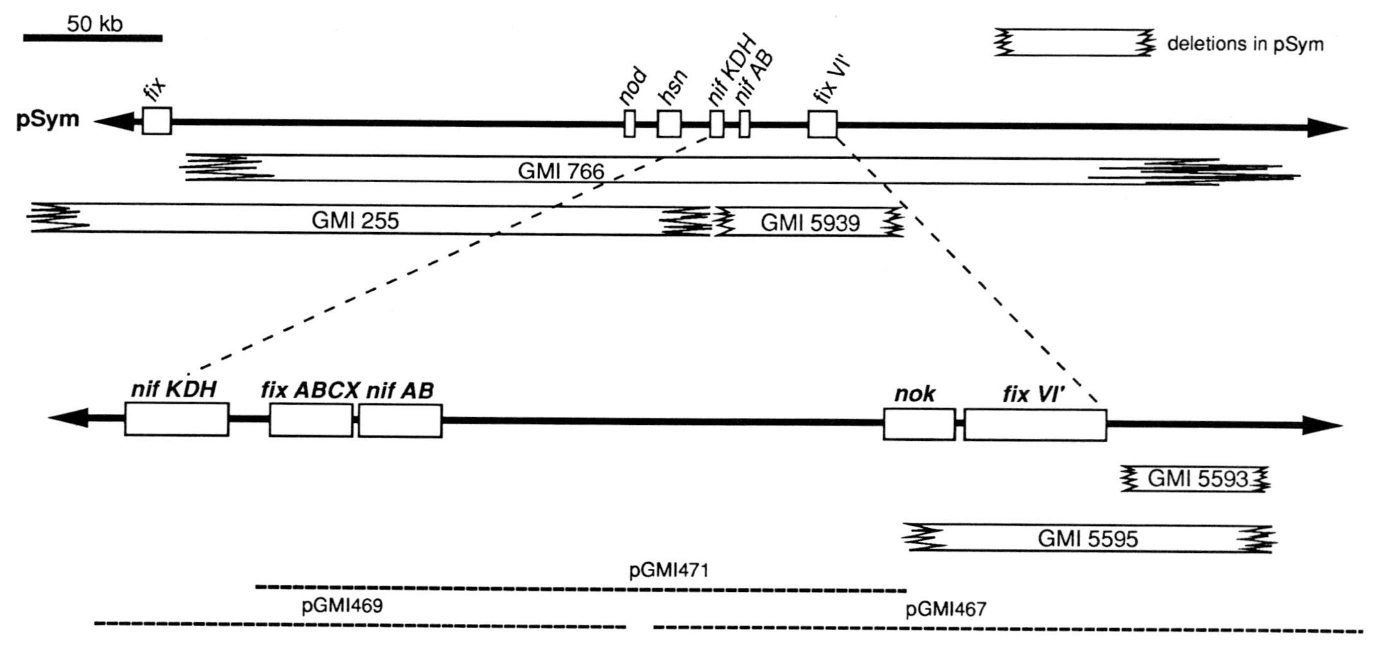

Fig. 5.3. Partial genetic map of pSym from *R. meliloti* strain RCR2011, showing regions containing known genetic loci in boxes with unbroken borders. Deletions are shown below by boxes with broken borders indicating uncertainty in the boundaries. Plasmids used in this study are indicated by dashed lines. *nok* (nodulation kinetics) refers to the *nod* locus defined by Renalier *et al*. (1987). *fixVI'* is a functional *fix* region repeat (Renalier *et al*. 1987).

catabolic genes. Complete carnitine catabolism was not restored, thus this plasmid encodes functions governing only part of the degradative pathway required for its efficient catabolism. Catabolism of choline and glycine betaine could not be restored by these plasmids. Thus, at least some information essential to their utilization is situated on pSym, but less closely linked to the symbiotic genes (located more than 70 kb to the right of *nifKDH*) (Goldmann *et al.* 1991). We conclude that the megaplasmid pSym of *R. meliloti* RCR2011 carries functions essential to the catabolism of the betaines produced by its plant-host *Medicago sativa*.

Biological significance of trigonelline and stachydrine catabolism

The presence of the trigonelline, stachydrine, and carnitine catabolism genes on the pSym plasmid, in a region closely surrounded by symbiotic genes, suggests a relationship between betaine catabolism and nitrogen-fixing symbiosis, thus insertional mutagenesis was used to identify and characterize the genetic loci controlling trigonelline and stachydrine catabolism. Insertions of Tn5-B20, a derivative of Tn5 able to generate transcriptional fusions, showed that the trigonelline catabolism region is a continuous DNA segment of 9 kb located 4 kb downstream of symbiotic genes *nifAB* and *fdxN* genes (Boivin *et al.* 1991). No significant defect in nodule formation or nitrogen fixation was detected with the mutants unable to grow on trigonelline as the sole carbon and nitrogen source, indicating that trigonelline catabolism is not essential for symbiotic effectiveness, at least under laboratory conditions. However, the trigonelline catabolism genes are expressed during all of the stages of symbiosis in alfalfa (Boivin *et al.* 1990*a*). Interestingly, trigonelline induces *nod* functions (Schmidt *et al.* 1986), suggesting that trigonelline catabolism and symbiosis may retain common modes of regulation.

Using Tn5-B20 mutagenesis, the stachydrine genes were localized to a region of approximately 3 kb to the right of *nodD2*. Two insertions were recovered that are induced by stachydrine (unpublished results). Under osmotic stress, catabolism is negatively osmoregulated and stachydrine accumulates in the bacteria (Gloux and Le Rudulier 1989); the authors are currently examining the possibility that genes necessary for stachydrine catabolism are stress regulated. In the future it may be possible to alter stachydrine catabolic functions and improve the bacterium's ability to withstand drought.

Conclusions

Attempting to identify substances produced by plant roots that might act as nutritional mediators of specific plant–bacterium relationships led the authors to discover new compounds, the calystegins, and to re-examine already described substances, the betaines. Genes implicated in these catabolic relations are plasmid-borne. Interestingly, genes controlling the catabolism of metabolites of legumes are carried by the symbiotic plasmid, and genes involved in the catabolism of the compounds of non-hosts are encoded on a different replicon that is not essential for symbiosis. Calystegins have the properties we would ascribe to nutritional mediators. In contrast to the calystegins, whose synthesis is rare in the plant kingdom, trigonelline, stachydrine, carnitine, choline, and glycine betaine are produced by numerous plant species, indicating that if they were nutritional mediators, specificity would have to depend on differences in the degree and timing of their production by host and non-host plants. Although the betaines cannot at present be clearly identified as nutritional mediators, the localization of genes encoding their catabolism in the proximity of genes directly involved in nodulation and nitrogen fixation raises the hypothesis that catabolism genes have a role in nitrogen-fixing symbiosis or were important in its evolution.

If nutritional mediation is a major source of plant–microorganism specificity, manipulation of the synthetic genes in plants and the catabolic genes in microorganisms could lead to the improvement of existing plant–microorganism interactions or to the creation of new ones. Currently, inoculation of plants in the field with a beneficial microorganism, such as *Rhizobium*, is often unsuccessful because the laboratory-grown organism cannot compete in the wild. We would imagine giving it genes to catabolize secondary metabolites specific to the host-plant with which we want it to associate, thereby increasing its chances of survival. Conversely, it may be possible to introduce genes into plants that encode the synthesis and release of novel metabolites into the rhizosphere. Thus new plant/microorganism interactions may someday be created to the benefit of agriculture.

Acknowledgements

We wish to thank C. Boivin, C. Rosenberg, and J. Dénarié (INRA, Toulouse) for general support and helping us with the bacterial genetics. We are indebted to J.-Y. Lallemand and C. Descoins (and their collaborators) for working out the structures of the calystegins. Parts of this work were funded by the French Ministry for Research and Technology.

References

Armitage, J. P., Gallagher, A., and Johnston, A. W. B. (1988). Comparison of the chemotactic behaviour of *Rhizobium leguminosarum* with and without the nodulation plasmid. *Mol. Microbiol.*, **2,** 743–8.

Bernard, T., Pocard, J. A., Perroud, B., and Le Rudulier, D. (1986). Variations in the response of salt-stressed *Rhizobium* strains to betaines. *Arch. Microbiol.*, **143,** 359–64.

Boivin, C., Camut, S., Malpica, C., Truchet, G., and Rosenberg, C. (1990*a*). *Rhizobium meliloti* genes encoding catabolism of trigonelline, a legume secondary metabolite, are induced in the alfalfa rhizosphere, infection threads and nodules. *Plant Cell*, **2,** 1157–70.

Boivin, C., Malpica, C., Rosenberg, C., Goldmann, A., Fleury, V., Maille, M., *et al.* (1990*b*). Catabolism of the plant secondary metabolites calystegins and trigonelline by *Rhizobium meliloti. Symbiosis*, **9,** 147–54.

Boivin, C., Barran, L., Malpica, C., and Rosenberg, C. (1991). Genetic Analysis of a region of the *Rhizobium meliloti* pSym plasmid specifying catabolism of trigonelline, a secondary metabolite present in legumes. *J. Bacteriol.*, **173,** 2809–17.

Csonka, L. N. (1989). Physiological and genetic responses of bacteria to osmotic stress. *Microbiol. Rev.*, 121–7.

Elbein, A. D. and Molyneux, R. J. (1987). The chemistry and biochemistry of simple indolizidine and related polyhydroxy alkaloids. In *Alkaloids: Chemical and biological perspectives* (ed. S. W. Pelletier), Vol. 5, pp. 1–55, Wiley, New York.

Evans, L. S. and Tramontano, W. A. (1984). Trigonelline and promotion of cell arrest in G2 of various legumes. *Phytochemistry*, **23,** 1837–40.

Firmin, J., Wilson, K., Rossen, L., and Johnston, A. W. B. (1986). Flavonoid activation of nodulation genes in *Rhizobium* reversed by other compounds present in plants. *Nature*, **324,** 90–2.

Gloux, K. and Le Rudulier, D. (1989). Transport and catabolism of proline betaine in salt-stressed *Rhizobium meliloti. Arch. Microbiol.*, **151,** 143–8.

Goldmann, A., Milat, M. L., Ducrot, P. H., Lallemand, J. Y., Maille, M., Lepingle, A., *et al.* (1990). Tropane derivatives from *Calystegia sepium. Phytochemistry*, **29,** 2125–7.

Goldmann, A., Boivin, C., Fleury, V., Message, B., Lecoeur, L., Maille, M., and Tepfer, D. (1991). Betaine use by rhizosphere bacteria: genes essential for trigonelline, stachydrine and carnitine catabolism in *Rhizobium meliloti* are located on pSym in the symbiotic region. *Mol. Plant-Microbe Interactions*, **4,** 571–8.

Ikegami, F., Kuo, Y. H., and Lambein, F. (1982). Release of amino acids, trigonelline and related compounds from imbibing *Lathyrus odoratus* seeds. *Arch. Int. Physiol. Biochim.*, **90,** B35–6.

Murphy, P. J., Heycke, N., Banfalvi, Z., Tate, M. E., de Bruijn, F., Kondorosi, A., *et al.* (1987). Genes for the catabolism and synthesis of an opine-like compound in *Rhizobium meliloti* are closely linked and on the Sym plasmid. *Proc. Natl. Acad. Sci. USA*, **84**, 493–7.

Nutman, P. S. (1965). The relation between nodule bacteria and the legume host in the rhizosphere and in the process of infection. In *Ecology of soil-borne pathogens* (ed. K. F. Baker and W. C. Synder), pp. 231–47. University of California Press, Berkeley.

Peters, A., Frost, J., and Long, S. (1986). A plant flavone, luteolin, induces expression of *Rhizobium meliloti* nodulation genes. *Science*, **233**, 977–9.

Renalier, M. H., Batut, J., Ghai, J., Terzaghi, B., Ghérardi, M., David, M., *et al.* (1987). A new symbiotic cluster on the pSym megaplasmid of *Rhizobium meliloti* 2011 carries a functional *fix* gene repeat and a *nod* locus. *J. Bacteriol.*, **169**, 2231–6.

Rossignol, G. and Dion, P. (1985). Octopine, nopaline and octopinic acid utilization in *Pseudomonas. Can. J. Microbiol.*, **31**, 68–74.

Schmidt, J., John, M., Wieneke, U., Krussmann, H. D., and Schell, J. (1986). Expression of the nodulation gene *nod A* in *Rhizobium meliloti* and localization of the gene product in the cytosol. *Proc. Natl. Acad. Sci. USA*, **83**, 9581–5.

Stachel, S., Messens, E., Van Montagu, M., and Zambryski, P. (1985). Identification of the signal molecules produced by wounded plant cells that activate T-DNA transfer in *Agrobacterium tumefaciens. Nature*, **318**, 624–9.

Tempé, J. and Goldmann, A. (1982). Occurence and biosynthesis of opines. In *Molecular biology of plant tumors* (ed. G. Kahl and J. Schell), pp. 451–9. Academic Press, New York.

Tempé, J., Petit, A., and Farrand, S. K. (1984). Induction of cell proliferation by *Agrobacterium tumefaciens* and *Agrobacterium rhizogenes*: a parasite point of view. In *Genes involved in microbe–plant interactions* (ed. D. P. S. Verma and T. H. Hohn), pp. 271–86. Springer-Verlag, New York.

Tepfer, D., Goldmann, A., Fleury, V., Maille, M., Message, B., Pamboukdjian, N., *et al.* (1988*a*). Calystegins, nutritional mediators in plant–microbe interactions. In *Molecular genetics of plant/microbe interactions* (ed. R. Palacios and D. Verma), pp. 139–44. APS Press, St Paul.

Tepfer, D., Goldmann, A., Pamboukdjian, N., Maille, M., Lépingle, A., Chevalier, D., *et al.* (1988*b*). A plasmid of *Rhizobium meliloti* 41 encodes catabolism of two compounds from root exudate of *Calystegia sepium. J. Bacteriol.*, **170**, 1153–61.

Tramontano, W. A., McGinley, P., Ciancaglini, E., and Evans, L. (1986). A survey of trigonelline concentrations in dry seeds of dicotyledoneae. *Environ. Exp. Botany*, **26**, 197–205.

Wyn Jones, R. G. and Storey, R. (1981). Betaines. In *The physiology and biochemistry of drought resistance in plants* (ed. L. G. Paleg and. D. Aspinall), pp. 171–204. Academic Press, Sydney.

6. Practical significance of tritrophic interactions for crop protection

PETER W. PRICE

Department of Biological Sciences, Northern Arizona University, Flagstaff, Arizona 86011–5640, USA

Introduction

Three-trophic-level interactions can be defined as interactions between plants, herbivores and carnivores that pass up and down food webs spanning more than direct linkage between two species. General awareness of the commonness and importance of such interactions was recognized over 10 years ago (Bergman and Tingey 1979; Lawton and McNeill 1979; Price *et al.* 1980). These interactions may involve physical factors such as a trichome layer on the leaf that modifies herbivore behaviour and thus affects the herbivore–carnivore interaction, or chemical factors. The chemical effects may result from constitutive defenses such as alkaloids or tannins, or they may involve volatile compounds that become used as cues or repellents by higher trophic levels. Chemical messages between different species are called **semiochemicals**, and have been divided, for between-species interactions, as follows: **allomones** which are defensive against another species, **kairomones** which act as positive cues to food or prey for the receiver, **synomones** which are mutually beneficial to the sender and receiver, and **apneumones** in which emission of a signal is from a non-living source such as rotting fruit or carcasses that attracts the enemies of inhabitants currently living there (Nordlund 1981). Examples to help clarify these differences would include, for an allomone, a chemical defence emitted by a plant that is repellent to an insect herbivore. A kairomone may be a chemical in a plant that attracts a herbivore to that plant. A synomone could be a chemical released by a plant that acts as a beneficial cue to a predator or parasitoid attacking a herbivore, such that when the herbivore is killed both the emitter of the chemical and the receiver benefit. Apneumones are chemicals emitted from dead organic matter, but as a result predators or parasitoids can more easily find their hosts which attack and live in, for example, dead animals or plant material. Parasitoids of fly larvae in a rotting carcase may find the host larvae in the carcase first by responding to a smell emitted by the rotting body. The specific cue in the smell would then act as an apneumone.

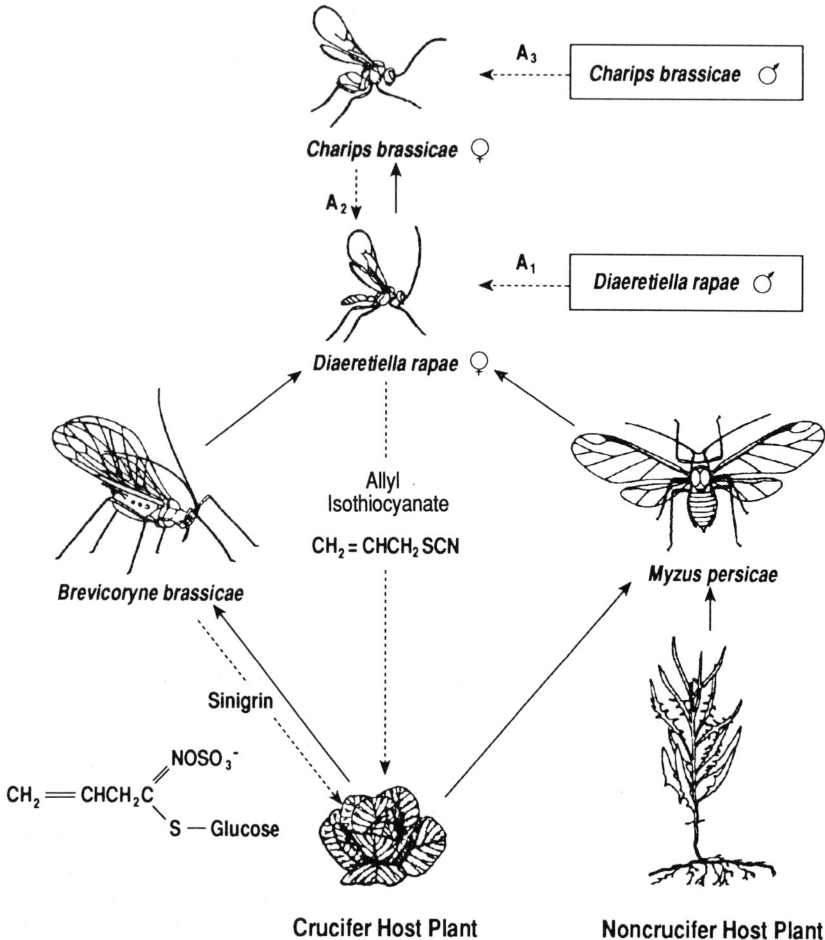

Fig. 6.1. Chemically mediated linkage in a trophic system based on crucifer host-plants in the genus *Brassica*. The specialized aphid *Brevicoryne brassicae* is attracted to the host-plant by the mustard oil glucoside or glucosinolate, sinigrin. The hydrolysis product allyl isothiocyanate is used by the primary parasitoid, *Diaeretiella rapae*, to find the host-plant first, and its aphid host secondarily. Aphids on non-crucifer plants, such as *Myzus persicae*, are not easily discovered by *Diaeretiella* females and are therefore less parasitized, even when close to crucifer hosts. Male *Diaeretiella* locate females via pheromones (A_1), the hyper-parasitoid, *Charips*, uses body odours to locate *Diaeretiella* (A_2, a kairomone), and male *Charips* are attracted to females by pheromones (A_3). Based on Read *et al.* (1970), and reproduced from Price (1984). Copyright © 1984 John Wiley and Sons, Inc. Reprinted by permission of John Wiley and Sons, Inc.

As a result of this terminology semiochemicals can perform two different functions at the same time. Sinigrin may attract the specialist aphid, *Brevicoryne brassicae* to its host-plant, a *Brassica* species, acting as a kairomone (Fig. 6.1). However, when sinigrin is hydrolyzed to allyl isothiocyanate it becomes an attractant to the female primary parasitoid, a parasitic wasp, *Diaeretiella rapae*. When this parasitic wasp causes the death of an aphid, this is beneficial to the host-plant and the semiochemical thus becomes a synomone, with benefits to the first and third trophic levels. In this context the carnivores of herbivores may be regarded as the **bodyguards** of the plant and it is possible to envision the evolution of plants to increase defensive roles by bodyguards (Bentley 1977; Dicke and Sabelis 1988, 1989; Dicke *et al.* 1990*b*; Takabayashi *et al.* 1991*a*).

In these tritrophic systems we see the possibility for complex ecological and evolutionary interactions (Fig. 6.2). There are direct step-by-step

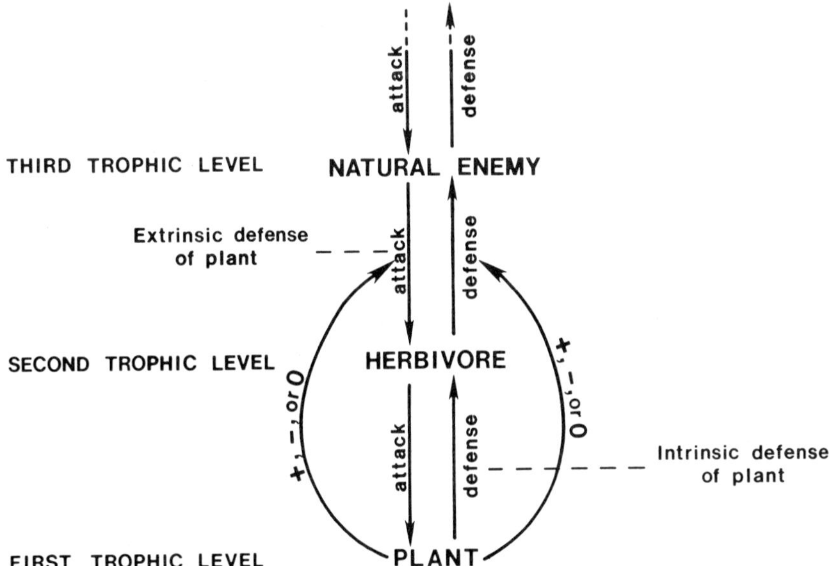

Fig. 6.2. The kinds of tritrophic interactions involving plants, herbivores, and natural enemies. Plant defense against herbivores may involve intrinsic defense or extrinsic defense. Plants may have positive (+), negative (–), or no (0) effect on the interaction between herbivores and their natural enemies. Reproduced with permission from Price (1986). In *Interactions of plant resistance and parasitoids and predators of insects* by D. J. Boethel and R. D. Eikenbary (eds), published in 1986 by Ellis Horwood Ltd, Chichester.

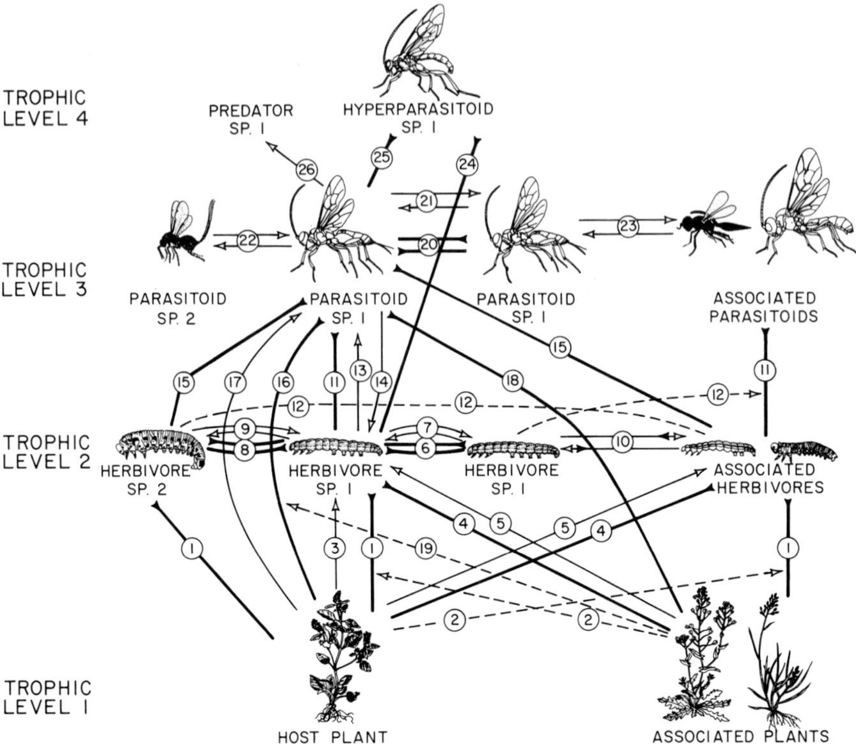

Fig. 6.3. Examples of most of the known cases of chemically mediated interactions in terrestrial trophic systems based on plants. Specific examples of each numbered interaction are provided in the original article. Arrows are placed against the responding organism. Thick solid lines and solid arrows illustrate attraction to a stimulus (e.g. 1, 4, 11, and 24). Thin solid lines and open arrows show repulsion (e.g. 3, 5, 13, 17 and 26). Thin dashed lines indicate indirect effects such as interference with another response (e.g. 2, 12, and 19). Reproduced from Price (1981). Copyright © 1981 John Wiley and Sons, Inc. Reprinted with permission of John Wiley and Sons, Inc.

interactions of attack and defense between trophic levels, but there are also indirect effects of plants, either positive or negative, on the carnivores feeding on herbivores. The extent to which a plant evolves its **intrinsic defenses**, or benefits from **extrinsic defense** from the third trophic level, will depend on many ecological variables, and we have been slow to develop hypotheses on the patterns to be expected in nature.

When tritrophic interactions are involved with semiochemicals, we can think of them as ecological aspects of body odour or BO (Price 1981).

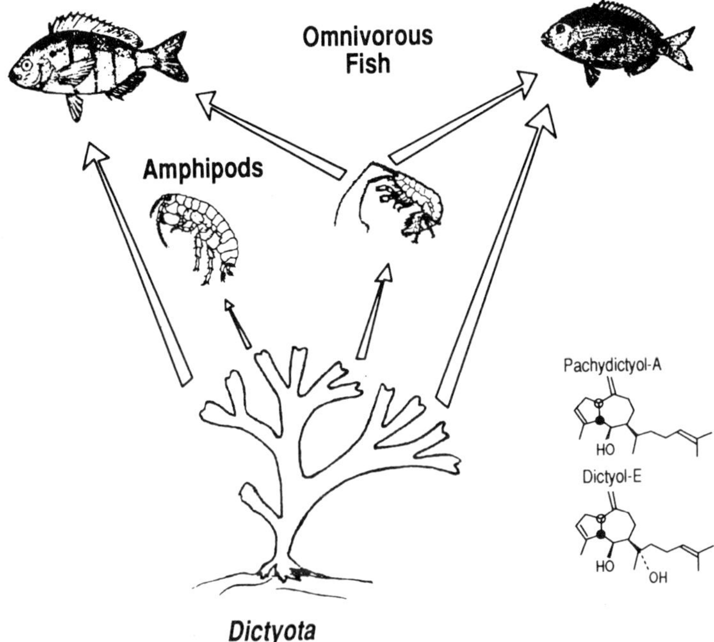

Fig. 6.4. An example of marine tritrophic interactions between a brown seaweed, *Dictyota* amphipod herbivores and omnivorous fish that eat both seaweeds and herbivores (based on Hay *et al.* 1987, 1988). The structures of the defensive compounds in *Dictyota* are also illustrated.

Humans tend to suppress BO or hide it with fragrances, but BO is utilized extensively in natural systems. Specific chemical messages are transmitted and received and such codes may be deciphered by some species and not by others. Codes may be broken, for example by carnivorous species, which aids detection of prey, as for the female's BO utilized by the hyperparasitoid *Charips brassicae* to find *Diaeretiella rapae* (Fig. 6.1).

The ubiquity of three-trophic-level interactions is not adequately appreciated. Almost all food webs based on plants may well involve tritrophic interactions. The range of semiochemical interactions alone is very extensive in terrestrial systems (Fig. 6.3), and may involve plants associated with the host-plant. Such associated plants may act as repellents to herbivores, providing **associational resistance**, or they may be attractive, acting as **lures or decoys** (Tahavanainen and Root 1972; Atsatt and O'Dowd 1976). Similar interactions are being discovered rapidly in shallow marine systems by Hay, Fenical, and associates (e.g. Hay *et al.* 1983, 1987, 1989; Hay 1985, 1986; Hay and Fenical 1988). For example, the diterpene alcohols pachydictyol-A and dictyol-E are synthesized by the brown seaweed *Dictyota dichotoma* (Fig. 6.4). These are unpalatable to herbivorous fish and sea urchins, but not to amphipods. Although the amphipods do not sequester the dictyols, they gain protection from predatory fish by feeding on the unpalatable plant (Hay *et al.* 1987).

Importance in plant protection

Many aspects of the plants we wish to protect are changed by our husbandry. In agriculture, we remove plant species from their natural ecological setting and grow them in environments to which they are aliens (Feeny 1976). We may well select for increased resistance, greater palatability, or productivity, and, almost inevitably, change in ecological setting, and plant characteristics will influence the three-trophic-level interactions based on the crop plant, and change the balance of intrinsic and extrinsic defenses evolved in the natural setting over thousands of generations. We disrupt 'the balance of nature'; we need to be aware of this, and attempt restoration of strong linkages if extrinsic defense has been significant in the plant's ecology.

Restoration and enhancement of extrinsic plant defenses is of pressing concern, for in many localities chemical control of pests has broken down or is not viable in the short term. Pesticide resistance, contamination of the environment, emergence of pests induced by humans, high costs of pesticide development, and long delays in certification all contribute to an obvious conclusion that enhancing natural extrinsic defenses through natural enemies of herbivores, for example, is one of the few long-term solutions.

The study of tritrophic interactions in relation to crop protection is important for many reasons.

1. Understanding the basic biology of the species and their interaction is fundamental to an understanding of how to manipulate them. Such understanding will ultimately reveal hypotheses and theories on patterns and predictability in these detailed, complex, and idiosyncratic systems (e.g. Vet and Dicke 1992).
2. In managed systems, phytochemistry is influenced by artificial selection, fertilizer application, planting density, toxicants, etc., and so we need to understand the ramifications up the trophic system.
3. It is possible to manipulate the behaviour of herbivores by selecting for repellents, antifeedants, toxins, and digestibility reducers.

4. In a tritrophic level context it is most important that we integrate all this information into an understanding of how to enhance the role of carnivores as bodyguards of the plants we cultivate. For this we need to know the physical and chemical properties of plants that foster or impede natural enemy attack rates. We need to know the innate, genetically based responses of the carnivores to body odours of plants and herbivores, and we need to understand the learning capacities of the carnivores to better influence their behaviours for crop protection. We must learn to reinstate, as well as possible, the mutually beneficial linkages between crop-plants and their bodyguards, even though we grow plants as aliens in man-made environments.

Examples and complexities

Plant breeding

The plant breeder, in the absence of concern for three-trophic-level interactions, may wish to increase plant resistance by selecting for, say, higher levels of the alkaloid tomatine in the tomato plant. The vegetative parts would have higher intrinsic defense without impact on fruit quality. However, Campbell and Duffey (1979) found that tomato cultivars with high tomatine were not better defended against the corn earworm, *Heliothis zea*, but its parasitoid *Hyposoter exiguae*, was impacted heavily, by the higher tomatine levels in the host caterpillar. Hence a breeding programme may not increase intrinsic defense while at the same time reducing severely extrinsic defense.

These kinds of unsuspected effects are found commonly in tritrophic interactions. An early case involved corn, *Zea mays*, breeding trials and resistance to the European corn borer, *Ostrinia nubilalis* (Chiang and Holdaway 1960). In this case a more resistant strain to this herbivore, actually conferred on the herbivore more resistance to a protozoan disease caused by *Perezia pyraustae*, so the net result of intrinsic and extrinsic defenses was about the same in the two corn strains. As is so common in these cases, the active compounds and their modes of action were not investigated. As biological control attempts increase using bacterial and fungal pathogens of insect pests it will become increasingly important to select strains of the host-plant compatible with extrinsic defense. For example, the entomopathogenic fungus *Beauveria bassiana* actually grows within the corn plant and provides season-long suppression of European corn borer larvae (Bing and Lewis 1991; Lewis and Bing 1991). Finding mutually highly compatible strains of host-plant and fungal pathogen appears to hold excellent promise for development of an extrinsic defense reliably associated with the crop-plant.

Table 6.1 Ten examples of kairomones, synomones, and apneumones involved with tritrophic interactions in insects and mites

Parasitoid (Pa) or predator (Pr)	Hosts or prey	Stimulant(s)	Action	Source
1. *Microplitis croceipes* (Pa)	*Heliothis virescens*, *H. zea*	13-Methyl hentriacontane	Host-searching stimulant	Jones *et al.* 1971
2. *Trichogramma evanescens* (Pa)	Eggs of many herbivores	Tricosane	Host-searching stimulant	Jones *et al.* 1973
3. *Orgilus lepidus* (Pa)	*Phthorimaea operculella*	*n*-Heptanoic acid	Host-searching stimulant	Hendry *et al.* 1973
4. *Cardiochiles nigriceps* (Pa)	*Heliothis virescens*	12-Methyl hentriacontane	Host-searching stimulant	Vinson *et al.* 1975
		13-Methyl dotriacontane	Host-searching stimulant	
		13-Methyl tritriacontane	Host-searching stimulant	
5. *Diaeretiella rapae* (Pa)	*Brevicoryne brassicae*	Allyl isothiocyanate	Kairomone	Read *et al.* 1970
6. *Cotesia marginiventris* (Pa)	*Spodoptera exigua*	(*Z*)-3-hexanal and many others	Associative learning	Turlings *et al.* 1990

Table 6.1 (*continued*)

Parasitoid (Pa) or predator (Pr)	Hosts or prey	Stimulant(s)	Action	Source
7. *Phytoseiulus persimilis* (Pr)	*Tetranychus urticae*	Linalool, methyl salicylate, (*E*)-β-ocimene etc.	Kairomone	Dicke 1988; Dicke *et al.* 1990*a*,*b*
Phytoseiulus persimilis (Pr)	*Tetranychus urticae*	Linalool, methyl salicylate, (*E*)-β-ocimene etc.	Associative learning	Dicke *et al.* 1990*d*
8. *Biosteres longicaudatus* (Pa)	Tephritid fruit fly	Fermentation products of fruit	Apneumone	Greany *et al.* 1977
9. *Asobara tabida* (Pa)	*Drosophila* spp.	Fresh fermenting sugar/yeast odours	Associative learning	Vet and Opzeeland 1985
10. *Leptopilina heterotoma* (Pa)	*Drosophila* spp.	Later fermenting sugar/yeast odours	Associative learning	Vet and Opzeeland 1985; Vet and Groenewold 1990

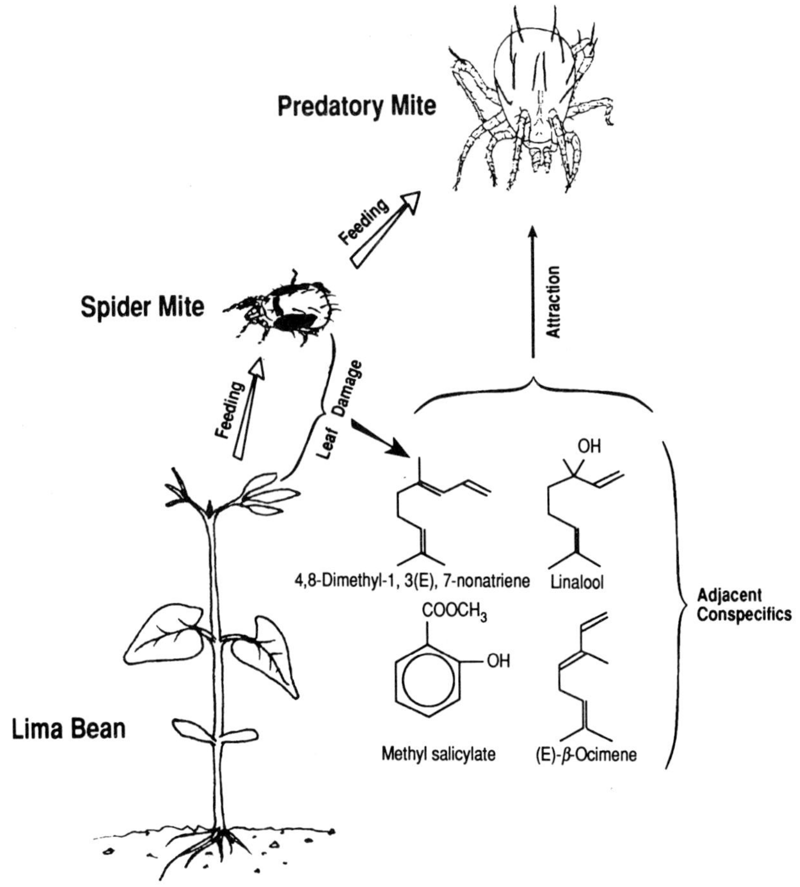

Fig. 6.5. The roles of induced phytochemicals from damaged lima bean leaves caused by spider mites in attracting predatory mites and inducing chemical defense in conspecifics downwind (based on Dicke 1988; Dicke *et al.* 1990*a*,*b*).

Semiochemicals and behavioural responses

In 1975 only four cases were known of the chemical identity of kairomones for parasitoids searching for hosts (Vinson 1975). Many more have been discovered since and a sampling is provided in Table 6.1. Synomones are also included. Both insects and mites provide examples.

The sources of kairomones may derive from the constitutive host-plant chemicals such as for tricosane, heptanoic acid, and the triacontanes, and appear unchanged in body parts, mandibular secretions, or excrement of herbivorous larvae. For example, corn contains tricosane and the corn earworm incorporates the chemical unchanged into its egg. The egg parasitoid, *Trichogramma evanescens*, then uses tricosane as a kairomone to find the host (Lewis *et al.* 1972). In a similar way heptanoic acid in potato tubers appears in the excrement or frass of the potato tuberworm, *Phthorimaea opercullela*, and serves as a kairomone for the parasitoid *Orgilus lepidus* (Hendry *et al.* 1973, 1976).

More recently important discoveries show that phytochemicals induced by damage by spider mites act as synomones attractive to predatory mites. Both mite damage and artificial damage, in experiments conducted by Dicke and associates (Dicke 1988; Dicke *et al.* 1990*a*,*b*), resulted in release of volatile phytochemicals from Lima bean plants (Fig. 6.5). The compounds included terpenes and phenolics. In closed experimental systems some of these chemicals also acted as pheromones inducing heightened defensive phytochemicals in conspecifics downwind (Dicke *et al.* 1990*b*). Plants contribute more to the volatiles than the herbivores (Takabayashi *et al.* 1991*b*).

Some of the pioneering research and some of the fascinating recent developments have been emphasized here, but in all systems in which kairomones and synomones have been identified the potential exists for behavioural manipulation of the plant's extrinsic defenses for the benefit of crop protection. Specific examples will be discussed later in this chapter.

Chemical cues and learning by carnivores

Repeatedly, researchers have discovered that arthropod carnivores of pest herbivores associate chemical cues with host or prey availability, and learn to respond again to such cues. This **associative learning** is of immense importance for applications to crop protection. Phytochemicals, herbivore odours, and even chemicals strange to nature may be active in such learning. Even chocolate mousse and vanilla ice cream could probably induce associative learning.

The study of learning in tritrophic systems is advancing rapidly. Lewis and Tumlinson (1988) showed that the parasitoid *Microplitis croceipes* learns to associate non-volatile chemicals in the faeces of its host, *Heliothis zea*, with volatile odours in the close vicinity, either in the frass, or completely novel odours such as vanilla. Subsequently, Lewis *et al.* (1991) discovered that naive females do not orient to faecal odours, but on contact with a host, antennation of faeces provides a learned stimulus to antennate frass. The chemical identity of the stimulus is 13-methylhentriacontane, derived from the host-plant. This non-volatile host-specific

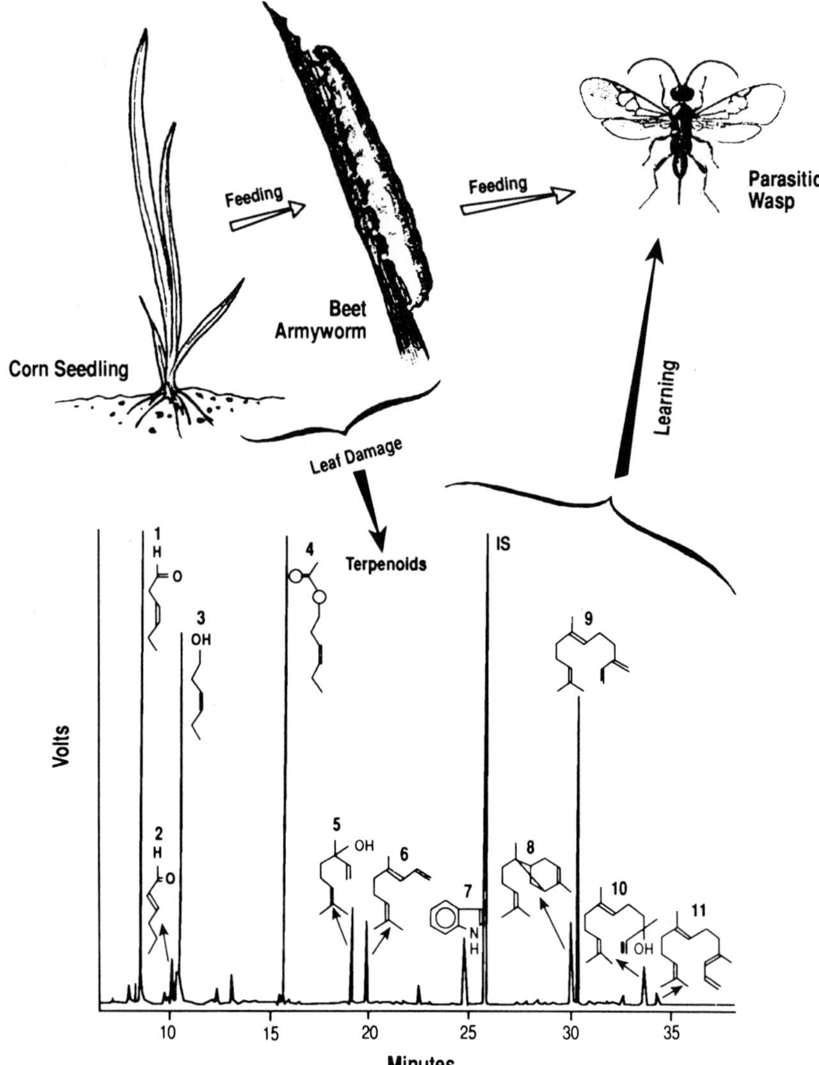

Fig. 6.6. The role of semiochemicals in the associative learning of a parasitoid wasp, *Cotesia marginiventris*. Corn, *Zea mays*, seedlings when damaged by the beet armyworm, *Spodoptera exigua*, which liberates mandibular secretions on to leaves while feeding, release a complex bouquet of volatiles. Female parasitoids learn to associate the odours with the presence of hosts (based on Turlings *et al.* 1990). The numbered compounds are: 1, (*Z*)-3-hexenal; 2, (*E*)-2-hexanal; 3, (*Z*)-3-hexen-1-ol; 4, (*Z*)-3-hexen-1-yl acetate; 5, linalool; 6, (3*E*)-4,8-dimethyl-1,3,7-nonatriene; 7, indole; 8, α-trans-bergamotene; 9, (*E*)-β-farnesene; 10, (*E*)-nerolidol; and 11, (3*E*, 7*E*)-4,8,12-trimethyl-1,3,7,11-tridecatetraene.

recognition signal may be associated with volatiles such as vanilla extract as a longer range chemical cue for host discovery. This parasitoid can also learn to associate odours with adult food sources such as nectar, and can respond appropriately to different odours associated with food for adults and hosts for larvae (Lewis and Takasu 1990). Similar studies have shown the importance of experience and learning also in the parasitoid *Cotesia marginiventris* which parasitizes the beet armyworm, *Spodoptera exiguae* (Turlings *et al.* 1990) and the fall armyworm, *Spodoptera frugiperda* (Dmoch *et al.* 1985, Turlings *et al.* 1990). Parasitoids respond to and learn to associate hosts with contact stimulants from hosts and terpenoid volatiles from damaged leaves. Only leaves damaged by caterpillars, or artificially damaged leaves with the caterpillars' oral secretions applied, provided a host-seeking stimulus for the parasitoids. The terpenoids included hexenals, hexenyl acetate, linalool, dimethylnonatriene, indole, α-*trans*-bergamotene, β-farnesene, and trimethyltridecatetraene (Turlings *et al.* 1990) (Fig. 6.6).

Damaged Lima bean plants, *Phaseolus lunatus*, produce a similar bouquet of volatile terpenoids when damaged by the two-spotted spider mite, *Tetranychus urticae*, which are attractive to a major predatory mite, *Phytoseiulus persimilis* (Dicke *et al.* 1990*a*,*b*). Plants recruit bodyguards. Not only this, but the odours are repellent to conspecific herbivorous mites (Dicke 1986), and the herbivore repellency and carnivore attraction effects are transferred to neighbouring plants either by absorption of bouquet components on the undamaged plant or by induced production of compounds *in situ* (Dicke *et al.* 1990*b*). 'This can be interpreted as a strategy of 'overhearing' other plants' cries for help and subsequently sounding the alarm' (Dicke *et al.* 1990*b*, p. 3109).

For applications to crop protection we therefore have a cornucopia of chemicals of potential value, and a large range of interactions that may be manipulated. Some of this potential is being explored and will be discussed in the next section.

Applications in crop protection

In a tritrophic context relevant to crop protection we have two basic options. First, we can manipulate or select plants in ways that maximize extrinsic defenses, while hopefully not losing any intrinsic defense relevant to the direct plant–herbivore interaction. Second, we can modify the behaviour of natural enemies of herbivores by using synomones, kairomones, and associative learning, to boost the efficacy of the plant's bodyguards. Hopefully, an integrated approach would recognize the need for combining the bottom-up effects of plant quality with the top-down effects from carnivores to jam the herbivorous pests between 'the devil and the deep blue sea' as so aptly put by Lawton and McNeill (1979).

We must recognize that the ecology of all species in crop systems is likely to be radically different from that in which the species evolved, and in which they are probably best adapted. In addition, many techniques are disruptive to the delicate and intricate tritrophic relationships we wish to foster: plowing, fertilizer application, harvesting, watering, and so on. To restore tritrophic linkage requires great knowledge and sophisticated husbandry, not only of the crop, but of the tritrophic relationship. This is a challenge of intense interest because other means of pest control are deteriorating rapidly.

Bottom-up approaches

In crop systems there are at least five levels of variation in the plantings relevant to tritrophic interactions that can play a role in improving use of synomones and kairomones to escalate efficacy of carnivores. From the larger scale to the smaller they are:

1. Plant species mixtures, as in polycultures that promote attraction and maintenance of the herbivores' natural enemies. This involves associational resistance (Nordlund *et al.* 1988).
2. Plant species selection, if choices are available, between plants attractive to carnivores and those that are unattractive (e.g. Nordlund *et al.* 1985).
3. Varieties or cultivars of plant species that differ in synomone production effective in extrinsic defense (e.g. Elzen *et al.* 1985, 1986; Dicke *et al.* 1990*b*).
4. Genotypes of plants within varieties or cultivars no doubt vary in allomone, synomone, and kairomone production, with differing effects up the trophic system, which could be matched to individual variation in populations of natural enemies of herbivores (Lewis *et al.* 1990).
5. Plants may be damaged or undamaged, or induced or uninduced, with different effects on the tritrophic systems, as we have seen in this chapter.

The last example concerns a change in phenotype resulting from damage, in contrast to the other four examples that deal with plant species and genotypic variation. However, the response of plants to damage is no doubt influenced by the genotype, which may contribute more variation than expected or investigated to date.

The potential for using these aspects of extensive variation have not been adequately explored although the evidence of important benefits in agriculture is clear from many reviews (e.g. Lewis *et al.* 1981; Nordlund *et al.* 1981; Boethel and Eikenbary 1986; Barbosa and Letourneau 1988; Dicke *et al.* 1990*c*; Lewis and Martin 1990; Vet and Dicke 1992). Use in pest control has been very limited but the potential for major advances in their use is high (Lewis and Martin 1990).

Top-down approaches

Beyond the need to consider carefully plant variation and its role in crop protection strategies, this must be coupled with equal finesse in husbandry of the third trophic level. Actual tactics may involve four general approaches:

1. Use of attractants and food supplements for natural populations of carnivores (e.g. Hagen *et al.* 1971; Hagen 1986).
2. Release of mass-reared carnivores into crops coupled with chemicals that arrest behaviour and stimulate host seeking (e.g. Lewis *et al.* 1975*a*,*b*, 1979, 1982).
3. Prerelease conditioning of mass-reared carnivores by associative learning with relevant synomones or kairomones (Lewis and Martin 1990).
4. Artificial selection for traits in carnivore populations that improve the interaction of bottom-up and top-down approaches (Lewis *et al.* 1990).

I think we should be optimistic that the use of coupled bottom-up and top-down approaches to crop protection will become widely used within the decade. Practical problems with pesticide resistance and development, environmental contamination, and the increasing economic incentives for effective biological control of pests combine to force funding agencies, research institutions, research scientists, and crop growers to develop the full potentials of three-trophic-level systems in crop protection.

Acknowledgements

I am grateful to the organizers of this symposium for their invitation to participate, and for their financial support. Additional support was provided by the US National Science Foundation grants BSR-8705302 and BSR-9020317.

References

Atsatt, P. R. and O'Dowd, D. J. (1976). Plant defense guilds. *Science*, **193**, 24–9.

Barbosa, P. and Letourneau, D. K. (ed.) (1988). *Novel aspects of insect–plant interactions*. Wiley, New York.

Bentley, B. L. (1977). Extrafloral nectaries and protection by pugnacious bodyguards. *Annu. Rev. Ecol. Syst.*, **8**, 407–27.

Bergman, J. M. and Tingey, W. M. (1979). Aspects of interaction between plant genotypes and biological control. *Bull. Entomol. Soc. Am.*, **25**, 275–9.

Bing, L. A. and Lewis, L. C. (1991). Suppression of *Ostrinia nubilalis* (Hübner) (Lepidoptera: Pyralidae) by endophytic *Beauveria bassiana* (Balsamo) Vuillemin. *Environ. Entomol.*, **20**, 1207–11.

Boethel, D. J. and Eikenbary, R. D. (ed.) (1986). *Interactions of plant resistance and parasitoids and predators of insects*. Ellis Horwood, Chichester.

Campbell, B. C. and Duffey, S. S. (1979). Tomatine and parasite wasps: potential incompatability of plant antibiosis with biological control. *Science*, **205**, 700–2.

Chiang, H. C. and Holdaway, F. G. (1960). Relative effectiveness of resistance of field corn to the European corn borer, *Pyrausta nubilalis*, in crop protection and in population control. *J. Econ. Entomol.*, **53**, 918–24.

Dicke, M. (1986). Volatile spider-mite pheromone and host-plant kairomone, involved in spaced-out gregariousness in the spider mite *Tetranychus urticae*. *Physiol. Entomol.*, **11**, 251–62.

Dicke, M. (1988). *Infochemicals in tritrophic interactions*. Ph.D. thesis. Agricultural University, Wageningen.

Dicke, M. and Sabelis, M. W. (1988). How plants obtain predatory mites as body guards. *Neth. J. Zool.*, **38**, 148–65.

Dicke, M. and Sabelis, M. W. (1989). Does it pay plants to advertize for bodyguards? Towards a cost-benefit analysis of induced synomone production. In *Causes and consequences of variation in growth rate and productivity of higher plants* (ed. H. Lambers, Konings, H., Cambridge, M. L., and Pons, T. L.), pp. 341–58. Academic Press, The Hague.

Dicke, M., van Beek, T. A., Posthumus, M. A., Ben Dom, N., van Bokhoven, H., and de Groot, Æ. (1990*a*). Isolation and identification of volatile kairomone that affects acarine predator–prey interactions: Involvement of host plant in its production. *J. Chem. Ecol.*, **16**, 381–96.

Dicke, M., Sabelis, M. W., Takabayashi, J., Bruin, J., and Posthumus, M. A. (1990*b*). Plant strategies of manipulating predator–prey interactions through allelochemicals: Prospects for application in pest control. *J. Chem. Ecol.*, **16**, 3091–118.

Dicke, M., van Lenteren, J. C., Minks, A. K., and Schoonhoven, L. M. (ed.). (1990*c*). Semiochemicals and pest control—Prospects for new applications. *J. Chem. Ecol.*, **16**, 3017–212.

Dicke, M., van der Maas, K. J., Takabayashi, J., and Vet, L. E. M. (1990*d*).

Learning affects response to volatile allelochemicals by predatory mites. *Proc. Exp. Appl. Entomol.*, **1**, 31–6.

Dmoch, J., Lewis, W. J., Martin, P. B., and Nordlund, D. A. (1985). Role of host-produced stimuli and learning in host selection behavior of *Cotesia* (= *Apanteles*) *marginiventris* (Cresson). *J. Chem. Ecol.*, **11**, 453–63.

Elzen, G. W., Williams, H. J., Bell, A. A., Stipanovic, R. D., and Vinson, S. B. (1985). Quantification of volatile terpenes of glanded and glandless *Gossypium hirsutum* L. cultivars and lines by gas chromatography. *J. Agric. Food Chem.*, **33**, 1079–82.

Elzen, G. W., Williams, H. J., and Vinson, S. B. (1986). Wind tunnel flight responses by humenopterous parasitoid *Campoletis sonorensis* to cotton cultivars and lines. *Entomol. Exp. Appl.*, **42**, 285–9.

Feeny, P. (1976). Plant apparency and chemical defense. In *Biochemical interaction between plants and insects* (ed. J. W. Wallace and R. L. Mansell), pp. 1–40. Plenum, New York.

Greany, P. D., Tumlinson, J. H., Chambers, D. L., and Boush, G. M. (1977). Chemically mediated host finding by *Biosteres* (*Opius*) *longicaudatus*, a parasitoid of tephritid fruitfly larvae. *J. Chem. Ecol.*, **3**, 189–95.

Hagen, K. S. (1986). Ecosystem analysis: Plant cultivars (HPR), entomophagous species and food supplements. In *Interactions of plant resistant and parasitoids and predators of insects* (ed. D. J. Boethel and R. D. Eikenbary), pp. 151–97. Ellis Horwood, Chichester.

Hagen, K. S. Sawall, E. F., and Tassan, R. L. (1971). The use of food sprays to increase effectiveness of entomophagous insects. In *Proc. Tall Timbers Conf. Ecol. Anim. Control by Habitat Management*, Vol. 2, pp. 59–81. Tall Timbers Research Station, Tallahassee, Florida.

Hay, M. E. (1985). Spatial patterns of herbivore impact and their importance in maintaining algal species richness. In *Proc. 5th Int. Coral Reef Cong., Tahiti* Vol. 4, 29–34. Antenne Museum-Ephe, Moorea.

Hay, M. E. (1986). Associational plant defenses and the maintenance of species diversity: Turning competitors into accomplices. *Am. Natur.*, **128**, 617–41.

Hay, M. E. and Fenical, W. (1988). Marine plant–herbivore interactions: The ecology of chemical defense. *Annu. Rev. Ecol. Syst.*, **19**, 111–45.

Hay, M. E., Colburn, T., and Downing, D. (1983). Spatial and temporal patterns in herbivory on a Caribbean fringing reef: The effects of plant distribution. *Oecologia*, **58**, 299–308.

Hay, M. E. Duffy, J. E., Pfister, C. A., and Fenical, W. (1987). Chemical defence against different marine herbivores: Are amphipods insect equivalents? *Ecology*, **68**, 1567–80.

Hay, M. E., Renaud, P. E., and Fenical, W. (1988). Large mobile versus small sedentary herbivores and their resistance to seaweed chemical defenses. *Oecologia*, **75**, 246–52.

Hay, M. E., Pawlik, J. R., Duffy, J. E., and Fenical, W. (1989). Seaweed–herbivore–predator interactions: Host-plant specialization reduces predation on small herbivores. *Oecologia*, **81**, 418–27.

Hendry, L. B., Greany, P. D., and Gill, R. J. (1973). Kariomone mediated host-finding behavior in the parasitic wasp *Orgilus lepidus. Entomol. Exp. Appl.*, **16**, 471–7.

Hendry, L. B., Wickmann, J. K., Hindenlang, D. M., Weaver, K. M., and Korzeniowski, S. H. (1976). Plant—The origin of kairomones utilized by parasitoids of phytophagous insects? *J. Chem. Ecol.*, **2**, 271–83.

Jones, R. L., Lewis, W. J., Bowman, M. C., Beroza, M., and Bierl, B. A. (1971). Host-seeking stimulants for parasite of corn earworm: Isolation, identification, and synthesis. *Science*, **173**, 842–3.

Jones, R. L., Lewis, W. J., Beroza, M., Bierl, B. A., and Sparks, A. N. (1973). Host-seeking (kairomones) for the egg parasite *Trichogramma evanescens. Environ. Entomol.*, **2**, 593–5.

Lawton, J. H. and McNeill, S. (1979). Between the devil and the deep blue sea: On the problem of being a herbivore. In *Population dynamics* (ed. R. M. Anderson, B. D. Turner, and L. R. Taylor), pp. 223–44. Blackwell, Oxford.

Lewis, L. C. and Bing, L. A. (1991). *Bacillus thuringiensis* Berliner and *Beauveria bassiana* (Balsamo) Vuillimen for European corn borer control: Program for immediate and season-long suppression. *Can. Entomol.*, **123**, 387–93.

Lewis, W. J. and Martin, W. R. (1990). Semiochemicals for use with parasitoids: Status and future. *J. Chem. Ecol.*, **16**, 3067–89.

Lewis, W. J. and Takasu, K. (1990). Use of learned odours by a parasitic wasp in accordance with host and food needs. *Nature*, **348**, 635–6.

Lewis, W. J. and Tumlinson, J. H. (1988). Host detection by chemically mediated associative learning in a parasitic wasp. *Nature*, **331**, 257–9.

Lewis, W. J., Jones, R. L., and Sparks, A. N. (1972). A host-seeking stimulant for the egg parasite *Trichogramma evanescens*: Its source and a demonstration of its laboratory and field activity. *Ann. Entomol. Soc. Am.*, **65**, 1087–9.

Lewis, W. J., Jones, R. L., Nordlund, D. A., and Sparks, A. N. (1975*a*). Kairomones and their use for management of entomophagous insects I. Evaluation for increasing rates of parasitization by *Trichogramma* spp. in the field. *J. Chem. Ecol.*, **1**, 343–7.

Lewis, W. J., Jones, R. L., Nordlund, D. A., and Gross, H. R. (1975*b*). Kairomones and their use for management of entomophagous insects: II. Mechanism causing increase in rate of parasitization by *Trichogramma* spp. *J. Chem. Ecol.*, **1**, 349–60.

Lewis, W. J., Beevers, M., Nordlund, D. A., Gross, H. R., and Hagen, K. S. (1979). Kairomones and their use for management of entomophagous insects: IX. Investigations of various kairomone-treatment patterns for *Trichogramma* spp. *J. Chem. Ecol.*, **5**, 673–80.

Lewis, W. J., Nordlund, D. A., and Gueldner, R. C. (1981). Semiochemicals influencing behaviour of entomophages: Roles and strategies for their employment in pest control. *Les Colloques de L'INRA*, **7**, 225–42.

Lewis, W. J., Nordlund, D. A., Gueldner, R. C., Teal, P. E. A., and Tumlinson, J. H. (1982). Kairomones and their use for management of entomophagous insects: XIII. Kairomonal activity for *Trichogramma* spp. of abdominal tips, excretions and a synthetic sex pheromone blend of *Heliothis zea* (Boddie) moths. *J. Chem. Ecol.*, **8**, 1323–31.

Lewis, W. J., Vet, L. E. M., Tumlinson, J. H., van Lenteren, J. C., and Papaj, D. R. (1990). Variations in parasitoid foraging behavior: Essential element of a sound biological control theory. *Environ. Entomol.*, **19**, 1183–93.

Lewis, W. J., Tumlinson, J. H., and Krasnoff, S. (1991). Chemically mediated associative learning: An important function in the foraging behavior of *Microplitis croceipes* (Cresson). *J. Chem. Ecol.*, **17**, 1309–25.

Nordlund, D. A. (1981). Semiochemicals: A review of the terminology. In *Semiochemicals: Their role in pest control* (ed. D. A. Nordlund, R. L. Jones, and W. J. Lewis), pp. 13–28. Wiley, New York.

Nordlund, D. A., Jones, R. L., and Lewis, W. J. (ed.) (1981). *Semiochemicals: Their role in pest control.* Wiley, New York.

Nordlund, D. A., Chalfant, R. B., and Lewis, W. J. (1985). Response of *Trichogramma pretiosum* females to extracts of two plants attacked by *Heliothis zea. Agric. Ecosyst. Environ.*, **12**, 127–33.

Nordlund, D. A., Lewis, W. J., and Altieri, M. A. (1988). Influences of plant produced allelochemicals on the host and prey selection behavior of entomophagous insects. In *Novel aspects of plant–insect interactions* (ed. P. Barbosa and D. K. Letourneau), pp. 65–90. Wiley, New York.

Price, P. W. (1981). Semiochemicals in evolutionary time. In *Semiochemicals: Their role in pest control* (ed. D. A. Nordlund, R. L. Jones, and W. J. Lewis), pp. 251–79. Wiley, New York.

Price, P. W. (1984). The concept of the ecosystem. In *Ecological entomology* (ed. C. B. Huffaker and R. L. Rabb), pp. 19–50. Wiley, New York.

Price, P. W. (1986). Ecological aspects of host plant resistance and biological control: Interactions among three trophic levels. In *Interactions of plant resistance and parasitoids and predators of insects* (ed. D. J. Boethel and R. D. Eikenbary), pp. 11–30. Ellis Horwood, Chichester.

Price, P. W., Bouton, C. E., Gross, P., McPheron, B. A., Thompson, J. N., and Weis, A. E. (1980). Interactions among three trophic levels: Influence of plants on interactions between insect herbivores and natural enemies. *Annu. Rev. Ecol. Syst.*, **11**, 41–65.

Read, D. P., Feeny, P. P., and Root, R. B. (1970). Habitat selection by the aphid parasite *Diaeretiella rapae* (Hymenoptera: Braconidae) and hyperparasite *Charips brassicae* (Hymenoptera: Cynipidae). *Can. Entomol.*, **102**, 1567–78.

Tahavanainen, J. O. and Root, R. B. (1972). The influence of vegetational diversity on the population ecology of a specialized herbivore, *Phyllotreta cruciferae* (Coleoptera: Chrysomelidae). *Oecologia*, **10**, 321–46.

Takabayashi, J., Dicke, M., and Posthumus, M. A. (1991*a*). Induction of indirect defence against spider mites in uninfested Lima bean leaves. *Phytochemistry*, **30**, 1459–62.

Takabayashi, J., Dicke, M., and Posthumus, M. A. (1991*b*). Variation in composition of predator-attracting allelochemicals emitted by herbivore-infested plants: Relative influence of plant and herbivore. *Chemoecology*, **2**, 1–6.

Turlings, T. C. J., Tumlinson, J. H., and Lewis, W. J. (1990). Exploitation of herbivore-induced plant odors by host-seeking parasitic wasps. *Science*, **250**, 1251–3.

Vet, L. E. M. and Dicke, M. (1992). Ecology of infochemical use by natural enemies in a tritrophic context. *Annu. Rev. Entomol.*, **37**, 141–72.

Vet, L. E. M. and Groenewold, A. W. (1990). Semiochemicals and learning in parasitoids. *J. Chem. Ecol.*, **16**, 3119–35.

Vet, L. E. M. and van Opzeeland, K. (1985). Olfactory microhabitat selection in *Leptopilina heterotoma* (Thomson) (Hym.: Eucoilidae), a parasitoid of Drosophilidae. *Neth. J. Zool.*, **35**, 497–504.

Vinson, S. B. (1975). Biochemical coevolution between parasitoids and their hosts. In *Evolutionary strategies of parasitic insects and mites* (ed. P. W. Price), pp. 14–48. Plenum, New York.

Vinson, S. B., Jones, R. L., Sonnet, P. E., Bierl, B. A., and Beroza, M. (1975). Isolation, identification and synthesis of host-seeking stimulants for *Cardiochiles nigriceps*, a parasitoid of the tobacco budworm. *Entomol. Exp. Appl.*, **18**, 443–50.

7. Cyanogenesis and foodplants

ADOLF NAHRSTEDT

Institut für Pharmazeutische Biologie und Phytochemie der Westf. Wilhelms-Universität, D-48149 Münster, Germany

Introduction

Plants constitute a considerable proportion of the human daily diet, with most foodplants providing non-toxic carbohydrates, proteins, or fats; others contain more or less toxic constituents which belong to the so-called secondary metabolites. One class of potentially toxic constituents are the cyanogenic glycosides. Several reviews already exist on their structure variation and distribution in the plant kingdom (Nahrstedt 1992; Seigler 1992) and in insects (Nahrstedt 1988); on their biology (Nahrstedt 1987), biochemistry and metabolism (Conn 1991), and toxicity (Conn 1979; Montgomery 1979; Oke 1980); and on their occurrence in and significance for foodplants of man and animals (Oke 1979; Nartey 1981; Poulton 1983, 1988*a*, 1990). Reviews concerning isolation and structure elucidation (Nahrstedt 1981; Seigler and Brinker, 1992) as well as liberation, detection, and quantification of hydrogen cyanide as the most important product of degradation have been published (Nahrstedt *et al.* 1981; Brinker and Seigler 1989).

The following review will concentrate on the main source of hydrogen cyanide, the cyanogenic glycosides, and will exclude the cyanogenic lipids (Mikolajcak 1977; Seigler 1992) and the pseudocyanogens such as the *N*-methylazoxy-glycosides and β-nitropropionic acid glycosides (Hegnauer 1986; Seigler 1992), both of which also give rise under certain conditions to hydrogen cyanide release from plant material and plant-derived foods. This article will also not refer to the cyclohexanecyano-methylene glucosides (Seigler 1992) which are said to be toxic but do not release hydrogen cyanide and are thus not cyanogenic compounds (Fig. 7.1).

The cyanogenic glycosides have received much attention during the past 190 years, as it was discovered very early (in 1803) that plant tissues can liberate hydrogen cyanide:

1. The first isolation of amygdalin from bitter almonds 160 years ago, stimulated organic chemistry and enzyme chemistry (Nahrstedt 1973 for review).

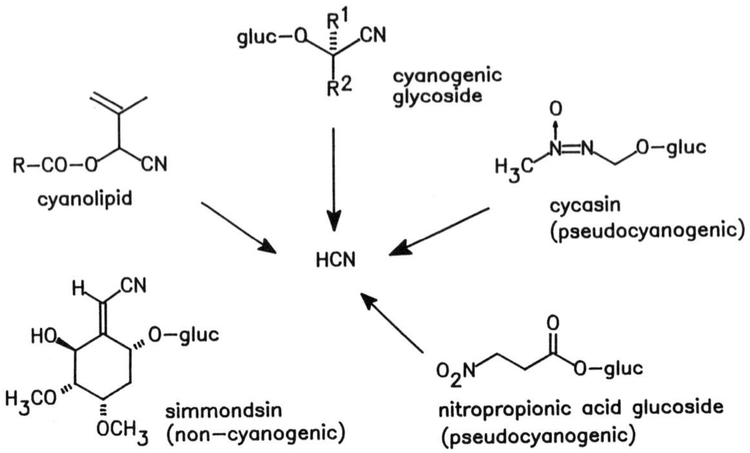

Fig. 7.1. Cyanogenic glycosides and other cyanogenic compounds such as cyanolipids and the so-called pseudocyanogenic substances (the latter release HCN upon alkaline degradation followed by acidification). Nitriles not bearing an α-hydroxy group are usually not cyanogenic.

2. The investigation of the channelled biosynthetic pathways producing the cyanoglucosides dhurrin and linamarin in *Sorghum*, flax (Conn 1991), and *Manihot* (Koch *et al.* 1992) stimulated plant biochemistry and physiology.

3. The use of cyanogenic glycosides as chemical markers for taxonomy (Hegnauer 1962–1992) stimulated plant chemotaxonomy.

4. The cyanogenic glycosides as potentially toxic plant constituents stimulated the research on plant/insect interactions (Nahrstedt 1985).

5. They stimulated the comparison of plant and insect biochemical pathways, as some Lepidoptera produce the same cyanoglycosides as do their host-plants (Nahrstedt 1988).

6. Last but not least, cyanoglycosides are constituents of several foodplants such as cassava or *Sorghum*. This fact has stimulated research on acute and chronic toxicity of hydrogen cyanide, processing of cyanogenic foodplants, biosynthesis and metabolism of these compounds, and also their advantage for producing special kinds of foodstuff. Some of these topics will be discussed in this chapter.

Characteristics of cyanogenic glycosides

The aglycones of the cyanoglycosides are hydroxynitriles (cyanohydrins) which are stabilized by glycosylation; the aglycone-bound sugar is, without exception, glucose with the β-configuration of the glycosidic linkage in all known cyanoglycosides (Seigler 1992). R^1 and R^2 of the aglycone can be aliphatic or aromatic: an asymmetric carbon at position 2 of the nitrile, resulting from different substituents, gives (*R*) and (*S*) configurated epimeric pairs that occur for several cyanogenic glycosides such as the (*R*)-configurated taxiphyllin and the (*S*)-configurated dhurrin, or the (*R*)-configurated lotaustralin and the (*S*)-configurated epilotaustralin; only linamarin, linustatin, and triglochinin do not possess a chiral carbon at the C-2 position (Fig. 7.2).

Most cyanogenic glycosides are fairly stable in cold and hot solvents such as ethanol or methanol, while others isomerize and/or decompose depending on their structure in boiling aqueous solvents or water itself (Nahrstedt 1981). This is especially true for *p*-hydroxylated mandelonitrile glucosides such as dhurrin or taxiphyllin and the dicarboxylic acid glucoside triglochinin which is a constituent of the aerial parts of *Alocasia* and *Colocasia* spp. (Araceae) whose rhizomes are used for starch production (Heywood 1978). These compounds quickly decompose when

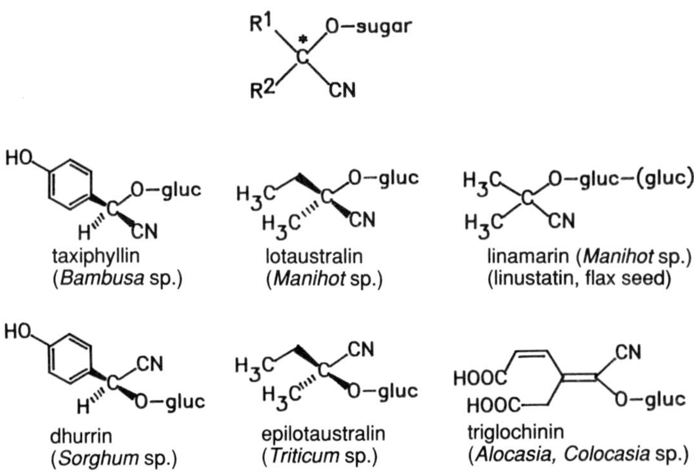

Fig. 7.2. The basic structure of the cyanogenic glycosides. Those of relevance for foodplants are shown as epimeric pairs (with respect to C-2 of the nitrile). Those on the right do not form epimeric pairs.

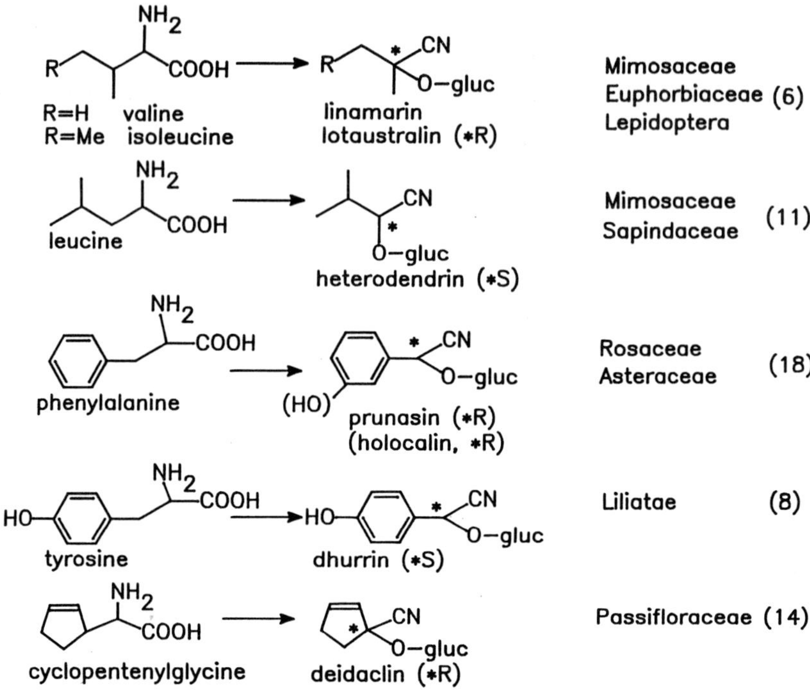

Fig. 7.3. Precursors and some products of the biogenetic pathway leading to cyanogenic glycosides. Taxa of higher plants and insects which contain these groups of cyanoglycosides are shown. The number in parenthesis indicates the number of different known structures belonging to the various biogenetic types (Seigler and Brinker 1992). The putative nicotinic acid derivative, acalyphin, is not shown here.

boiled in water. Taxiphyllin, for instance, is the cyanoglucoside of bamboo, which is used as a vegetable in Far-East countries; bamboo is edible because of taxiphyllin's instability during cooking (Schwarzmaier 1976).

To date about 60 cyanoglycosides are known from several hundred plant species (Seigler 1992)—on the other hand more than 2500 species are known to be cyanogenic in that they release hydrogen cyanide (Hegnauer 1962–1992). All these structures are biogenetically derived from a few amino acids (Fig. 7.3), namely valine, isoleucine, leucine, phenylalanine, and tyrosine, from the non-proteinogenic amino acid cyclopentenylglycin and, probably, from nicotinic acid (for review see Seigler 1992). Growing plant tissues are usually high in cyanoglycosides, although the tissue in which accumulation occurs is not generally the site of biosynthesis, and transport may occur within the entire plant (Selmar and Lieberei 1988; Koch *et al.* 1992).

Intact cyanogenic glycosides in general are bitter but not acutely toxic to higher animals. For instance, during the treatment of cancer patients with pure amygdalin (so-called laetrile), doses up to *c.* 9 g per person (4.5 $g/m^2/day$) were injected intravenously without severe intoxication phenomena (Moertel *et al.* 1981). When the cyanogenic glycosides, however, were ingested orally at doses of 1.5 g/day together with hydrolysing enzymes (and this is usually the case when cyanogenic foodstuffs are consumed), more or less severe intoxications including death are reported (Humbert *et al.* 1977; Sadoff *et al.* 1978; Rubino and Davidoff 1979; Moertel *et al.* 1981).

Liberation of HCN and its toxicity.

What makes cyanogenic glycosides toxic? These compounds often occur together with degrading enzymes, β-glucosidases, in the same plant. Many examples show that enzyme and substrate are separated from each other by being localized in different tissues (Kojima *et al.* 1979), or the β-glucosidases are located in the apoplastic space and the substrate within the cells (Frehner and Conn 1987; Mkpong *et al.* 1990). After disruption of the plant material this compartmentation is destroyed and the substrate can be hydrolysed by more or less specific and effective β-glucosidases to give the corresponding cyanohydrin (for review see Poulton 1990). For the hydrolysis two different mechanisms exist: (a) the simultaneous mechanism in which a β-glucosidase cleaves off the entire diglycoside chain in one step and (b) the sequential mechanism in which two different glucosidases hydrolyse the glycosidic bonds step by step (Poulton 1988*a*). The cyanohydrins themselves are unstable and decompose, depending on the pH (the more alkaline the quicker), to yield a carbonyl component and hydrogen cyanide. Many organisms possess a hydroxynitrile lyase that accelerates this step up to 20-fold in comparison to the spontaneous decomposition (Selmar *et al.* 1989). This effective and usually entirely enzymatically catalysed catabolic pathway makes the organisms cyanogenic (Fig. 7.4). Consequently, hydrogen cyanide is generally the component responsible for intoxications; plants containing more than 20 mg potential cyanide per 100 g fresh weight are considered as toxic (Kingsbury 1964). It is yet not known to what extent carbonyl compounds, aldehydes and ketones, may also contribute to toxicity.

The β-glucosidases usually stem from the ingested plant material; on the other hand reports exist on the involvement of the microflora of the

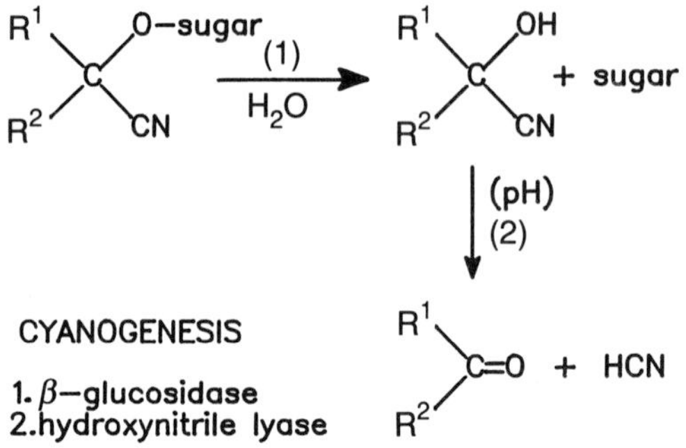

Fig. 7.4. Liberation of hydrogen cyanide from cyanogenic glycosides. Step (1) catalysis by β-glucosidases; step (2) catalysis by hydroxynitrile lyases. Step (2) also occurs spontaneously, being faster the more alkaline the pH.

gastro-intestinal tract of laboratory animals in such processes (Carter *et al.* 1980). In the GI-tract of humans, the activity of β-glucosidases, obviously present in the intestinal mucosa, is limited (Conn 1979). Pure amygdalin, given orally to cancer patients at a dose of three times 500 mg per day (that represents *c.* 75 mg HCN per day), caused a cyanide blood level of *c.* 0.5–2 μg/ml, that is *c.* 2–8 mg HCN/person, and, thus, a subacute level (Moertel *et al.* 1981). It should be pointed out, however, that longer periods of intake of such amounts will lead to chronic cyanide intoxication, which will be described below.

Cyanogenic monoglucosides are absorbed from the GI-tract, whereas diglycosides are not. After consumption of insufficiently processed cassava (see below) by humans its monoglucoside linamarin has been observed in the urine (Rosling 1987). The oral bioavailability of the monoglucoside prunasin in dogs, for example, is approximately 50 per cent (Rauws *et al.* 1982/83), whereas amygdalin is hardly absorbed (Rauws *et al.* 1982*a*). In the GI-tract of the dog, however, and probably also in humans, diglucosides such as amygdalin are hydrolysed to a certain extent to the corresponding monoglucosides (Rauws *et al.* 1982*b*) following the sequential mechanism mentioned above. The monoglucoside then undergoes a far better absorption. This is clearly of toxicological relevance as several tissues of higher animals possess β-glucosidase activity (e.g. the liver, kidney; for review see Schliemann 1983) which may be able to liberate HCN from absorbed cyanoglucosides; no quantitative data, however, are known.

In connection with plant-derived food both the acute and the chronic toxicities of hydrogen cyanide are important (Montgomery 1979; Egekeze and Oehme 1980). Acute toxicity results from oral ingestion of 0.5–3.5 mg cyanide/kg body weight (normally $\geq 5\ \mu g/ml$ blood) and often leads to fatal poisoning when immediately incorporated by inhibiting cellular respiration; in sublethal doses the symptoms are headache, hyperventilation, vomiting, and a general weakness (Egekeze and Oehme 1980). Severe intoxications from foodstuff, however, occur seldom for two reasons:

(1) a high amount of plant material must be ingested in a sufficiently short time together with active β-glucosidases to accumulate lethal cyanide levels, and

(2) the detoxifying mechanisms of the human body are very effective (Egekeze and Oehme 1980).

Nevertheless, consumption of bitter almonds or apricot kernels have indeed caused severe intoxications especially in young children.

Foodplants more often lead to chronic cyanide intoxication resulting from long term exposure to low cyanide levels (Conn 1979; Montgomery 1979). What happens from a pathological point of view? Hydrogen cyanide, once absorbed, is detoxified in several processes (Fig. 7.5), the most important of which is the transfer of sulphur from thiosulphate to form rhodanide or thiocyanide (Egekeze and Oehme 1980). This process

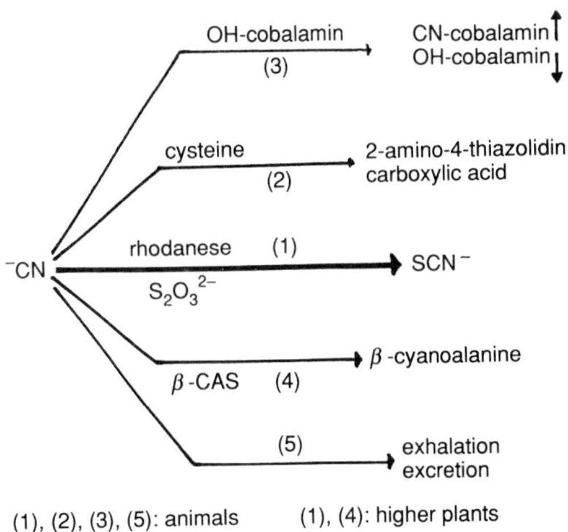

Fig. 7.5. Main detoxication pathways and elimination routes of the cyanide ion. The main route in higher animals is formation of thiocyanate (1), the main route in higher plants is formation of β-cyanoalanine (4) catalysed by β-cyanoalanine synthase (β-CAS).

depends on a good supply of sulphur-containing amino acids and elevates the rhodanide plasma level; this in turn causes endemic goitre, that is also observed after ingestion of brassicacean plants containing glucosinolates.

A second process with pathological relevance is binding of cyanide to hydroxycobalamin to form the physiologically inactive cyanocobalamin. This process is at least partly assumed to cause the degeneration of peripheral nerves leading to neuropathies, that include severe illnesses such as optic nerve atrophy, perceptive deafness, or myelopathy (Conn 1979; Montgomery 1979; Rosling 1987). It is clear from this that cyanide levels and, thus, cyanogenic glycosides should be avoided or diminished in human foodstuff at least for long term ingestion.

Cyanogenic glycosides in foodplants and their variability

Table 7.1 summarizes some foodplants which accumulate considerable amounts of cyanogenic glycosides. These are cassava, the lima bean, and *Sorghum*, used for starch and/or protein supply in large areas of the world. The others are used as delicacies, as crude drugs, as spices, or for production of jam, juices, or brandies. The data show that enzymatically liberated hydrogen cyanide can reach fairly high concentrations up to approximately 1 per cent. For example, 100 g of tops of young flax seedlings contain the amount of bound HCN with a potential of killing 10 human adults.

The data also show that in different parts of a given plant the concentration of hydrogen cyanide varies considerably; thus accumulation of cyanogenic compounds shows a distinct organ to organ variation. For example, in *Sorghum* the fruits are edible whereas young leaves and shoot tips, sometimes used as vegetables, are potentially hazardous (Panasiuk and Bills 1984). The data of *Sorghum* are representative of many cultivated cereals, in that corn, barley, or wheat have fruits almost free of cyanoglycosides, whereas the young shoots contain detectable amounts; the old leaves, however, also do not contain cyanoglycosides (Lehmann *et al.* 1979; Erb *et al.* 1981*a*,*b*). In general cyanogenic glycosides show a remarkable ontogenetic variation with respect to their accumulation in given organs and in the whole plant (Dement and Mooney 1974; Vetter and Haraszti 1977; McDiarmid and Hall 1981; Kaplan *et al.* 1983; Lieberei *et al.* 1986), and even intraindividual heterogeneity occurs as observed for *Trifolium repens* (Till 1987).

Table 7.1 shows that varieties exist with different potencies to accumulate cyanoglycosides as shown for the lima bean or *Sorghum*. Such intraspecific chemical variation resulting in chemical polymorphism was often observed for cyanogenic glycosides (Seigler 1992). Besides genetic factors the cyanoglycoside content of a given plant population depends on diurnal, seasonal, environmental, and nutritional factors which have been more or less discussed in the literature (e.g. for *Sorghum*, Kim 1987; Wheeler *et al.* 1990; Saskatoon serviceberry, Majak *et al.* 1980*a*; arrowgrass, Majak *et al.* 1980*b*; *Holcus* spp., Zandee 1983; *Lotus* spp., Ellis *et al.* 1977; *Trifolium* spp., Foulds and Young 1977). In some populations of a single species cyanogenesis may differ quantitatively between individuals ranging from non-cyanogenic to strongly cyanogenic. Chemical polymorphism has been thoroughly described for *Trifolium repens* (Collinge and Hughes 1982; Fraser and Nowak 1988,) and *Lotus corniculatus* (Jones 1962, 1977; Ramnani and Jones 1984); in both cases investigations on the genetic background have been performed (see below). Polymorphism with respect to cyanogens is also found for many edible cyanogenic plants such as *Sorghum*, *Prunus* spp. or *Manihot*.

In principle in young growing tissues of cyanogenic plants the amount of HCN that can be liberated from cyanoglycosides is potentially high. Many plants accumulate cyanoglycosides in flowers (e.g. *Prunus* spp.), seeds (e.g. *Linum* spp., *Phaseolus* spp., *Prunus* spp., Frehner *et al.* 1990), young shoots (e.g. sprouts of *Sorghum* spp., flax seedlings, especially young leaves of some members of the Araceae), or the outer parts of their organs (bark of cassava tuber, Table 7.1). This suggests one of the ecological functions of these toxic compounds is to confer a relative protection on tissues of high value (generative organs, growing tissues) or the first line of invasion such as the epidermis against herbivores (Nahrstedt 1985).

The Cassava problem and detoxication procedures

One of the most important foodplants is *Manihot esculenta*, whose tubers are rich in starch and are an energy supply for an estimated 300–800 million people in tropical areas of the world. Products prepared from *Manihot esculenta* are cassava, manioc, and tapioca. A comprehensive report exists on cassava from an interdisciplinary workshop held in London in 1973 (Nestel and MacIntyre 1973); a recent short overview was given by Rosling (1987). The daily uptake of cyanide in cassava is responsible for nutritional neuropathies mentioned above.

The cyanide potential of cassava is based on two cyanogenic glucosides, linamarin and lotaustralin, whose concentrations are high in bitter varieties and low in sweet varieties (Sundaresan *et al.* 1987); *c.* 70 per cent of the glucosides are located in the bark of the tubers (Kojima *et al.* 1983). A well investigated β-glucosidase, so-called linamarase, hydrolyses

Table 7.1 Amounts of hydrogen cyanide that are released from cyanogenic glycosides (source), from various organs of some foodplants, by the catalytic action of β-glucosidases (modified after Conn 1979; Poulton 1983)

Foodplant	Source of HCN	Organ	mg HCN/100g f.w.
Cassava	Linamarin	Leaves	77–104
(*Manihot esculenta*)	Lotaustralin	Bark of tuber	69–84
		Tuber: inner part	7–33
Lima bean	Linamarin	Seed Americ. white	10
(*Phaseolus lunatus*)	Lotaustralin	Java coloured	312
		Puerto Rico bl.	400
Sorghum	Dhurrin	Fruits (seeds)	0.6
(*Sorghum bicolor*)		Shoot tips (etiol.)	240
		young green leaves	60
Bamboo	Taxiphyllin	Unripe stem	300
(*Bambusa vulgaris*)		Tops of sprouts	800
Flax	Linustatin	Linseed	21–54
(*Linum usitatissimum*)	Neolinustatin	Linseed cake	50
	Linamarin	Seedling tops	910
	Lotaustralin		
Bitter almond	Amygdalin	Seed	290
(*Prunus amygdalus*)	Prunasin	Young leaves	20

Table 7.1 (*continued*)

Foodplant	Source of HCN	Organ	mg HCN/100g f.w.
Apricot (*Prunus armeniaca*)	Amygdalin Prunasin	Seed	40–400
Cherry (*Prunus avium*)	Amygdalin Prunasin	Seed	*c*. 100
Laurelcherry (*Prunus laurocerasus*)	Prunasin	Leaves	150

substrates to give the cyanohydrins (Eksittikul and Chulavatnatol 1988; Yeoh 1989; Mkpong *et al.* 1990); the hydroxynitrile lyase of the second step (*c.* Fig. 7.4) has also been investigated (Carvalho 1981).

The traditional procedures performed in order to detoxify the tubers of cassava (Nestel and MacIntyre 1973) will be discussed in more detail, as the entire process or parts of it, in principle, are useful to understand how to eliminate cyanoglycosides from other cyanogenic foodstuff, such as linseed cake, lima beans (Ologhobo *et al.* 1984), seeds of rosacean plants, or others. The first step is selection of low cyanoglucoside-producing strains as chemical variation also occurs in *Manihot*. The genetics, however, have not been well investigated; we know from the pioneering work of M. Hughes with *Trifolium repens* that cyanogenesis is controlled by two independent loci, one responsible for production of linamarin and lotaustralin, the other for production of the β-glucosidase (Collinge and Hughes 1982; Hughes *et al.* 1984). Only individuals with a functional allel for cyanoglucoside production produce linamarin and lotaustralin, with homozygotes producing more than double the amount of heterozygotes (Hughes and Stirling 1982). A second step to reduce potential cyanide is peeling; however, most of the β-glucosidase is also located in the living bark tissue, so that degrading activity is also removed (Kojima *et al.* 1983). Initial soaking in water followed by discarding the water is a principle step for reducing cyanoglycosides and break-down products. The glycosides and their products are thus hydrolysed and leached. Subsequent heating evaporates most or almost all toxic components resulting from the glycosides.

Processing of cassava performed mainly in Africa consists of peeling, soaking the whole or more or less sliced tubers in water for a certain period, and then air drying (Monroy-Rivera *et al.* 1990; Tylleskär *et al.* 1992). The procedure, however, does not eliminate most of HCN and cyanogens; this may be the reason why neuropathies occur more often in African areas after cassava consumption. In South America traditional processing today includes: grinding to a fine wet powder that allows optimized contact between degrading enzymes and substrates; expression and discarding of the juice with an appreciable amount of glucosides, cyanohydrins, and HCN; and storage of the residue overnight. Cooking removes residual cyanohydrins and HCN which were produced by the overnight storage (Seigler and Pereira 1981). This procedure allows production of an acceptably secure material; scale up for industrial purpose is possible. Nevertheless, suitable modern procedures for a quick and effective elimination of cyanogens from cassava, which thereby allow large amounts to be processed, are still required.

Many microorganisms are able to degrade cyanogenic glycosides and to detoxify the resulting hydrogen cyanide (Knowles 1976, 1988), which is mainly converted by rhodanese or β-cyanoalanine synthase to rhodanide or β-cyanoalanine, respectively; the latter is responsible for lathyrism (Rosenthal and Bell 1979), the former is goitrogenic (Poulton 1983). In Mozambique and Rwanda cassava is processed by fermentation using an undefined mould, but the result seems to be a high exposure to aflatoxins (Rosling 1987). The fungus *Rhizopus oryzae* infects harvested cassava roots and its spoilage produces linamarase and rhodanese activities (Padmaja and Balagopal 1985). In principle such microorganisms can be used for detoxication of cyanogenic foodstuff (Nartey 1981). The usefulness of microorganisms was again recently investigated in order to reduce the cyanogens, and thereby the bitterness, of apricot kernels and their seed cake via fermentation with the fungus *Rhizopus oligosporus* (Tuncel *et al.* 1990). It was shown that up to 70 per cent of the cyanogens were removed during the entire fermentation process that includes soaking, boiling, and fermentation. However, no information is given on whether the metabolites of HCN produced by the fungus include rhodanide or β-cyanoalanine. Nevertheless, such procedures need further investigation as they may help in the use of large amounts of apricot kernels (a by-product of jam and juice production) as a protein-rich food.

Desired and undesired compounds derived from cyanogenic glycosides

In the context of apricot kernels another toxic compound that arises from cyanogenic glycosides under certain conditions should be mentioned. Apricots, cherries, plums, and related rosacean fruits are used, after fermentation for the production of several types of brandy. Whisky is produced from malted barley after fermentation. The rosacean seeds and germinating barley contain cyanogenic glycosides, the former the well-known amygdalin and prunasin, the latter the leucin-derived epi-heterodendrin (Erb *et al.* 1979). The glycosides are degraded during fermentation to give cyanide, which is distilled and forms, dependent on the ethanol concentration, ethyl carbamate (Wucherpfennig *et al.* 1987; Aylott *et al.* 1990; Cook *et al.* 1990; MacKenzie *et al.* 1990). A postulated formation scheme (Aylott *et al.* 1990) includes the oxidation of cyanide to cyanate, which, via addition of H_2O, forms ethyl carbamate in the presence of ethanol (Fig. 7.6), up to 300 ppb (μg/litre) in some cherry brandies (Wucherpfennig *et al.* 1987), but not exceeding 20 ppb in whisky (Aylott *et al.* 1990). Ethyl carbamate is a carcinogenic chemical; in rats daily application up to 100 μg/kg in drinking water did not lead to tumors, but 500 μg/kg did (Classen *et al.* 1987). Drinking 1 litre

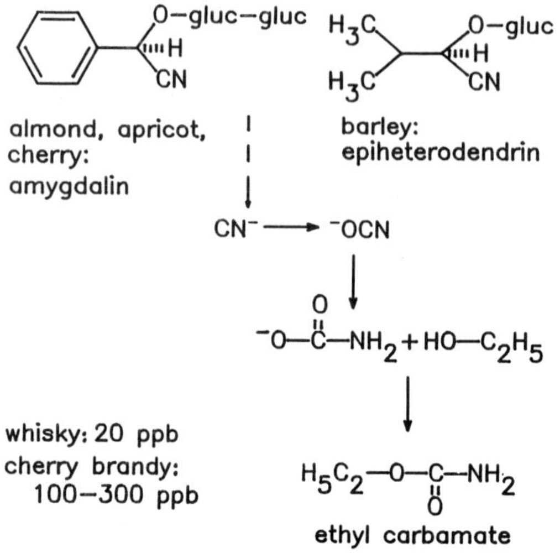
O–gluc–gluc
H
CN
H3C
H3C
O–gluc
H
CN
almond, apricot, cherry: amygdalin
barley: epiheterodendrin
CN⁻ → ⁻OCN
⁻O–C(=O)–NH2 + HO–C2H5
whisky: 20 ppb
cherry brandy: 100–300 ppb
H5C2–O–C(=O)–NH2
ethyl carbamate

Fig. 7.6. Possible formation of carcinogenic ethyl carbamate from cyanide released from cyanogenic glycosides during fermentation for production of cherry brandies and whisky (after Aylott *et al.* 1990).

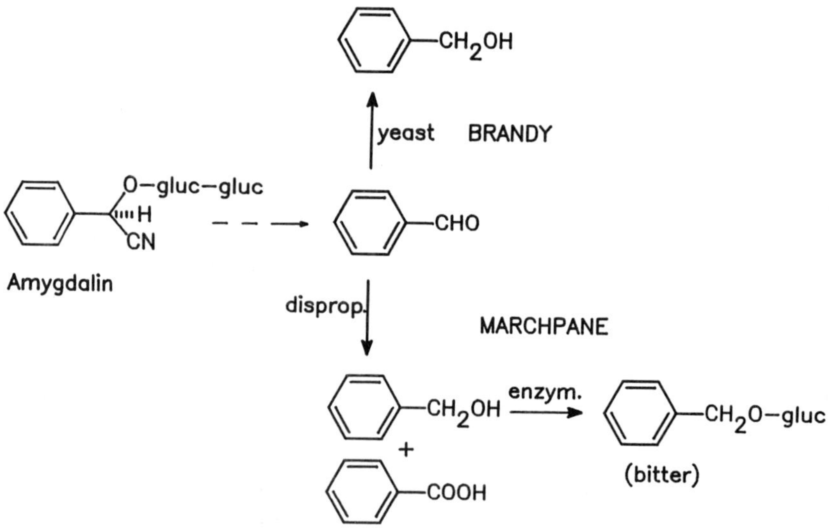
CH2OH
yeast
BRANDY
O–gluc–gluc
H
CN
CHO
Amygdalin
disprop.
MARCHPANE
CH2OH
enzym.
CH2O–gluc
+
COOH
(bitter)

Fig. 7.7. Formation of benzyl alcohol and its bitter β-D-glucoside from amygdalin and prunasin during fermentation processes and marchpane processing.

cherry brandy of high ethyl carbamate level daily, gives 4–5 $\mu g/kg$ to an adult human, which is beyond the safety limit from the carcinogenic risk alone.

In processed seeds of cherry and apricot cyanogenic glycosides such as amygdalin not only contribute to toxicity but also to aromatic compounds and bitterness (Fig. 7.7). It was shown several years ago that benzaldehyde, which arises from amygdalin during fermentation of cherry brandy is reduced by yeast to give benzyl alcohol which is part of the aroma of cherry brandy at 0.15 mmol/litre corresponding to *c.* 17 mg/litre (Tuttas and Nahrstedt 1977). Rjaer and coworkers have shown that benzyl alcohol may also stem from disproportionation of benzaldehyde, which then is glucosylated during the processing of almonds and apricots to produce marchpane (probably by catalysis of the β-glucosidase of emulsin) to its β-glucoside which is particularly bitter (Godtfredsen *et al.* 1978). At this point it should be realized that benzaldehyde itself is the aroma component of marchpane and of laurel cherry used as a spice in the kitchen. Thus, in addition to the important acute and chronic risks which cyanogenic glycosides and their break-down products cause in foodplants, there also is an effect, although less important, on the taste of plant-derived food.

Breeding and its consequences

As mentioned above variability and polymorphism with respect to cyanoglycosides is a characteristic of higher plants. This allows selective breeding for low cyanogenic strains which has been successfully achieved over the centuries for almonds, cassava, and lima beans. Two examples will indicate that edible low cyanogenic plants may suffer from enhanced pressure from herbivores. The larvae of the African insect *Zonocerus variegatus* feed on the leaves of *Manihot esculenta*, the cassava plant. Bernays *et al.* (1977) were able to show that cyanogenesis of the leaves correlates inversely with the nymph's feeding. Thus, low HCN liberation rates cause intense feeding of the late instars of the insect. It is of interest in this context that *Manihot* tubers respond to wounding of the roots with accumulation of cyanogenic compounds, suggesting that linamarin and lotaustralin are involved in a defense reaction (Kojima *et al.* 1983). In an early paper Nayar and Fraenkel (1963) showed that linamarin and lotaustralin, both present in the lima bean (*Phaseolus lunatus*) and in the bush bean (*P. vulgaris*), elicited biting responses in the Mexican bean beetle larvae (*Epilachna varivestis*) at low concentrations, but inhibited biting at higher concentrations. The authors conclude that the high level of cyanogenic compounds in the lima bean

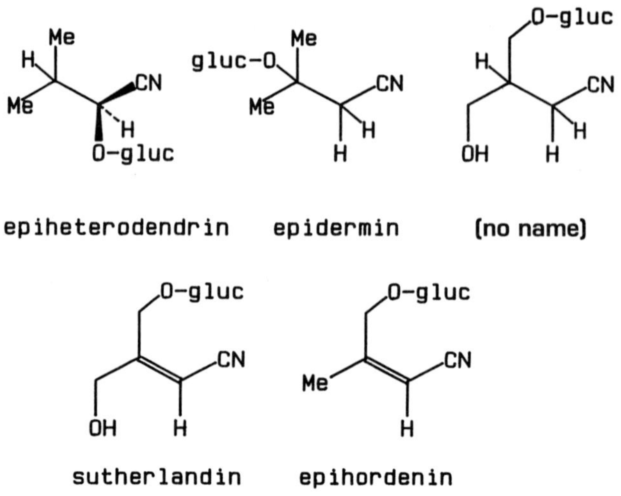

epiheterodendrin epidermin (no name)

sutherlandin epihordenin

Fig. 7.8. Cyanogenic (epiheterodendrin) and non-cyanogenic nitriles located in the epidermis of barley leaves responsible for the compatible reaction between fungi and the barley plant.

causes resistance to the bean beetle; this resistance is not found in the low-cyanogenic bush bean.

Both these examples taken from foodplants highlight problems arising from selection for low-cyanogenic varieties. Experience, however, over many centuries with low-producing strains of cassava, lima bean, and sweet almond has shown that such ecological pressure can well be counteracted by human cultivating practices such as biological plant protection or, if necessary, the controlled use of insecticides.

In some cases, however, highly cyanogenic strains of certain plants seem to be more susceptible to fungal infection than low producers. This was shown for *Hevea brasiliensis*, the rubber tree, where high cyanogenesis correlates with high infection rates of its pest *Microcyclus ulei* (Lieberei 1988). A recent paper presents evidence that HCN, produced during the fungal attack, suppresses the production of scopoletin that acts as a phytoalexin for the rubber plant (Lieberei *et al.* 1989). Thus HCN in this case represents a susceptibility factor in the plant's competition with a fungal parasite. A similar situation may be true for the leaves of the barley plant, *Hordeum vulgare*, in which the cyanogenic glucoside epiheterodendrin (that was mentioned above in combination with ethyl carbamate production in whisky) and related nitriles, including some non-cyanogenic compounds (Fig. 7.8), are responsible for the susceptibility to its pest *Erysiphe graminis* DC f.sp. *hordei* (Pourmohseni 1989). The nitriles account for 90 per cent of the glycosidic fraction in the epidermis of compatible barley leaves, whereas non-compatible ones contain much less. These results corroborate earlier findings obtained for the flax plant (Lüdtke and Hahn 1953) and show that cyanogenic glycosides probably provide no chemical defense against fungi but, in contrast, in some examples favour infection. However, Pandey *et al.* (1981) have shown that high cyanogenic strains (*c.* 190 μg/g f.w.) of *Linum usitatissimum* are resistant, those of lower cyanogenesis (*c.* 150 μg/g) are moderately resistant, and low cyanogenic cultivars (*c.* 20 μg/g) are susceptible to the powdery mildew (*Oidium lini*).

Conclusion

The cyanogenic glycosides influence well-established foodplants and plant-derived food at various levels, as their acute and chronic toxicity requires adapted processing procedures, which also change taste and aroma development, hence they must be considered in breeding and have consequences for plant/herbivore interactions. Continuous work on this selected group of secondary plant constituents will further increase the safety of foodplants with respect to cyanogenesis.

Acknowledgements

Thanks are due to Dr Victor Wray (Braunschweig-Stöckheim) for linguistic help and to Dr Birgit Dräger (Münster) for valuable discussions.

References

Aylott, R. I., Cochrane, G. C., Leonard, M. J., MacDonald, L. S., MacKenzie, W. M., and McNeish, A. S. (1990). Ethylcarbamate formation in grain based spirits. Part I: Post-distillation ethyl carbamate formation in maturing grain whisky. *J. Inst. Brew.*, **96**, 213–21.

Bernays, E. A., Chapman, R. F., Leather, E. M., and McCaffery, A. R. (1977). The relationships of *Zonocerus variegatus* with cassava. *Bull. Ent. Res.*, **67**, 391–404.

Brinker, A. M. and Seigler, D. S. (1989). Methods for the detection and quantitative determination of cyanide in plant materials. *Phytochemical Bull.*, **21**, 24–31.

Carter, J. H., McLafferty, M. A., and Goldman, P. (1980). Role of the gastrointestinal microflora in amygdalin-induced cyanide toxicity. *Biochem. Pharmacol.*, **29**, 301.

Carvalho, F. J. P. C. (1981). Studies on the aliphatic ketone cyanohydrin hydroxynitrile lyase from *Manihot esculenta* Crantz. Ph.D. Dissertation, University of California Davis.

Classen, H.-G., Elias, P. S., and Hammes, W. P. (1987). *Toxikologisch-hygienische Beurteilung von Lebensmittelinhalts- und -zusatzstoffen sowie bedenklicher Verunreinigungen.* p. 120. P. Parey, Berlin.

Collinge, D. B. and Hughes, M. A. (1982). Developmental and physiological studies on the cyanogenic glycosides of white clover *Trifolium repens* L. *J. Expt. Bot.*, **33**, 154–61.

Conn, E. E. (1979). Cyanogenic glycosides. In *Biochemistry of Nutrition* (ed. A. Neuberger and T. H. Jukes), pp. 21–43. University Park Press, Baltimore.

Conn, E. E. (1991). The metabolism of a natural product: Lessons learned from cyanogenic glycosides. *Planta Medica*, **57**, S1–S9.

Cook, R., McCaig, N., McMillan, J. M. B., and Lumsden, W. B. (1990). Ethyl carbamate formation in grain-based spirits. Part III. The primary source. *J. Inst. Brew.*, **96**, 233–44.

Dement, W. A. and Mooney, H. A. (1974). Seasonal variation in the production of tannins and cyanoglucosides in the Chapparal shrub *Heteromeles arbutifolia. Oecologica*, **15**, 65.

Egekeze, J. O. and Oehme, F. W. (1980). Cyanides and their toxicity: a literature review. *The Veterinary Quarterly*, **2**, 104.

Eksittikul, T. and Chulavatnatol, M. (1988). Characterization of cyanogenic β-glucosidase (linamarase) from cassava (*Manihot esculenta* Crantz). *Arch. Biochem. Biophys.*, **266**, 263–9.

Ellis, W. M., Keymer, R. J., and Jones, D. A. (1977). The effect of temperature on the polymorphism of cyanogenesis in *Lotus corniculatus. Heredity*, **38**, 339.

Erb, N., Zinsmeister, H. D., Lehmann, G., and Nahrstedt, A. (1979). Epiheterodendrin a new cyanogenic glucoside from *Hordeum vulgare. Phytochemistry*, **18**, 1515.

Erb, N., Zinsmeister, H. D., Lehmann, G., and Michely, D. (1981*a*). Prussic acid content of cereals in temperate climate regions. *Zeitschrift für Lebensmitteluntersuchung und -Forschung*, **173**, 176–9.

Erb, N., Zinsmeister, H. D., and Nahrstedt, A. (1981*b*). Die cyanogenen Glykoside von *Triticum*, *Secale* und *Sorghum*. *Planta Medica*, **41**, 84–9.

Foulds, W. and Young, L. (1977). Effect of frosting, moisture, stress and potassium cyanide on the metabolism of cyanogenic and acyanogenic phenotypes of *Lotus corniculatus* and *Trifolium repens. Heredity*, **38**, 19.

Fraser, J. and Nowak, J. (1988). Studies on variability in white clover: growth, habits and cyanogenic glucosides. *Ann. Botany*, **61**, 311–18.

Frehner, M. and Conn, E. E. (1987). The linamarin β-glucosidase in Costa Rica wild Lima beans (*Phaseolus vulgaris*) is apoplastic. *Plant Physiol.*, **84**, 1296.

Frehner, M., Scalet, M., and Conn, E. E. (1990) Pattern of the cyanide-potential in developing fruits. *Plant Physiol.*, **94**, 28–34.

Godtfredsen, S. E., Kjaer, A., and Madsen, J. O. (1978). Bitterness in aqueous extracts of aprico kernels. *Acta Chem. Scand.*, **B32**, 5588–92.
Hegnauer, R. (1962–1992). *Chemotaxonomie der Pflanzen*, Vol. 1–10. Birkhäuser Verlag, Basel.
Hegnauer, R. (1986). *Chemotaxonomie der Pflanzen*, Vol. 7, p. 345. Birkhäuser Verlag, Basel.
Heywood, V. H. (1978). *Flowering plants of the world*, p. 309. Oxford University Press.
Hughes, M. and Stirling, J. D. (1982). A study of dominance at the locus controlling cyanoglucoside production in *Trifolium repens* L. *Euphytica*, **31**, 477–83.
Hughes, M. A., Stirling, J. D., and Collinge, D. B. (1984). The inheritance of cyanoglucosides content in *Trifolium repens. Biochem. Genetics*, **22**, 139–51.
Humbert, R. J., Jonathan, H. T., and Kathleen, T. B. (1977). Fatal cyanide poisoning: Accidental ingestion of amygdalin. *J. Am. Med. Assoc.*, **238**, 482.
Jones, D. A. (1962). Selective eating of the acyanogenic form of the plant *Lotus corniculatus* by various animals. *Nature*, **193**, 1109.
Jones, D. A. (1977). On the polymorphism of cyanogenesis in *Lotus corniculatus*. VII: Distribution in Western Europe. *Heredity*, **39**, 27.
Kaplan, M. A. C., Figueiredo, M. R., and Gottlieb, O. R. (1983). Variation in cyanogenesis in plants with season and insect pressure. *Biochemical Systematics and Ecology*, **11**, 367–70.
Kim, J. G. (1987). Effects of morphological characteristics and air temperature on the hydrocyanic potential in *Sorghum* and *Sorghum*–sudangrass hybrids. *Han'guk Ch'uksan Hakhoechi*, **29**, 462–8 (cited from *Chem. Abstr.*, **108**, 164882).
Kingsbury, J. M. (1964). *Poisonous plants of the United States and Canada*. Prentice-Hall, Englewood Cliffs, NJ.
Knowles, C. J. (1976). Microorganisms and cyanide. *Bacteriological Rev.*, **40**, 652.
Knowles, C. J. (1988). Cyanide utilization and degradation by microorganisms. In *Cyanide compounds in biology* (ed. CIBA FOUNDATION), pp. 3–15. Wiley, Chichester.
Koch, B., Nielsen, V. S., Halkier, B. A., Olsen, C. E., and Moller, B. L. (1992). The biosynthesis of cyanogenic glucosides in seedlings of cassava. (Manihot esculenta Crantz). *Arch. Biochem. Biophys.*, **292**, 141–50.
Kojima, M., Poulton, J. E., Thayer, S. S., and Conn, E. E. (1979). Tissue distribution of dhurrin and of enzymes involved in its metabolism in leaves of *Sorghum* bicolor. *Plant Physiol.*, **63**, 1022–8.
Kojima, M., Iwatsuki, N., Data, E. S., Dolores, C., and Villegas, V. (1983). Changes of cyanide content and linamarase activity in wounded cassava roots. *Plant Physiol.*, **72**, 186–9.
Lehmann, G., Zinsmeister, H. D., Erb, N., and Neunhoeffer, O. (1979). Über den Blausäuregehalt von Getreiden und Getreideprodukten. *Zeitschrift für Ernährungswissenschaft*, **18**, 16.

Lieberei, R. (1988). Relationship of cyanogenic capacity (HCN- p) of the rubber tree Hevea brasiliensis to susceptibility of *Microcyclus ulei*, the agent causing South American leaf blight. *J. Phytopathol.*, **122,** 54–67.

Lieberei, R., Nahrstedt, A., Selmar, D., and Gasparotto, L. (1986). Occurrence of lotaustralin in the genus *Hevea* and change of HCN-potential in developing organs of *Hevea brasiliensis*. *Phytochemistry*, **25,** 705–9.

Lieberei, R., Biehl, B., Giesemann, A., and Junqueira, N. T. V. (1989). Cyanogenesis inhibits active defense reactions in plants. *Plant Physiol.*, **90,** 33–6.

Lüdtke, M. and Hahn, H. (1953). Über den Linamaringehalt gesunder und von *Colletotrichum* befallener junger Leinpflanzen. *Biochem. Zeitung*, **324,** 433–42.

MacKenzie, W. M., Clyne, A. H., and MacDonald, L. S. (1990). Ethyl carbamate formation in grain based spirits. Part II: The identification and determination of cyanide related species involved in ethyl carbamate formation in Scotch grain whisky. *J. Inst. Brew.*, **96,** 223–32.

Majak, W., McDiarmid, R. E., and van Ryswyk, A. L. (1980*a*). Factors affecting triglochinin accumulation in arrowgrass. *Plant Physiol. Suppl.*, **65,** 145.

Majak, W., Quinton, D. A., and Broersma, K. (1980*b*). Cyanogenic glycoside levels in Saskatoon Serviceberry. *J. Range Manage.*, **33,** 197–9.

McDiarmid, R. E. and Hall, J. W. (1981). The cyanide potential of saskatoon serviceberry (*Amelanchier alnifolia*) and chokecherry (*Prunus virginiana*). *Can. J. Anim. Sci.*, **61,** 681–6.

Mikolajcak, K. L. (1977). Cyanolipids. *Prog. Chem. Fats and Lipids*, **15,** 97.

Mkpong, O. E., Yan, H., Chism, G., and Sayre, R. T. (1990). Purification, characterization and localization of linamarase in cassava. *Plant Physiol.*, **93,** 176–81.

Moertel, C. G., Ames, M. W., Kovach, J. S., Moyer, T. P., Rubin, J. R., and Tinker, J. H. (1981). A toxicological and pharmacologic study of amygdalin. *J. Am. Med. Assoc.*, **245,** 591.

Monroy-Rivera, J. A., Lebert, A., Marty, C., Muchnik, J., and Bimbenet, J. J. (1990). Determination of cyanogenic compounds in cassava during heated air drying. *Sciences des Aliments*, **10,** 647–58.

Montgomery, D. R. (1979). Cyanogenetic glycosides. In *Handbook of clinical neurology* (ed. P. J. Vinken and G. W. Bruyn), Vol. 36, pp. 515–27. Elsevier, Amsterdam.

Nahrstedt, A. (1973). Cyanogene Glykoside in höheren Pflanzen. *Pharmazie in unserer Zeit*, **2,** 147.

Nahrstedt, A. (1981). Isolation and structure elucidation of cyanogenic glycosides. In *Cyanide in biology* (ed. B. Vennesland, E. E. Conn, C. J. Knowles, J. Westley, and F. Wissing), pp. 145–81. Academic Press, London.

Nahrstedt, A. (1985). Cyanogenic compounds as protecting agents for organisms. *Plant Systematics and Evolution*, **150,** 35–47.

Nahrstedt, A. (1987). Recent developments in chemistry, distribution and biology of the cyanogenic glycosides. In *Biologically active natural products* (ed. K. Hostettmann and P. J. Lea), pp. 213–34. Clarendon Press, Oxford.

Nahrstedt, A. (1988). Cyanogenesis and the role of cyanogenic compounds in insects. In *Cyanide compounds in biology* (ed. CIBA FOUNDATION), Vol. 140, pp. 131–45. Wiley, Chichester.

Nahrstedt, A. (1992). The biology of the cyanogenic glycosides. New developments. In *Nitrogen metabolism of plants* (ed. D. J. Pilbeam and K. Mengel), pp. 249–69. Oxford University Press.

Nahrstedt, A., Erb, N., and Zinsmeister, H. D. (1981). Methods of liberating and estimating of HCN from cyanogenic plant material. In *Cyanide in biology* (ed. B. Vennesland, E. E. Conn, C. J. Knowles, J. Westley, and F. Wissing), pp. 461–71. Academic Press, London.

Nartey, F. (1981). Cyanogenesis in tropical feeds and foodstuffs. In *Cyanide in biology* (ed. B. Vennesland, E. E. Conn, C. J. Knowles, J. Westley, and F. Wissing), pp. 114–32. Academic Press,

Nayar, J. R. and Fraenkel, G. (1963). The chemical basis of the host selection in the mexican bean beetle, *Epilachna varivestis*. *Ann. Entomol. Soc. Am.*, **56**, 174.

Nestel, B. and MacIntyre, R. (1973). *Chronic cassava toxicity*. Proceedings of an interdisciplinary workshop. London 1973. Intern. Development Res. Centre, Box 8500, Ottawa, Canada.

Oke, O. L. (1979). Some aspects on the role of cyanogenic glycosides in nutrition. *Wld. Rev. Nutr. Diet. (Basel)*, **33**, 70–103.

Oke, O. L. (1980). Toxicity of cyanogenic glycosides. *Food Chem.*, **6**, 97–109.

Ologhobo, A. D., Fetuga, B. L., and Tewe, O. O. (1984). The cyanogenic glycoside content of raw and processed lima bean varieties. *Food Chem.*, **13**, 117–28.

Padmaja, G. and Balagopal, C. (1985). Cyanide degradation by *Rhizopus oryzae*. *Can. J. Microbiol.*, **31**, 663–9.

Panasiuk, O. and Bills, D. R. (1984). Cyanide content of *Sorghum* sprouts. *J. Food Sci.*, **49**, 791–3.

Pandey, R. N., Kush, A. K., and Misra, D. P. (1981). Hydrocyanic acid content, a biochemical marker for reactions to powdery mildew in linseed. *Curr. Sci.*, **50**, 902–4.

Poulton, J. E. (1983). Cyanogenic compounds in plants and their toxic effects. In *Handbook of natural toxins. Plant and fungal toxins* (ed. R. F. Keeler and A. T. Tu), Vol. 1, p. 117. Dekker, New York.

Poulton, J. E. (1988*a*). Localization and catabolism of cyanogenic glycosides. In *Cyanide compounds in biology* (ed. CIBA FOUNDATION), Vol. 140. Wiley, Chichester.

Poulton, J. E. (1988*b*). Toxic compounds in plant foodstuffs: cyanogens. *Food Proteins*, 381–401.

Poulton, J. E. (1990). Cyanogenesis in plants. *Plant Physiol.*, **94**, 401.

Pourmohseni, H. (1989). Cyanoglycoside in der Epidermis von Sommergerste und ihre Bedeutung für die quantitative Resistenz bezw. Anfälligkeit gegenüber dem Echten Mehltau (*Erysiphe graminis* DC f.sp *hordei* Marchal). Ph.D. thesis, Göttingen.

Ramnani, A. D. and Jones, D. A. (1984). Genetics of cyanogenesis, cyanoglycoside and linamarase production in the leaves of *Lotus corniculatus* L. *Pakistan J. Bot.*, **16**, 145–54.

Rauws, A. G., Olling, M., and Timmermann, A. (1982*a*). The pharmacokinetics of amygdalin. *Arch. Toxicol.*, **49**, 311–19.

Rauws, A. G., Gramberg, L. G., and Olling, M. (1982*b*). Determination of amygdalin and its major metabolite prunasin in plasma and urine by HPLC. *Pharmaceutisch Weekblad Sci. Ed.*, **4**, 172.

Rauws, A. G., Olling, M., and Timmermann, A. (1982/83). The pharmacokinetics of prunasin, a metabolite of amygdalin. *J. Toxicol.: Clin. Toxicol.*, **19**, 851.

Rosenthal, G. A. and Bell, E. A. (1979). Naturally occuring, toxic non-protein amino acids. In *Herbivores. Their interaction with secondary plant metabolites* (ed. G. A. Rosenthal and D. A. Janzen). Academic Press, New York.

Rosling, H. (1987). *Cassava toxicity and food security.* H. Rosling, Uppsala.

Rubino, M. J. and Davidoff, F. (1979). Cyanide poisoning from apricot seeds. *J. Am. Med. Assoc.*, **241**, 359.

Sadoff, L., Fuchs, K., and Hollander, J. (1978). Rapid death associated with laetrile ingestion. *J. Am. Med. Assoc.*, **239**, 1532.

Schliemann, W. (1983). β-Glucosidasen (EC 3.2.1.21) von Tieren. *Biologische Rundschau.*, **21**, 333–47.

Schwarzmaier, U. (1976). On the cyanogenesis of *Bambusa vulgaris* and *B. guadua. Chem. Ber.*, **109**, 3379.

Seigler, D. S. (1992). Cyanide and cyanogenic glycosides. In *Herbivores: Their interactions with secondary plant metabolites. The chemical participants* (ed. G. A. Rosenthal and M. R. Berenbaum), Vol. 1, pp. 35–77. Academic Press, San Diego.

Seigler, D. S. and Brinker, A. (1992) Characterization of cyanogenic glycosides, cyanolipids, nitroglycosides, organic nitro compounds and nitrile glucosides from plants. In *Methods of plant biochemistry* (ed. P. M. Dey and J. B. Harborne), Vol. 8, pp. 51–131. Academic Press, London.

Seigler, D. S. and Pereira, J. F. (1981). Modernized preparation of cassava in the Llanos of Venezuela. *Econ. Bot.*, **35**, 356.

Selmar, D. and Lieberei, R. (1988). Mobilization and utilization of cyanogenic glycosides—The linustatin pathway. *Plant Physiol.*, **86**, 711.

Selmar, D., Lieberei, R., Biel, B., and Conn, E. E. (1989). α-Hydroxynitrile lyase in *Hevea brasiliensis* and its significance for rapid cyanogenesis. *Physiologica Plantarum*, **75**, 97–101.

Sundaresan, S., Nambisan, B., and Amma, C. S. E. (1987). Bitterness in cassava in relation to cyanoglycoside content. *Ind. J. Agric. Sci.*, **57**, 37–40.

Till, I. (1987). Variability of expression of cyanogenesis in white clover (*Trifolium repens* L.). *Heredity*, **59**, 265.

Tuncel, G., Nout, M. J. R., Brimer, L., and Göktan, D. (1990). Toxicological, nutritional and microbiological evaluation of tempe fermentation with Rhizopus oligosporus of bitter and sweet apricot seeds. *Int. J. Food Microbiol.*, **11**, 337–44.

Tuttas, R. and Nahrstedt, A. (1977). Bedeutung der Cyanglykoside für den

Benzylalkoholgehalt in Kirschbränden. *Zeitschrift für Lebensmitteluntersuchung und -Forschung*, **163**, 257.

Tylleskär, T., Banea, M., Bikangi, N., Cooke, R. D., Poulter, N. H., and Rosling, H. (1992). Cassava cyanogens and konzo, an upper motoneuron disease found in Africa. *Lancet*, **339**, 208–11.

Vetter, J. and Haraszti, E. (1977). Changes in the hydrogen cyanide content of sudan grass (*Sorghum sudanense*) and broomcorn (*Sorghum bicolor* var. *technicum*) during growing season. *Acta Agronomica Academiae Scientiae Hungaricae*, **26**, 15–22.

Wheeler, J. L., Mulcahy, C., Walcott, J. J., and Rapp, G. G. (1990). Factors affecting the hydrogen cyanide potential of forage sorghum. *Aust. J. Agric. Res.*, **41**, 1093–100.

Wucherpfennig, K., Clauss, E., and Konja, G. (1987). Beitrag zur Entstehung des Ethylcarbamats in alkoholischen Getränken auf der Basis der Sauerkirsche "Maraska". *Deutsche Lebensmittel Rundschau*, **83**, 344.

Yeoh, H.-H. (1989). Kinetic properties of β-glucosidase from cassava. *Phytochemistry*, **28**, 721–4.

Zandee, M. (1983). Studies on the *Holcus lanatus–Holcus mollis* complex. Cyanogenesis in the pentaploid cytodeme. *Proceedings of the Koninklijke Nederlandse Akademie van Wetenschappen (Amsterdam)*, **86c**, 255.

8. Digestibility-inhibiting substances from alfalfa

G. MASSIOT and C. LAVAUD

Laboratoire de Pharmacognosie (URA CNRS 492), Faculté de Pharmacie, Université de Reims Champagne-Ardenne, 51, rue Cognacq-Jay, 51096 Reims Cedex, France

---

The main sources of proteins for animal feed are cereals and soya. Only polygastric animals are able to partly degrade cellulose and to make use of non-cereal vegetable proteins. At the present time, soya-cake is mainly imported from the USA and Argentina, and in a quest for alternative and independent sources of proteins, Europe is trying to promote its own cultures of sunflower, rape, lupin, pea, and alfalfa.

Alfalfa is the best source of protein under temperate climatic conditions. Yields of proteins are in the range of 2 tons per hectare, two or three times more than wheat and soya. The land surface cultivated with alfalfa is less than it used to be at the beginning of the century, but it is still important: in the order of 500 000 hectares in 1990. A fifth of the surface is industrial alfalfa, the rest of it is used for making hay. Most of industrial alfalfa is planted in the Champagne area, which is one of the reasons why the authors work on alfalfa. Industrial alfalfa is processed in small factories which make pellets of dried forage from fresh plants. They also prepare what is called Leaf Protein Concentrates (PX) by coagulation of the juice obtained by pressing of the plant. These protein concentrates are well balanced in amino acids, especially rich in lysine, and they contain vitamins and pigments which are greatly appreciated in poultry feeding (Gastineau and de Mathan 1981). The alfalfa industry represents 1000 jobs in the authors' region.

The situation would be ideal—good yields, good quality—if alfalfa and the protein concentrates did not contain deleterious principles referred to as 'digestibility-inhibiting substances'. Their presence at high doses limits the growth of the animals and in case of hens decreases egg laying (Heywang and Bird 1954). As a result, it is not possible to incorporate more than 10 per cent of alfalfa in poultry feed. For years, people have been trying to improve their knowledge of these chemicals and to breed new varieties of alfalfa more suited to the poultry feed market. Chemists have played an important role in this multidisciplinary research and the purpose of this paper is to briefly report on our techniques and results.

In the early days, the relative toxicity of alfalfa was assigned to coumestrol, a substance with oestrogenic properties, however, this was not proven. The presence of an unusual amino acid—L-canavallin—has been mentioned, but this is mostly a seed constituent which is also present in soya (Albourine *et al.* 1991). As everybody knows and must keep in mind, soya is not toxic! The third category of suspected compounds are the saponins and there are good reasons to think that they are biologically active (Cheeke 1980). The main characteristic of saponins is their affinity for cholesterol, which can lead to stable one-to-one complexes. This property is responsible for the behaviour of saponins towards membranes; an extreme example is the interaction of saponins with red blood cells which leads to total disruption of the membrane, a phenomenon known as haemolysis. Formation of cholesterol complexes is a means to separate saponins from complex mixtures. Other properties such as the formation of foams are linked to their amphiphilic character, lipophilic in the terpene part and hydrophilic in the sugar part.

Chemistry of alfalfa saponins dates back to the early 1960s, one of the first major achievements was the structural elucidation of medicagenic acid by Djerassi *et al.* (1957). Medicagenic acid owes its name to the systematic name of alfalfa: *Medicago sativa*. It is a triterpene with two acid and two alcohol functions (Fig. 8.1). Other important triterpenes from alfalfa are hederagenin (Shany *et al.* 1972) which usually belongs to ivy (*Hedera*) and the soyasapogenols from soya (Fig. 8.1).

A 1974 article (Berrang *et al.* 1974) summarized the knowledge on alfalfa sapogenins including at least three unknown triterpenes, one of which was luzernic acid. This was first isolated in 1959 (Livingston 1959), but its structure was elucidated 28 years later (Dimbi *et al.* 1984; Massiot

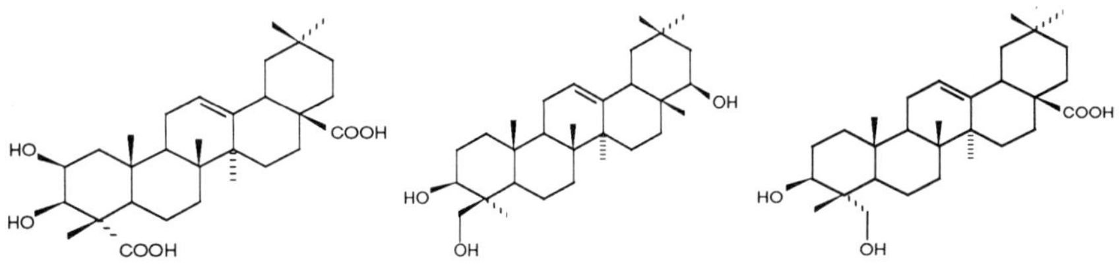

Medicagenic acid Soyasapogenol B Hederagenin

Fig. 8.1. Known sapogenins from alfalfa.

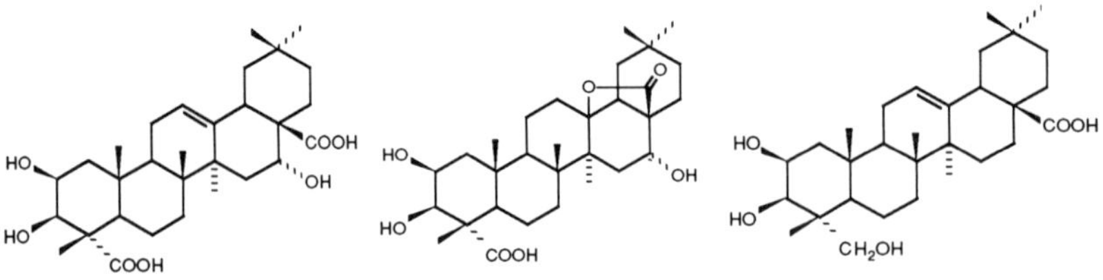

Zahnic acid Luzernic acid Bayogenin

Fig. 8.2. Recently identified sapogenins from alfalfa.

et al. 1988*b*; Fig. 8.2). For the first time, Berrang *et al.* (1974) pointed out differences in composition between alfalfa cultivars.

Morris was the first to isolate and identify an intact saponin from alfalfa. It was the 3-*O*-glucoside of medicagenic acid (**1**) and at the time, the sugar was placed at position 3 to justify resistance to hydrolysis; a partial synthesis later supported this hypothesis (Morris *et al.* 1961; Morris and Tankersley 1963). Morris later isolated another saponin with medicagenic acid, but its structure was more complex (three sugars) and could not be fully elucidated (Morris and Hussey 1965).

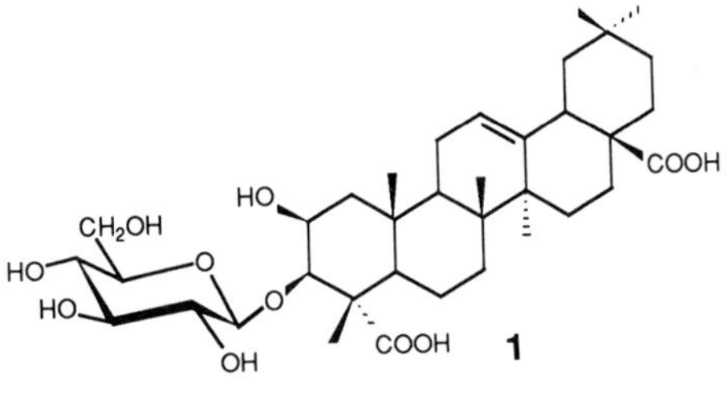

1

The first really complex saponin isolated from alfalfa was a triglucoside of medicagenic acid (**2**) identified by Gestetner in 1971 (Gestetner 1971). The structural elucidation was based on derivatization, partial and total hydrolysis, and subsequent GC-MS identification of the fragments. This follows the techniques developed by Tschesche (Tschesche

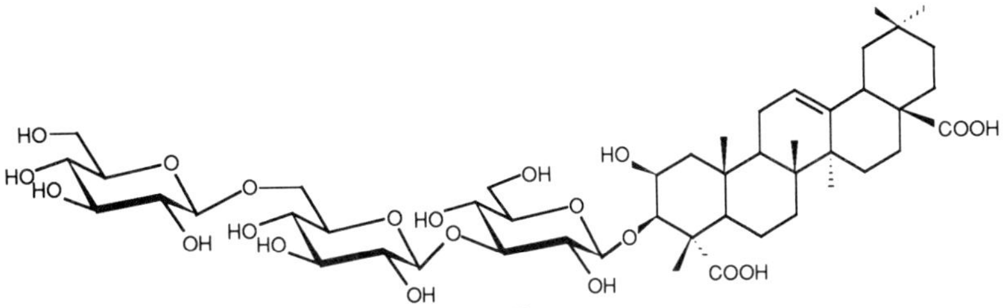

2

and Wulff 1973) and Kitagawa (Kitagawa *et al.* 1976), which seem to be the safest in the field of glycoconjugates. We shall see that new tools have emerged which complement these chemical methods.

The structural elucidation work on alfalfa saponins had a rapid start with a few major compounds being identified, but it seems that activity in the domain ceased around 1975. One of the reasons might have been economical: alfalfa might not have been of interest anymore in the USA where soya was becoming the most important protein source. The second and sad reason was the decease of Gestetner in one of the Middle East wars. He had published five articles and would certainly have published more.

The authors started looking at alfalfa saponins in 1985, in a joint effort with French agronomists and nutritionists from INRA, under the impulsion of the France-Luzerne company. At the same time, but independently, groups in Tashkent under Abubakirov (Timbekova and Abubakirov 1987*a*,*b*; Timbekova *et al.* 1990) and in Poland under Jurzysta (Jurzysta 1973) had resumed the work on alfalfa saponins. These groups used degradation techniques while the authors worked with NMR, the results however, were in accordance. Later on, some activity in the field was developed in Israel (Levy *et al.* 1986) and in Japan with Kitagawa in person abandoning soya for alfalfa (Kitagawa *et al.* 1988).

Our method relies on NMR studies of peracetylated saponins. This philosophy can be summarized in a few points (Massiot *et al.* 1986, 1988*a*, 1991*a*):

1. A peracetylated saponin yields a high resolution ^1H NMR spectrum in organic solutions. Protons are identified by decoupling or 2D COSY experiments, analysis of which starts with the easily identifiable doublets of anomeric protons.

2. The chemical shift of an anomeric proton tells whether a sugar is an ether or an ester, and α or β when its absolute configuration is known.

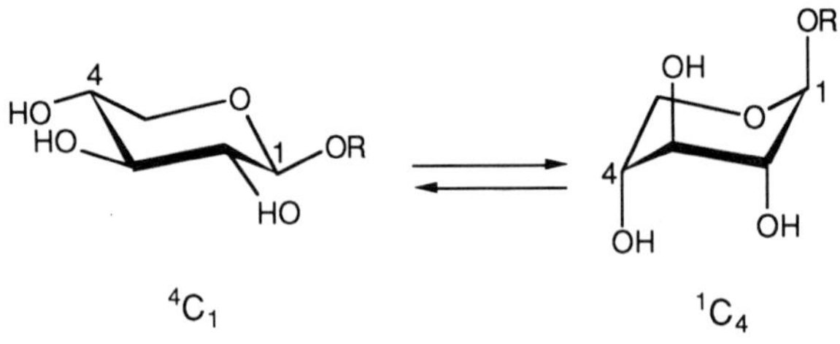

$^4C_1$ $\rightleftharpoons$ $^1C_4$

Fig. 8.3. The $^1C_4$, $^4C_1$ equilibrium in xylose.

3. A sugar may be identified through the analysis of its proton spin–spin couplings. Difficulties arise when there is strong coupling because of overlap or when $^1C_4$ to $^4C_1$ equilibrium modifies coupling constants (Fig. 8.3). In this respect, it must be kept in mind that the anomeric effect is powerful enough to force the C-1–oxygen bond into an axial position whatever the other substituents are. The similarities between the values of $^3J_{ax,eq}$ and $^3J_{eq,eq}$ do not make the identification easy.
4. Branching points on a sugar are characterized by shielded α protons. This feature is the key to our reasoning, and the safest way to detect ether substituents in a sugar.
5. Sequencing requires observation of interactions between distinct elements of two sugars (or sugar and triterpene), this can be:
   (a) long range (^4J) couplings;
   (b) through space interactions (NOE, ROE);
   (c) heteronuclear ^3J C–H couplings (two possibilities per linkage).

   The problem of ester links is best solved with this latter technique. Natural acetates must obviously be treated in a different fashion, i.e. on the native saponin or on a perdeutero-acetylated derivative.

To make these points more clear, we shall now take the major saponin of alfalfa roots as an example of a structural determination. The 300 MHz ^1H NMR spectrum of the peracetylated saponin (**3**) shows good resolution but few directly identifiable resonances (Fig. 8.4). At high field (right part of the spectrum), one sees most of the resonances for the protons of the triterpene. This is the lipophilic part of the molecule and the protons are far away from polar groups and thus 'shielded'. The sharp signals between 0.6 and 1.5 ppm correspond to the angular methyls. Experts are able to recognize a triterpene through the pattern of resonances for the methyls. The set of sharp peaks at 2 ppm are the acetate methyls; these 'non-natural' signals only obscure a small zone of the spectrum where few resonances of the triterpene are hidden. The 3–6 ppm area is the most important since it contains all the sugar renances, the triterpene CH—O resonances, functionalities (such as double bonds) and all the information on the sequence. Anomeric protons of the sugars appear as doublets since these protons have only one neighbour. Three such resonances at 5.7, 4.7, and 4.6 ppm may be directly assigned. The deshielded resonance at 5.7 ppm corresponds (most probably) to a sugar linked to an acid function of the triterpene. Anomeric protons usually resonate in the vicinity of 4.5 ppm. In our example, there is a fourth anomeric proton at 5 ppm; we shall see later how it is detected.

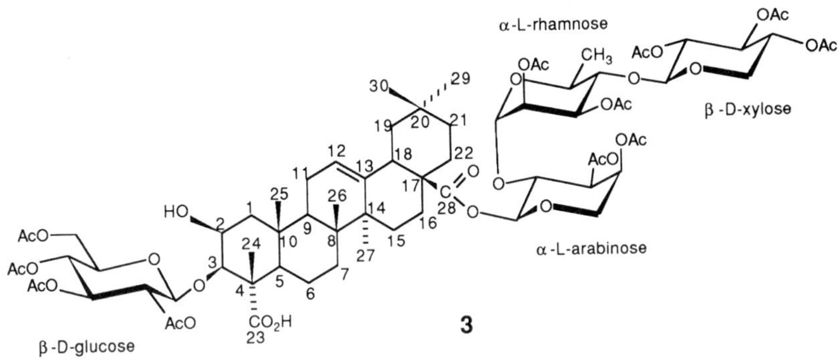

Regular spin–spin analysis cannot be performed on the entangled 3–5 ppm area, not because of strong coupling but because of overlap. These systems are now routinely analysed by means of COSY experiments. A COSY yields ($\omega \times \omega$) maps with two sorts of peaks. The diagonal peaks (coordinates: ω_i, ω_i) correspond to the resonances of the 1-D-spectrum (peaks at frequency ω_i). Off-diagonal peaks of coordinates ω_i, ω_j only occur when there is scalar coupling between nuclei resonating at ω_i and ω_j, i.e. when protons are neighbours of one another. They are important and provide information on connectivities. The analysis of the COSY experiment for a saponin starts from anomeric resonances. On the row or column corresponding to an anomeric resonance, one finds the cross-peak corresponding to the H_1–H_2 coupling (rows and columns are usually equivalent since the maps are made symmetrical). From H-2, one determines H-3 and so on. The solid lines correspond to the full analysis of the resonances of one sugar (Fig. 8.5).

When two coupling resonances have close chemical shifts, they yield off-diagonal peaks (cross-peaks) near the diagonal, and these are not

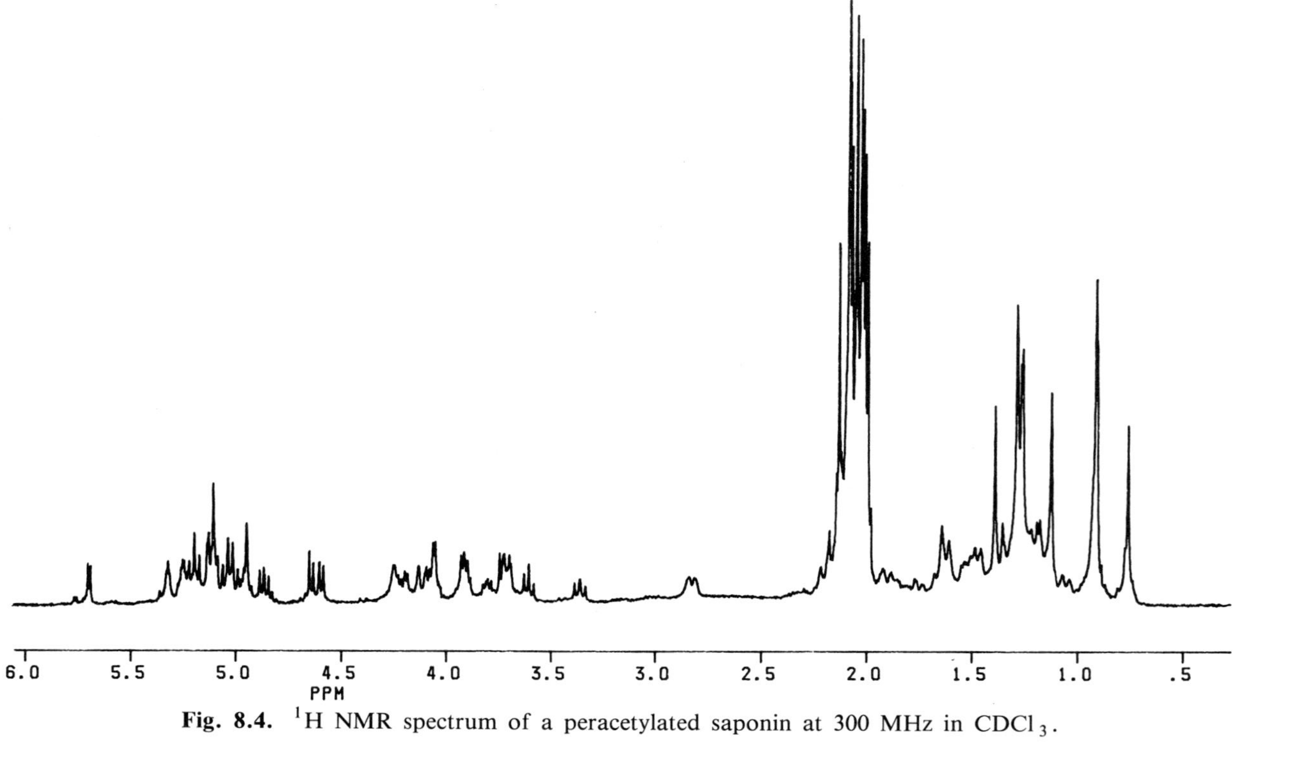

Fig. 8.4. ^{1}H NMR spectrum of a peracetylated saponin at 300 MHz in $CDCl_3$.

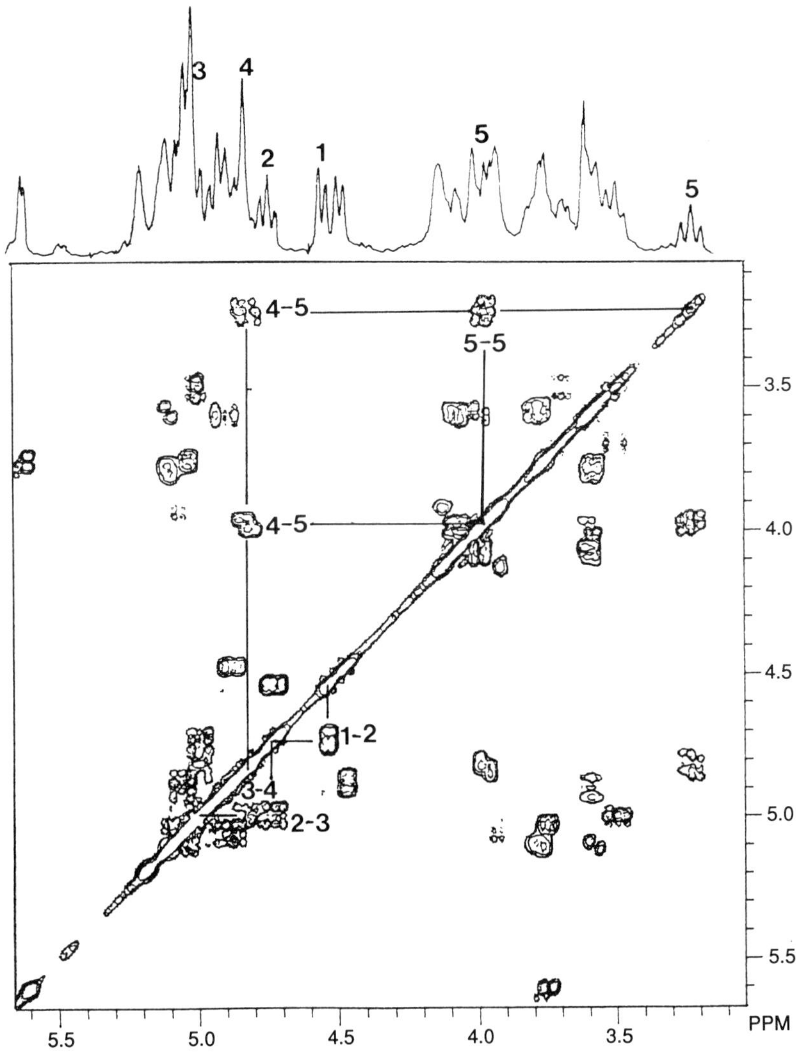

Fig. 8.5. COSY H–H of a peracetylated saponin (sugar part); correlations of the β-D-xylose are indicated by solid lines.

easily detectable. Many experiments have been devised to overcome these assignments problems. The relayed COSY experiment (Fig. 8.6) yields cross-peaks with the 'neighbour of the neighbour' and for example, in the case of sugars H-1 gives cross-peaks with H-2 and H-3 (via a relay). Full

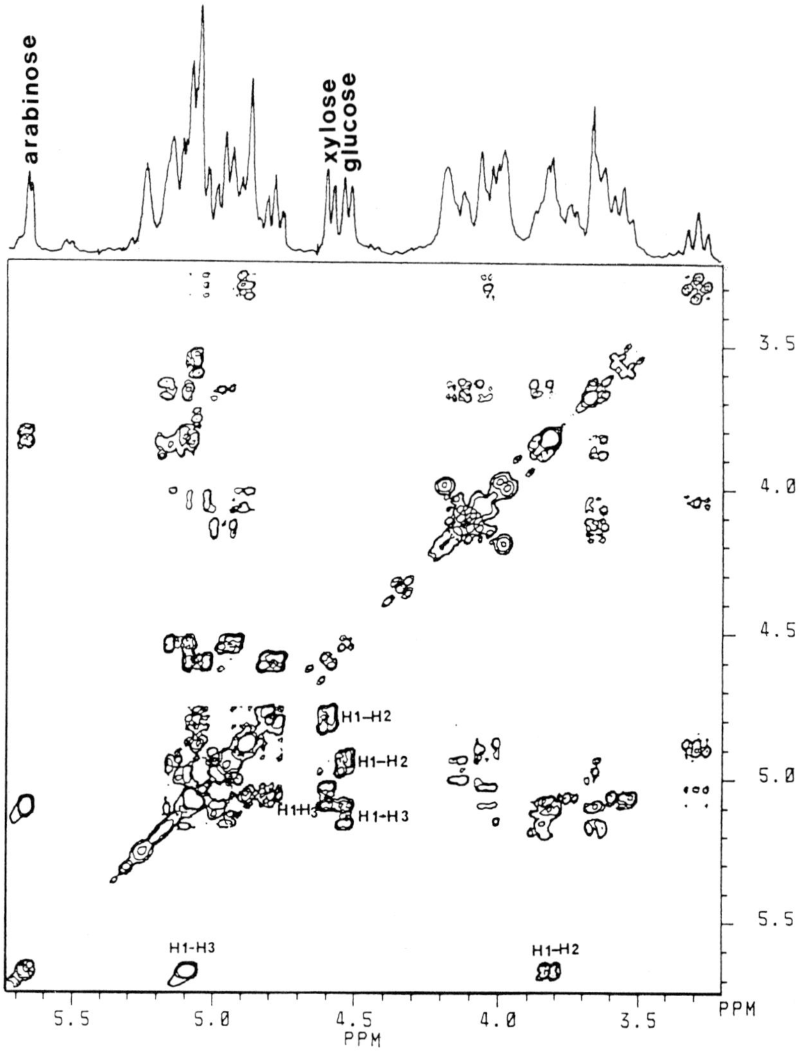

Fig. 8.6. Relayed COSY H–H of a peracetylated saponin (sugar part); correlations from H-1 of β-D-glucose, β-D-xylose, and α-L-arabinose with H-3 via a relay are indicated.

spin systems are detected by means of HOHAHA (Fig. 8.7) or TOCSY experiments and when the coupling path is not interrupted, one sees cross-peaks for all the protons of the same sugar in a single column (or row).

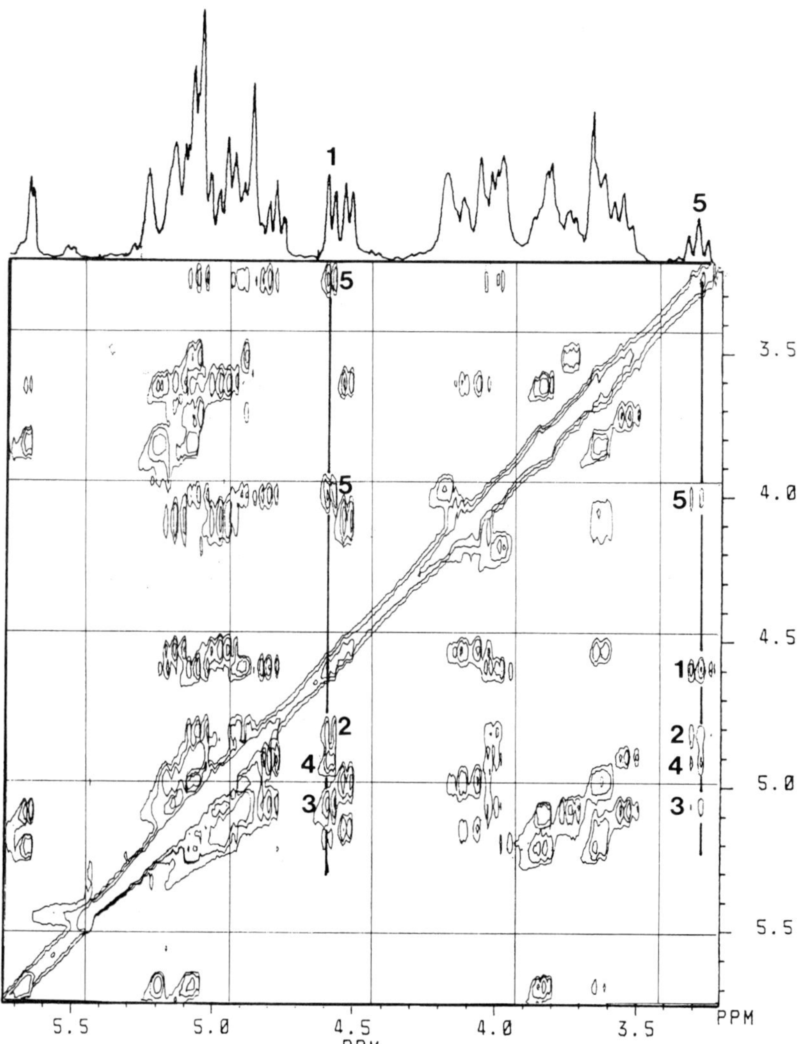

Fig. 8.7. HOHAHA of a peracetylated saponin (sugar part); correlations for H-1 and H-5 of the β-D-xylose are indicated (note that the data are presented without symmetrization).

Sugars are initially identified by the number of protons they have (five for uronic acids, six for pentoses, and seven for hexoses). The shape of the cross-peaks gives an idea of the values of the coupling constants

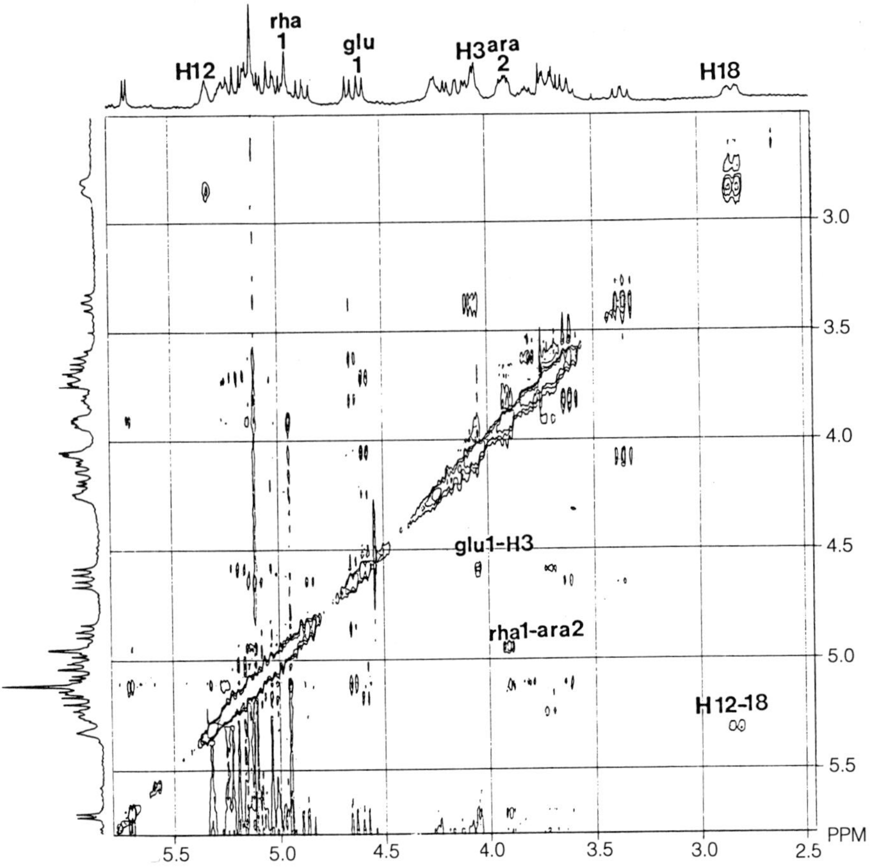

Fig. 8.8. ROESY of a peracetylated saponin (sugar part). Correlations useful for sequencing are H-3–glu-1 and rha-1-ara-2. The H-12–H-18 correlation gives the C-18 configuration.

which is in most cases enough to identify the sugars. In our specific example, there is glucose, rhamnose, arabinose, and xylose.

The shielding of H-2 in arabinose and of H-4 in rhamnose (3.9 and 3.6 ppm) tells that these positions in these sugars are not acylated. These correspond to branching points and this is probably the safest and simplest way to determine them! The two other sugars (glucose and xylose) are peracetylated and therefore terminal.

To sequence the five elements of the saponin, one must link them two by two. This is done here by the observation of through space nuclear

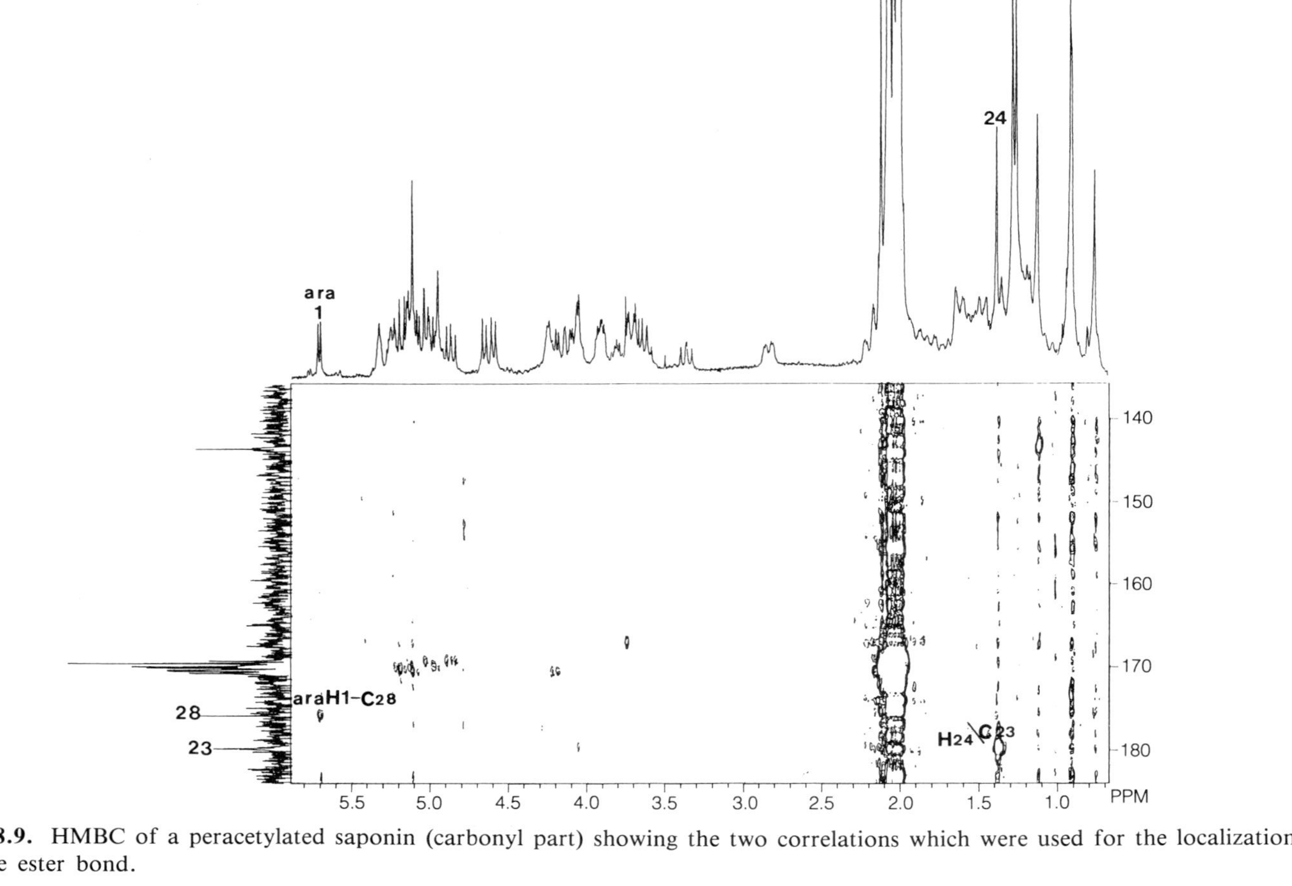

Fig. 8.9. HMBC of a peracetylated saponin (carbonyl part) showing the two correlations which were used for the localization of the ester bond.

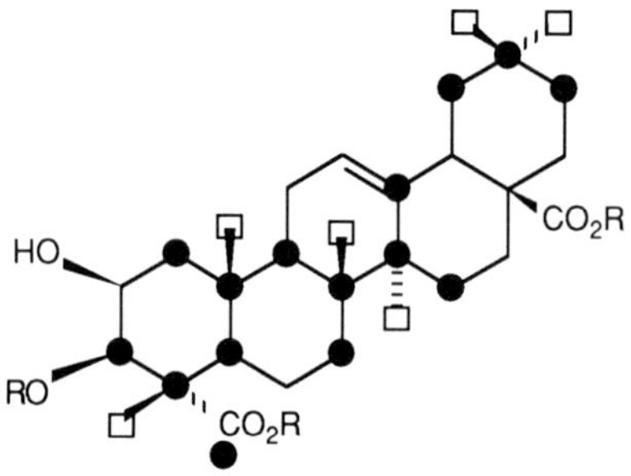

Fig. 8.10. HMBC assignments of the carbon skeleton through correlation with the methyl proton resonances.

Overhauser effects. There are also 2D NMR experiments allowing these measurements, namely NOESY and ROESY. The demonstration has been performed here with a ROESY experiment for sensitivity reasons (Fig. 8.8). Cross-peaks are observed between H-3 of medicagenic acid and H-1 of glucose, H-2 of arabinose and H-1 of rhamnose which is enough to sequence the sugar chain as Xyl$\xrightarrow{1\ 4}$Rha$\xrightarrow{1\ 2}$Ara. It remains to determine to which acid (C-23 or C-28) it is linked. This information is obtained by means of a long range C–H correlation (HMBC is used here; Fig. 8.9) which links C-28 to H-1 of arabinose (three bond coupling). C-28 is distinguished from C-23 because of the observation of a coupling between Me-24 and C-23.

There are two C-H correlation experiments which have been found useful in saponin structural elucidation: HMQC (one bond correlation) and HMBC (two or three bonds correlation). We have seen the use of HMBC for sequencing, but this experiment is also useful for assignments (Fig. 8.10). The angular methyls are linked by ^2J or ^3J coupling to four carbon atoms and with six or seven such methyls present in saponins one may assign almost all the resonances of the triterpene (Fig. 8.11).

The HMQC experiment is only used for assignments and as an example, anomeric protons can be detected through ^1J couplings with the anomeric carbons which resonate in a well-separated area of the spectrum (95–105 ppm) (Fig. 8.12).

It probably follows from the above discussion that sequencing is the most difficult problem to solve with NMR. This can be done more easily with mass spectrometry. Californium Plasma Desorption Mass Spectrometry (CPDMS) has two particularities which deserve attention: it is a cheap MS technique and it works with acetylated saponins. In our case (Fig. 8.13), it gives a quasi molecular ion $[M^* + 23]^+$ at

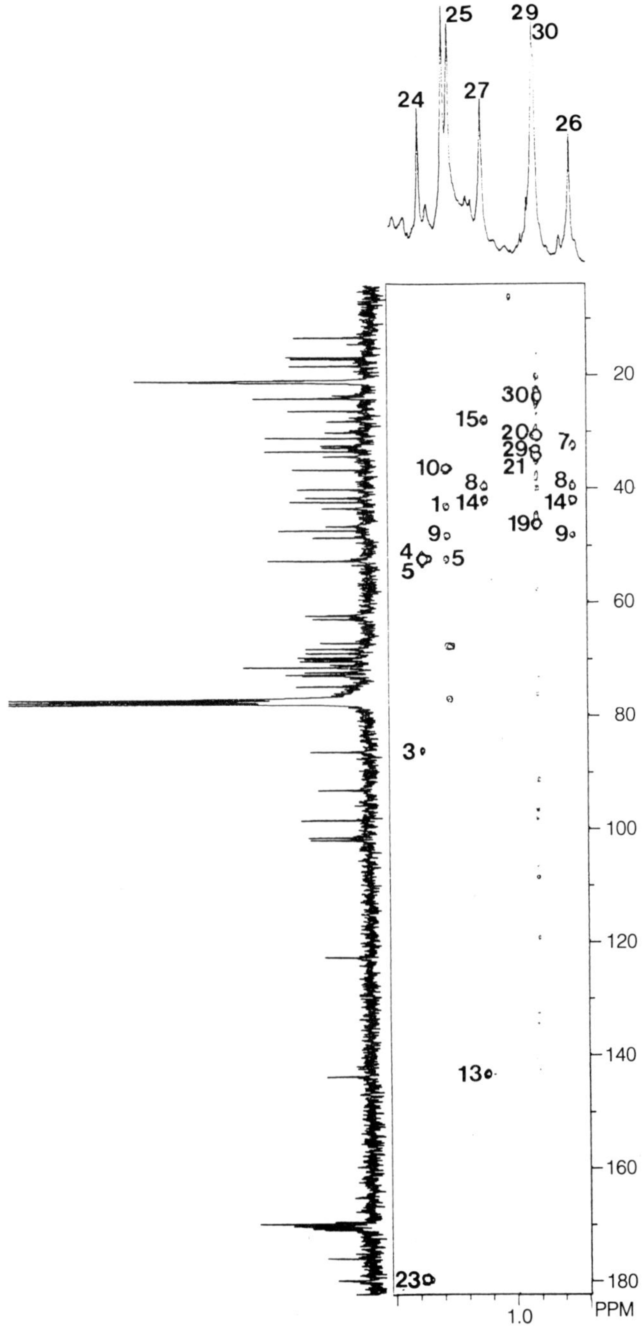

Fig. 8.11. HMBC of a peracetylated saponin (methyls part).

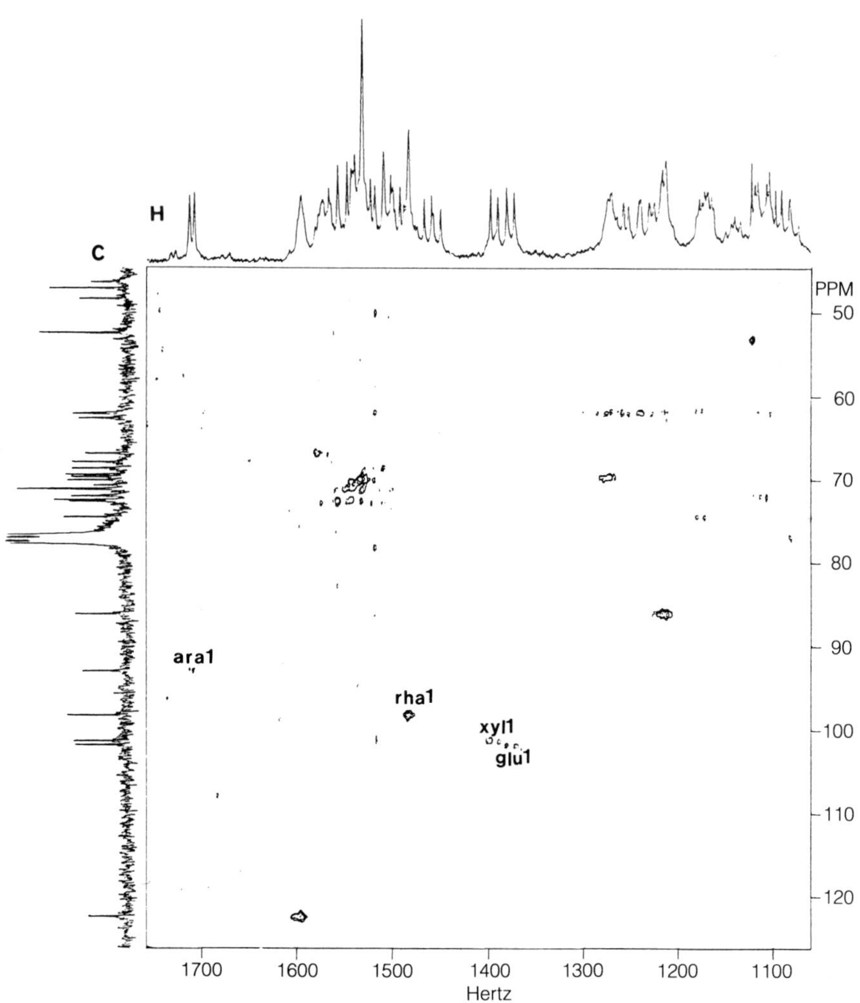

Fig. 8.12. HMQC of a peracetylated saponin (sugar part); identification of the anomeric resonances.

m/z 1560 and fragments corresponding to the sequential loses of the sugar of the side chain: 259 (peracetylated pentose), 489 (pentose + desoxyhexose), and 705 (pentose + desoxyhexose + pentose).

This set of techniques has allowed the structural elucidation of about ten saponins with medicagenic acid (Fig. 8.14), hederagenin, and soyasapogenol B. One of the saponins with rhamnose, glucose and

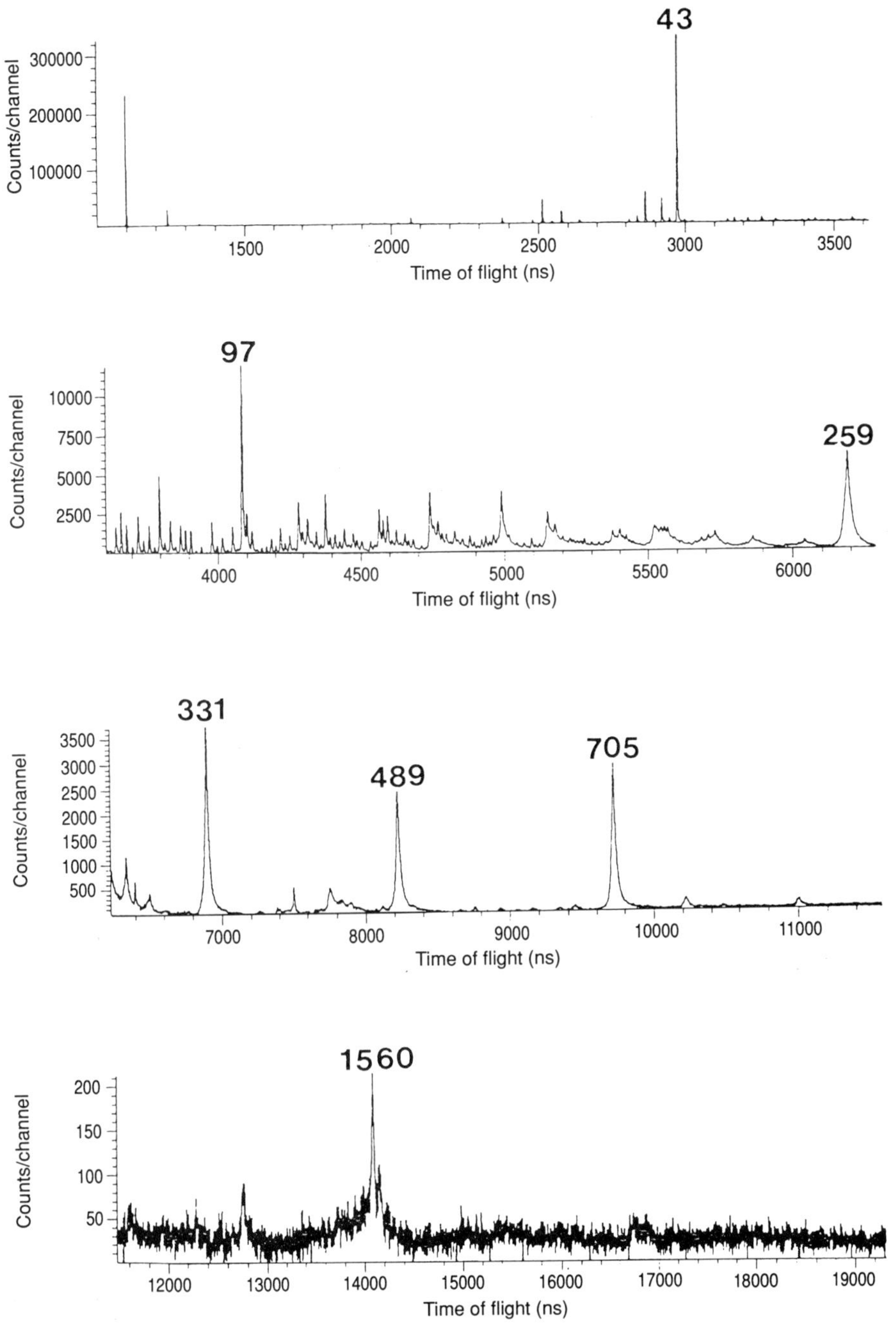

Fig. 8.13. CPDMS of a peracetylated saponin.

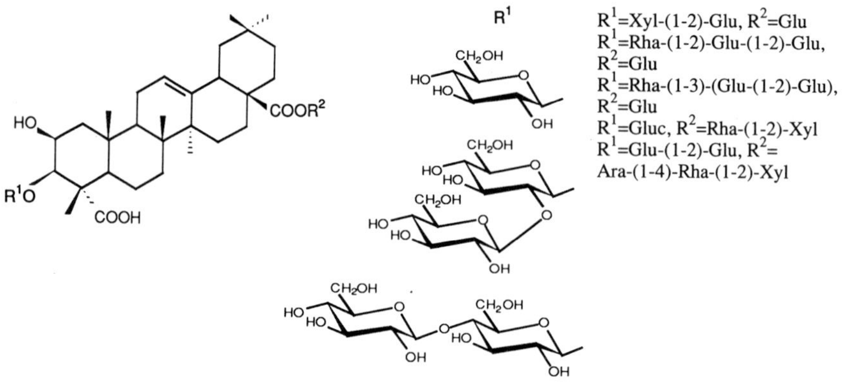

R^1=Xyl-(1-2)-Glu, R^2=Glu
R^1=Rha-(1-2)-Glu-(1-2)-Glu, R^2=Glu
R^1=Rha-(1-3)-(Glu-(1-2)-Glu), R^2=Glu
R^1=Gluc, R^2=Rha-(1-2)-Xyl
R^1=Glu-(1-2)-Glu, R^2= Ara-(1-4)-Rha-(1-2)-Xyl

Fig. 8.14. Some saponins with medicagenic acid as aglycone.

glucuronic acid, and soyasapogenol B as constituents illustrates some of the difficulties of the work with saponins. The structure was established by spectral means and published *bona fide* as new, before it was realized that the saponin was isolated and characterized earlier by Kitagawa under the name of azukisaponin. This shows some of the problems met when working with saponins which, unlike most small metabolites, cannot be identified on the basis of a molecular ion and of 5–6 typical NMR chemical shifts.

Preluzernic acid or zahnic acid was identified at least 25 years after luzernic acid (an artefact) was mentioned in the literature. The first saponins with zahnic acid were identified only recently in a joint Anglo–Polish effort (Oleszek *et al.* 1992). These saponins are among the most complex ever isolated and have apiose—a rare pentose—as a terminal residue (Fig. 8.15). They are the most bitter compounds from alfalfa.

The risk that one takes when working with peracetates is that the molecule may be irreversibly modified during derivatization. An example of such a degradation is soyasaponin VI (**4b**) whose dihydromaltol unit completely vanishes during regular isolation (when not maintaining a neutral pH) and during acetylation. One isolates instead soyasaponin I (**4a**) whereas soyasaponin VI is the natural product in alfalfa, soya bean, lentils, peanuts, etc. To alleviate this risk, new NMR techniques are currently being developed to obtain selective information on selected sugars (Massiot *et al.* 1991*b*). Without going into details, one can simply say that it is now possible to obtain 1D spectra of individual sugars in a saponin and the corresponding COSY spectra. This approach is however limited by the narrow range of resonances in sugar protons

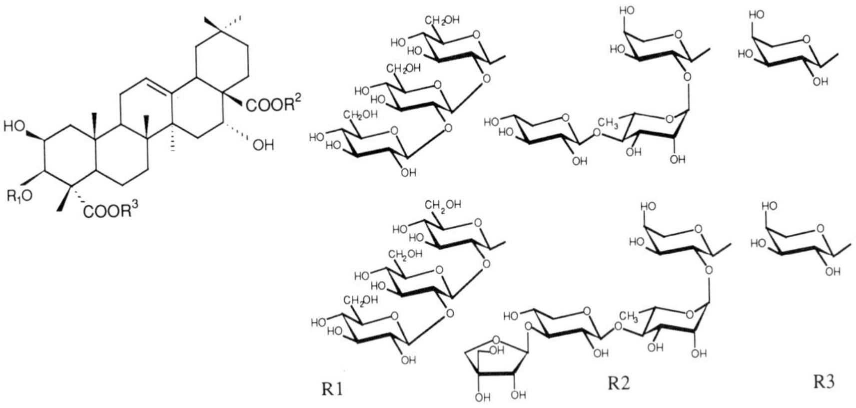

Fig. 8.15. Some saponins with zahnic acid as aglycone.

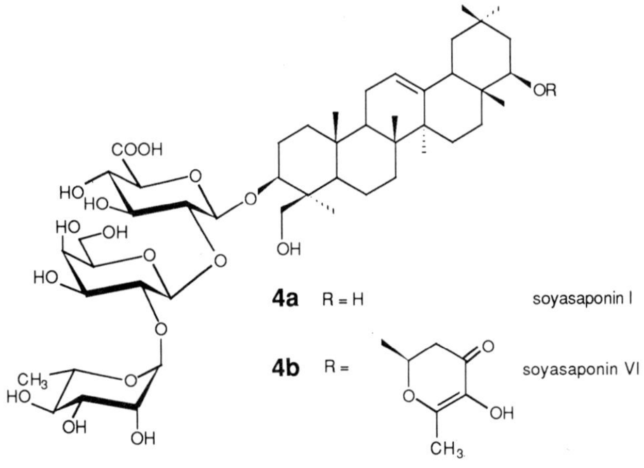

causing second order effects and overlap. Clearly the use of very high fields must be considered (600 MHz)!

To come back to the real problem which is the development of new alfalfa varieties with good digestibility properties, one must solve two problems:

(1) the determination of the real digestibility-inhibiting factors;

(2) their quantification.

Evaluation of a feed cannot be based on a single test such as 'take' or 'do not take'. The feed component must be incorporated into a well-balanced diet, given to animals, and measurements must be made on feed intake and animal weight gain (or possibly loss). On rare instances such experiments to test new protein concentrates were performed on chickens in batteries but these are costly experiments. The ideal test animal is, according to nutritionists at the INRA centre in Bordeaux, *Tenebrio molitor*, the meal worm (Pracros 1988). In 50 different experiments, this species (10 animals per experiment) was fed with protein concentrates for two weeks after which time the animal weights and food uptakes were measured. In parallel experiments, saponins were extracted from the protein concentrates and these saponins were hydrolysed. High Pressure Liquid Chromatography (HPLC) allowed quantification of medicagenic acid in each sample. Analysis of the results showed that saponins and medicagenic acid in particular had negative effects on gain of weight and food intake of the animals. For the genetic selection process, thousands of varieties are being produced each year. They come to maturity in a short period of time and the tests must be performed quickly. This cannot be done by HPLC or by the animal test and this is done by far infra-red measurements on small samples of ground leaves. The apparatus is first calibrated with the IR spectra of the 50 samples which were submitted to the full analysis, and comparisons are made between the IR spectrum of a new sample with those of the reference sample. These comparisons are done under computer control and a fraction of acceptable samples are kept for another round of selections.

With this technique however, the problem of rapid selection for low medicagenic acid content has not been satisfactorily solved and it might be better to address the selection problem in other terms, such as genetics. The selection is complicated because the ideal cultivar must have low saponin content, high productivity, resistance to lodging and early frost, etc.

This research led us to manipulate large quantities of saponins and it would be interesting to find new potential markets for these readily available compounds. One key property of saponins is their relationship with cholesterol, which is of interest in connection with work done by Malinow *et al.* (1980) in the USA, who were able to considerably lower fluid cholesterol level in human volunteers, using alfalfa in the diet. Tests on rats and monkeys displayed the same effects without noticeable toxicity (Malinow *et al.* 1982). The only difficulty met with humans was the bitter taste of the diet and several volunteers were not able to complete the test.

Acknowledgements

It is a pleasure to acknowledge the contributions of our colleagues whose names appear in the reference section and especially of Mrs Prof. Le Men-Olivier for many fruitful discussions. Our colleagues at the Lusignan INRA centre (Genier, Guy) and at France Luzerne are thanked for providing plant material, Mrs P. Pracros for performing biological tests, and INRA, CNRS, and URCA for financial support.

References

Albourine, A., Petit-Ramel, M., Thomas-David, C., and Vallon, J. J. (1991). Extraction de la L-canavalline à partir du soja et de la luzerne. *J. Pharm. Belg.*, **46**, 229–35.

Berrang, B., Davis, K. H., Wall, M. E., Hanson, C. H., and Pedersen, M. L. (1974). Saponins of two alfalfa cultivars. *Phytochemistry*, **13**, 2253–60.

Cheeke, P. R. (1980). Biological properties and nutritional significance of legume saponins. In *Leaf protein concentrates* (ed. L. Telek and H. D. Graham), pp. 396–414. AVI, Westport, CT.

Dimbi, M. Z., Warin, R., Delaude, C., Huls, R., Kapundu, M., and Lami, N. (1984). Structures de l'acide zahnique et de l'acide zahnique γ-lactone, triterpenoide nouveau isolé de *Zahna golugensis* et de *Ganophyllum giganteum*. *Bull. Soc. Chim. Belg.*, **93**, 323–8.

Djerassi, C., Thomas, D. B., Livingston, A. L., and Thompson, C. R. (1957). The structure and stereochemistry of medicagenic acid. *J. Am. Chem. Soc.*, **79**, 5292–7.

Gastineau, I. and de Mathan, O. (1981). La préparation industrielle de la protéine verte de luzerne. In *Protéines foliaires et alimentation* (ed. C. Costes), pp. 159–82. Gauthier-Villars-Bordas, Paris.

Gestetner, B. (1971). Structure of a saponin from lucerne. *Phytochemistry*, **10**, 2221–3.

Heywang, B. W. and Bird, H. R. (1954). The effect of alfalfa saponins on the growth, diet consumption and efficiency of diet utilisation of chicks. *Poultry. Sci.*, **33**, 239–41.

Jurzysta, M. (1973). Isolation and chemical characterization of saponins from lucerne seeds. *Acta Soc. Bot. Pol.*, **42**, 201–7.

Kitagawa, I., Yoshikawa, M., and Yosioka, I. (1976). Saponin and sapogenol XIII, Structures of three soybean saponins: soyasaponin I, soyasaponin II and soyasaponin III. *Chem. Pharm. Bull.*, **24**, 121–9.

Kitagawa, I., Taniyama, T., Murakami, T., Yoshihara, M., and Yoshikawa, M. (1988). On the constituents in aerial parts of American alfalfa. The structure of dehydrosoyasaponin I. *Yakugaku Zasshi*, **108**, 547–54.

Levy, M., Zehavi, U., Naim, M., and Polacheck, I. (1986). An improved procedure for the isolation of medicagenic acid 3-*O*-β-D-glucopyranoside from alfalfa roots and its antifungal activity on plants pathogens. *J. Agric. Food Chem.*, **34**, 960–3.

Livingston, A. L. (1959). Lucernic acid, a new triterpene from alfalfa. *J. Org. Chem.*, **24**, 1567.

Malinow, M. R., McLaughlin, P., and Stafford, C. (1980). Alfalfa seeds: effects on cholesterol metabolism. *Experientia*, **36**, 562–4.

Malinow, M. P., Mc Nulty, W. P., Houghton, D. C., Kessler, S., Stenzel, P., Goonight, S. H., *et al.* (1982). Lack of toxicity of alfalfa saponins in *Cynomolgus macaques. J. Med. Primatol.*, **11**, 106–18.

Massiot, G., Lavaud, C., Guillaume, D., Le Men-Olivier, L., and van Binst, G. (1986). Identification and sequencing of sugars in saponins using 2D 1H NMR Spectroscopy. *J. Chem. Soc., Chem. Commun.*, 1485–7.

Massiot, G., Lavaud, C., Le Men-Olivier, L., van Binst, G., Miller, S. P. F., and Fales, H. M. (1988*a*) Structural elucidation of alfalfa root saponins by mass spectrometry and NMR analysis. *J. Chem. Soc., Perkin Trans. 1*, 3071–9.

Massiot, G., Lavaud, C., Guillaume, D., and Le Men-Olivier, L. (1988*b*). Reinvestigation of the sapogenins and prosapogenins from alfalfa. *J. Agric. Food Chem.*, **36**, 902–9.

Massiot, G., Lavaud, C., Besson, V., Le Men-Olivier, L., and van Binst, G. (1991*a*). Saponins from aerial parts of alfalfa. *J. Agric. Food Chem.*, **39**, 78–82.

Massiot, G., Lavaud, C., and Nuzillard, J. M. (1991*b*). Revision des structures des chrysanthellines par R.M.N. *Bull. Soc. Chim. Fr.*, **127**, 100–7.

Morris, R. J. and Hussey, E. W. (1965). A natural glycoside of medicagenic acid. An alfalfa blossom saponin. *J. Org. Chem.*, **30**, 166–8.

Morris, R. J. and Tankersley, D. L. (1963). The synthesis of the β-D-glucoside of medicagenic acid, an alfalfa root saponin. *J. Org. Chem.*, **28**, 240–2.

Morris, R. J., Dye, W. B., and Gisler, P. S. (1961). Isolation, purification and structural identity of an alfalfa root saponin. *J. Org. Chem.*, **26**, 1241–3.

Oleszek, W., Jurzysta, M., Ploszynski, M., Colquhoun, I. J., Price, K. R., and Fenwick, G. R. (1992). Zahnic acid tridesmoside and other dominant saponins from alfalfa aerial parts. *J. Agric. Food Chem.*, **40**, 191–6.

Pracros, P. (1988). Mesure de l'activité des saponines de la luzerne par les larves du ver de farine *Tenebrio molitor. Agronomie*, **8**, 793–9.

Shany, S., Gestetner, B., Birk, Y., Bondi, A., and Kirson, I. (1972). Isolation of hederagenin and its saponin from alfalfa. *Isr. J. Chem.*, **10**, 881–4.

Timbekova, A. M. and Abubakirov, N. K. (1987*a*). Triterpenes glycosides of alfalfa. Medicoside I. *Khim. Prir. Soedin*, **5**, 571–3.

Timbekova, A. M. and Abubakirov, N. K. (1987*b*) Triterpenes glycosides of alfalfa. Medicoside J. *Khim. Prir. Soedin*, **5**, 574–7.

Timbekova, A. M., Vereschagin, A. L., Semenov, A. A., and Abubakirov, N. K. (1990). Triterpenes glycosides of alfalfa. Medicoside L. *Khim. Prir. Soedin.*, 178–83.

Tschesche, R. and Wulff, G. (1973). Chemie und biologie der saponine. *Chem. Org. Naturst.*, **30**, 461–606.

9. Toxic range plants and their constituents

RUSSELL J. MOLYNEUX

Western Regional Research Center, Agricultural Research Service, US Department of Agriculture, 800 Buchanan Street, Albany, California 94710, USA

Introduction

As settlers moved westward across North America they encountered an environment which became increasingly hostile to both themselves and their livestock. It was an environment with which they were completely unfamiliar, far removed in character from the 'European' type of grazing lands typical of the Eastern seaboard. The climate was generally arid or semi-arid, with diverse plant communities, and while many of the plants provided excellent forage they were intermixed with other species which proved to be extremely toxic to livestock.

Losses of livestock were often extremely large. For example, 25 000 cattle died from locoweed poisoning in a single county of southwestern Kansas during 1883 (Crawford 1908). As a consequence of such events, appeals for assistance were made to the federal government which initiated a research effort in the early years of this century. These investigations were conducted primarily by Dr C. Dwight Marsh, under the auspices of the Bureaus of Animal Industry and Plant Industry of the US Department of Agriculture (USDA), which resulted in a series of monographs describing the etiology of the major livestock poisoning problems and delineating the primary toxic plant species (Marsh 1909, 1924).

The USDA Bureaus eventually evolved into the Agricultural Research Service, which has continued to perform research on poisonous range plants until the present time. It is the purpose of this chapter to discuss current livestock toxicity problems caused by range plants; to present a coherent research approach aimed at these problems; to illustrate the role of phytochemistry in such investigations by a presentation of our research on locoweed poisoning; and to discuss some of the ancillary benefits which may derive from a thorough understanding of the chemistry and biochemistry of natural toxins.

Current rangeland toxicity problems

Livestock losses due to poisonous plants are still a major cost to producers in the 17 Western States of the US, equalling those due to infectious diseases. It has been estimated that the carrying capacity (stocking) of rangeland could be doubled if it were not for the presence of poisonous plants. Calculations based on an annual death loss of 1 per cent of all cattle and sheep, a 1 per cent decrease in calf crop, and a 3.5 per cent decrease in lamb crop, amount to a net reduction in income of $234 million, based on 1987 prices (Nielsen *et al.* 1988). These calculations do not account for poisoning of horses, goats, pigs, and wildlife. Nor do they include veterinary fees, poisonous plant control, labour costs of herding away from hazardous areas, fencing, and unusable grazing land. A current estimate, including the latter factors and adjusting for increased livestock prices, yields a realistic annual loss figure of $340 million.

On occasion losses may be catastrophic, leading to bankruptcy of the livestock operation. For example, during the 1960s more than 6400 sheep were poisoned by locoweed in the Uintah basin of eastern Utah and one operator lost $125 000 at that time. During the same period a single rancher lost 1200 sheep overnight due to *Halogeton* (oxalate) poisoning (Nielsen *et al.* 1988). More recently, 350 sheep died in southern Idaho through ingestion of death camas (*Zygadenus paniculatus*) which contains complex steroidal alkaloids (Panter *et al.* 1987). Another rancher lost 35 valuable Saler cattle, out of a herd of 43, due to their consumption of trimmings of the extremely poisonous ornamental plant, English yew (*Taxus baccata*), which were injudiciously discarded in an area to which the animals had access. We were able to confirm the ingestion of the plant by analysis of the rumen contents for the compound taxol, which is unique to *Taxus* species, using thin-layer chromatography (Panter *et al.* 1993).

While such losses are spectacular, of even greater impact overall are those in which a single animal or a small group of animals may be affected, leading to a significantly reduced profit margin for the producer. Such animals frequently die unobserved with no possibility of establishing the cause. At other times the period between ingestion of the plant and signs of poisoning may be considerable, leading to great difficulty in incriminating a particular species. We have shown this to be the situation with *Senecio* (groundsel) species which contain hepatotoxic pyrrolizidine alkaloids. Cattle have succumbed to stress induced by cold, malnutrition, and parturition at periods of 6 to 18 months after consumption of tansy ragwort (*S. jacobaea*), threadleaf groundsel (*S. longilobus*), and Riddell's groundsel (*S. riddellii*) (Molyneux *et al.* 1988*b*).

Additional losses may arise from birth defects (teratogenesis) due to fetal insult *in utero* by specific constituents when the mother consumes particular plants. The first documented occurrence of this phenomenon was the identification of the steroidal alkaloid jervine and two structurally related alkaloids as the cause of cyclopia in lambs (Keeler and Binns 1968). This defect, characterized by a single eye, and known as 'monkey-face lamb', was found to be produced only when the pregnant ewe ate the plant *Veratrum californicum* (Western false hellebore) on the 14th day of gestation (Binns *et al.* 1965). At other periods terata were minimal, and careful management practices have succeeded in reducing the frequency of cyclopic birth to insignificant levels. Congenital birth defects, including cleft palate and skeletal malformations of the limbs and spine, have been induced in pigs, goats, sheep, and cattle by poison hemlock (*Conium maculatum*), tree tobacco (*Nicotiana glauca*), and the lupines (*Lupinus formosus*, *L. caudatus*, and *L. sericeus*). All of these plants contain piperidine alkaloids or the quinolizidine alkaloid, anagyrine. Panter *et al.* (1990) have produced evidence, through ultrasound monitoring, that these constituents have a sedative effect upon the fetus. Reduced fetal movement of the tongue and limbs, with consequent obstruction of palate closure and inadequate extension of tendons and ligaments, accounts for the observed malformations.

Reproductive problems induced by poisonous plant consumption are frequent causes of reduced calf and lamb crops. Ingestion of Ponderosa pine needles (*Pinus ponderosa*) by cows in the third trimester of pregnancy causes premature parturition or abortion (James *et al.* 1989). While the calf may be viable, it is often too weak to survive without exceptional treatment. In certain Ponderosa pine endemic areas, abortion rates of 80–90 per cent have been reported. A similar syndrome is produced by consumption of broom snakeweed (*Gutierrezia* spp.). These abortions are characterized by a retained placenta which often leads to haemorrhage and infection, requiring veterinary treatment in order to save the mother.

Many additional poisonous plant toxicities are observed, including acute poisonings due to plants such as larkspurs (*Delphinium* spp.), milkweeds (*Asclepias* spp.), chokecherry (*Prunus virginiana*), poison hemlock (*Conium maculatum*), and waterhemlock (*Cicuta douglasii*). Several plants are photosensitizing, leading to swelling and inflammation of exposed skin which prevents the animal from feeding itself or from nursing its young. Species in this group are horsebrush (*Tetradymia glabrata* and *T. canescens*), St. Johnswort (*Hypericum perforatum*), and spring parsley (*Cymopterus watsonii*). Lack of weight gain (unthriftiness) and emaciation are frequent occurrences which may be caused by digestive disturbance and vomiting, or by prehension problems, muscular weakness, and irregular gait which prevent the animals from foraging. Plants in this category are bitter rubberweed (*Hymenoxys odorata*), pingue (*H. richardsonii*), rayless goldenrod (*Happlopappus heterophyllus*), sneezeweed (*Helenium hoopesii*), and starthistle (*Centaurea solstitialis*) (James *et al.* 1980).

Research approach

The exceptional diversity of plant species involved in livestock poisoning episodes and the great variation in syndromes and susceptibility of livestock groups prevents a consistent research approach from being adopted. Each particular problem requires a great deal of imagination and originality to unravel the complexities of the plant–animal interactions. In spite of a considerable body of knowledge regarding the majority of toxic plant species inhabiting rangelands, serious livestock losses regularly occur. *This is primarily due to a lack of understanding of the phytochemistry involved.* As a result, we too often lack the ability to predict a safe level of exposure. No Western US rangelands are completely safe. There are always poisonous plants present, but their toxin content varies greatly with plant species, stage of growth, and numerous environmental factors.

The author's research approach is firmly grounded upon the Axiom of Paracelsus: 'Sola dosis facit venenum' (Only the dose makes the poison). If the ingestion level of the toxin can be reduced to below its harmful level for each individual animal in the herd, then toxic plant species may be grazed with a reasonable degree of safety. An idealized sequence of research stages has been developed, shown in Fig. 9.1, which if successfully accomplished will lead to management recommendations designed to prevent episodes of poisoning.

The first step, determination of etiology of poisoning, is generally well established, the specific plants involved having been known for several decades. Obviously, the most difficult step is the isolation and characterization of toxins. The chances of success at this stage are greatly improved if a bioassay can be developed. This may be a large- or small-animal bioassay, or preferably an *in vitro* assay. If no bioassay is available, the toxin isolation process degenerates to a conventional phytochemical survey of the plant constituents, with considerable dependence upon experience and intuition as to the most likely classes of compounds involved in the poisoning. In such situations progress is usually seriously impeded. A valid bioassay often provides an ancillary benefit, once the specific toxin is identified, by providing insight into therapeutic potential or alternative systems in which the compound might exhibit bioactivity.

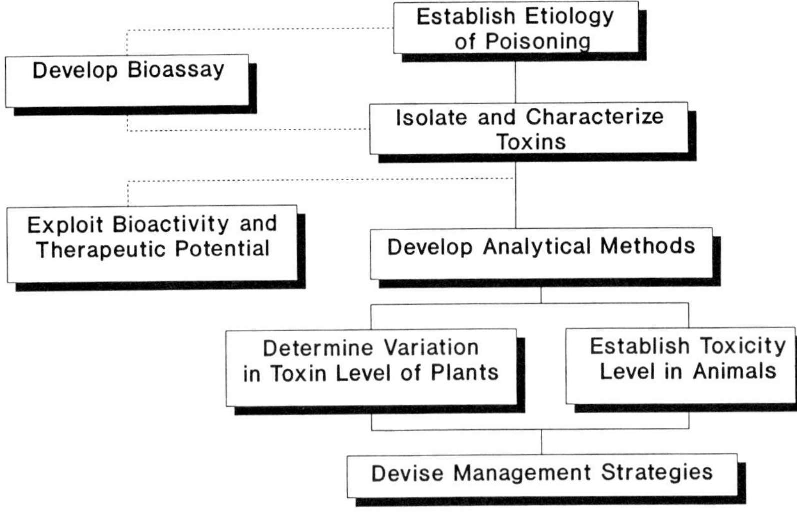

Fig. 9.1. Sequence of research stages essential for attacking the problem of livestock toxicity due to poisonous plants.

Isolation of the toxin(s) allows analytical methods to be developed, based upon the chemical nature of the compounds, which permits the variation of toxin level in the plants, and the toxicity level in animals, to be determined. It then remains only to devise practical methods of managing the rangelands and herds so that the animals have no possibility of ingesting harmful doses of the particular toxins of concern. It is obvious that no meaningful management recommendations can be developed unless the chemical structure of the toxin is completely defined.

Locoweed toxicity

The problem of livestock losses caused by the locoweeds presents the clearest illustration of a rational approach to the prevention of poisonous plant problems. Locoweeds are certain species of the genera *Astragalus* and *Oxytropis*, members of the plant family Leguminosae. These genera have circumglobal distribution but in North America alone there are 552 species and varieties of *Astragalus* and 35 of *Oxytropis* (Barneby 1952, 1964). A number of these are toxic due to their content of nitro-toxins (Williams and Barneby 1977) or because of their ability to accumulate high levels of selenium compounds (James 1983). However, these compounds

Table 9.1 Syndromes of toxicity produced in livestock by consumption of locoweeds (*Astragalus* and *Oxytropis* species)

Syndrome	Specific signs
Neurological (Locoism)	Depression, staggering gait, solitariness, nervousness and belligerence under stress, incoordination, difficulty in eating and drinking
Emaciation	Lack of weight gain despite normal feed intake
Habituation	Increased acceptance of normally unpalatable locoweed species
Congestive right-heart failure	Laboured respiration, fluid accumulation in neck and thorax, death under exertion
Abortion	High incidence at any stage of gestation
Reproductive failure	Reduced sexual activity, suppressed oestrus, reduced spermatogenesis
Teratogenesis	Malformation of forelimbs and joints, contracted tendons

are not the causative agents of the locoweed syndrome, which has had a romantic association with the settlement of the American West through its use in magazine articles, novels, and films.

About a dozen species have been definitely incriminated as 'locoweeds' through repeated, large-scale livestock poisoning episodes. Locoweed consumption produces a complex of syndromes (Table 9.1) which can serve as the epitome of virtually all poisonous plant problems. These signs of poisoning vary considerably with animal species, condition, and environment under which the plant is ingested. Neurological symptoms, or locoism, are the most dramatic, giving rise to problems of perception or proprioception and extreme sensitivity to sudden noise or movement ('spookiness'). More typical signs are depression and emaciation. Congestive right-heart failure is found only when locoweeds are consumed at high altitudes (~ 3000 m). Reproductive problems and birth defects are a major economic loss factor.

The similarity of locoweed toxicity to poisoning by *Swainsona* species (poison peas), a genus restricted to Australia, was noted in the early 1900s (Maiden 1901; Marsh 1909). However, it was not until 1970 that the microscopic and clinical pathology was recognized as being essentially identical to mannosidosis, a genetic disease of Angus cattle (Huxtable and Gibson, 1970). This disease, one of a group of lysosomal storage

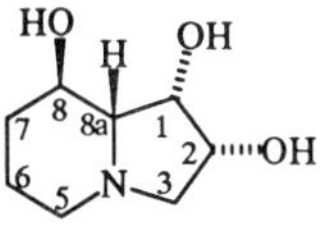

Swainsonine
Astragalus, Oxytropis, Swainsona spp.
Inhibits: α-mannosidase

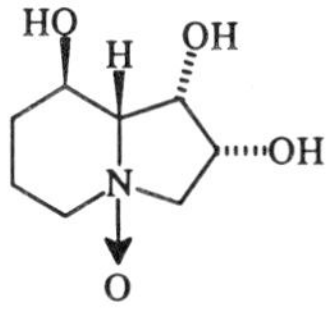

Swainsonine N-oxide
Astragalus, Oxytropis spp.

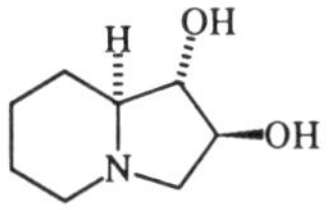

Lentiginosine
A. lentiginosus
Inhibits: α-glucosidase

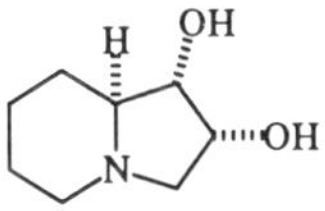

2-Epilentiginosine
A. lentiginosus
A. oxyphysus
Inhibits: None

Fig. 9.2. Structures of glycosidase inhibitory hydroxyindolizidine alkaloids present in *Astragalus*, *Oxytropis*, and *Swainsona* species.

diseases, also occurs in humans and is characterized by cellular vacuolation caused by cytoplasmic accumulation of incompletely processed glycoproteins. Lesions in the brain lead to neurological defects. Since the genetic disease results from an insufficiency of the enzyme α-mannosidase, Dorling *et al.* (1978) reasoned that *Swainsona* toxicosis was an induced mannosidosis produced by a compound in the plant which inhibits α-mannosidase. Using this enzyme as a biochemical probe for inhibitory activity, Colegate *et al.* (1979) were able to isolate and characterize the trihydroxyindolizidine alkaloid, swainsonine (Fig. 9.2), as the toxic constituent.

The earlier recognition of the similarity of locoweed poisoning to *Swainsona* toxicity led to the isolation of swainsonine from the spotted locoweed (*Astragalus lentiginosus*), in which it was discovered that it co-occurs with swainsonine *N*-oxide (Molyneux and James 1982). In contrast, none of the *N*-oxide can be detected in *Swainsona* species (Dorling *et al.* 1989). Swainsonine *N*-oxide is quite labile and prone to undergo thermal deoxygenation and decomposition. Consequently mass spectrometric analysis at normal probe temperatures gives a spectrum identical to that of swainsonine itself, and the molecular ion can only be observed when the measurements are made at probe temperatures of *c.* 140 °C (Molyneux *et al.* 1985). In addition to these alkaloids, two minor constituents have recently been isolated by careful thin-layer chromatographic separation of the residual alkaloid fraction after removal of swainsonine. These were shown by 2D-NMR techniques to be epimeric dihydroxyindolizidine alkaloids, which were named lentiginosine and 2-*epi*lentiginosine (Fig. 9.2). These probably arise via a divergent pathway from the biosynthesis of swainsonine. It is noteworthy that they differ from swainsonine in the absolute stereochemistry at the bridgehead position. Lentiginosine is a good inhibitor of α-glucosidase but epimerization of the hydroxyl group at the 2-position results in complete loss of inhibitory activity (Pastuszak *et al.* 1990).

It is particularly interesting that swainsonine has also been found to occur in the fungi, *Rhizoctonia leguminicola* and *Metarhizium anisopliae* (Schneider *et al.* 1983; Hino *et al.* 1985). The biosynthesis of swainsonine in Diablo locoweed (*Astragalus oxyphysus*) and *R. leguminicola* was subsequently shown to proceed by the same pathway (Harris *et al.* 1988). This raises an interesting teleological question as to whether the alkaloid first arose as a plant or a fungal metabolite.

Swainsonine analysis

The identification of swainsonine as the causative agent of locoweed poisoning provided the stimulus for development of analytical techniques applicable to prediction of the most toxic stages of plant development. Initial efforts concentrated upon thin-layer chromatography as a simple, rapid screening method for presence or absence of swainsonine. A spray reagent consisting of acetic anhydride, followed by Ehrlich's reagent, generated an intense purple colour with both swainsonine and swainsonine *N*-oxide which provided a minimum detection level of *c.* 0.5 μg. A large number of *Astragalus* and *Oxytropis* species were analysed and swainsonine was shown to be present in all those which had been historically associated with locoism, together with many species not previously identified as causative agents of the disease (Molyneux *et al.* 1991*a*).

In order to accurately assess locoweed toxicity it was essential to develop a quantitative analysis for swainsonine. This has been achieved by capillary gas chromatography of the tris-trimethylsilyl ether derivative, formed by treatment of the alkaloidal plant extract with MSTFA (*N*-methyl-*N*-trimethylsilyl-trifluoroacetamide). The detector response was calibrated with pure swainsonine derivatized under identical conditions.

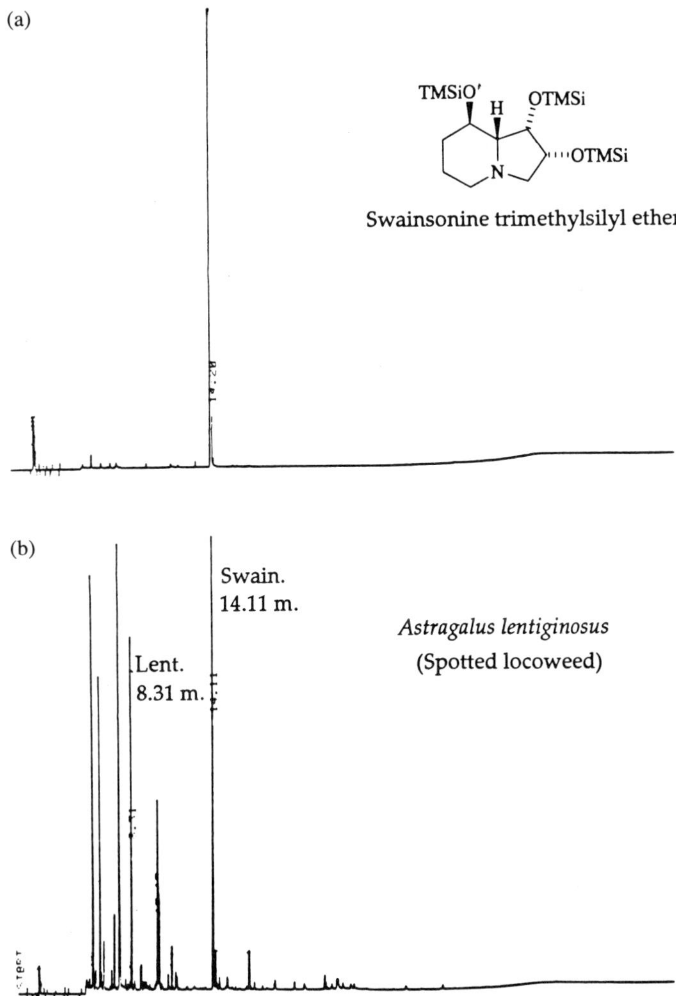

Fig. 9.3. Gas chromatographic analysis of trimethylsilyl derivatives of (a) pure swainsonine, and (b) *Astragalus lentiginosus* extract, showing presence of lentiginosine and swainsonine.

This technique provides a minimum detection level of 150 pg and analyses are routinely done on 500 mg plant samples. Analyses of a single locoweed seed weighing *c.* 2 mg have been successfully performed. A representative analysis of spotted locoweed (*A. lentiginosus*) is shown in Fig. 9.3.

Table 9.2 Swainsonine content of selected *Astragalus* and *Oxytropis* species from North America, South America, and Asia

Species	Collection site	Swainsonine content[a]
A. lentiginosus var. *lentiginosus*	Holbrook, Arizona	Seed: 0.36%
A. lentiginosus var. *wahweapensis*	Henry Mountains, Utah	Whole plant: 0.037% (green) Whole plant: 0.023% (senescent)
A. mollissimus	Gladstone, New Mexico	Flower: 0.284%
A. assymetricus	Corral Hollow, California	Pod/seed: 0.039%
A. oxyphysus	Diablo Range, California	Pod/seed: 0.051%
A. pycnostachyus	Pomponio Creek, California	Pod/seed: 0.081%
O. lambertii	Cache Co., Utah	Leaf: 0.031%
O. sericea	Raft River Mountains, Idaho/Utah (1984–1986)	Leaf: 0.007–0.106% Flower: 0.008–0.103% Pod/seed: 0.022–0.154%
A. pehuenches	Chubut Prov., Argentina	Seed: 0.048% Leaf: 0.032%
A. pehuenches	Mendoza Prov., Argentina	Seed: 0.098% Leaf: 0.025%
A. garbancillo	Cajamarca, Peru	Leaf: 0.005%
A. junin	Cajamarca, Peru	Leaf: 0.005%
A. strictus	Lhasa, Tibet	Leaf: 0.003%
A. variabilis	Inner Mongolia	Leaf: 0.005%
O. deflexa	Inner Mongolia	Leaf: 0.012%
O. ochrocephala	Ningxia Prov., China	Leaf: 0.012%

[a] Based on dry weight of plant material.

A sufficient number of quantitative analyses have now been performed for some general trends to be apparent (Table 9.2). Typically, highest levels of swainsonine are found in the pods and seeds of the plant, e.g. 0.35 per cent in *A. lentiginosus* seed. However, amounts approaching this level have been measured in the blooms of wooly locoweed

(*A. mollissimus*). Swainsonine is conserved in senescent plant material, which accords with field observations that animals, including wildlife such as elk, can be poisoned by eating dead locoweed stems. There is considerable variation in swainsonine content during the growing season, but it is our opinion that levels exceeding 0.005 per cent should be of concern, particularly if animals have grazing access for extended periods. This assessment has led to a recommendation for prevention of locoweed-induced right-heart failure. Since the syndrome is a function of swainsonine intake, in combination with time spent at high altitude, it has been proposed that the infested range be heavily stocked and grazed at an early plant growth stage, before the more toxic seed pods are formed. With such a regimen the swainsonine intake of individual animals is below the toxic dose and the herd is subjected to only a short period of high altitude stress (James *et al.* 1991).

A particularly interesting facet of the development of a quantitative analysis for swainsonine is that it has provided an opportunity to analyse *Astragalus* and *Oxytropis* species from areas of the world other than North America. Samples have been examined from Inner Mongolia, Tibet, Peru, and Argentina and the measured swainsonine levels are comparable to those present in the author's locoweed samples (Table 9.2). In some cases this has established the cause of livestock poisoning of unknown etiology as a manifestation of locoism (Molyneux and Gómez-Sosa 1991; Cao *et al.* 1992).

Related glycosidase inhibitors

The structural similarity of swainsonine to mannose suggested a mechanistic basis for its inhibitory properties towards α-mannosidase and the glycoprotein processing enzyme, mannosidase II (Dorling *et al.* 1980). It was hypothesized, therefore, that swainsonine might be the prototype for a family of polyhydroxy alkaloids having similar properties, inhibition of any particular glycosidase depending upon the stereochemistry of the individual alkaloid. A survey of the literature revealed that the only known related alkaloid at that time was castanospermine (Fig. 9.4), which had been isolated from the seeds of the Moreton Bay chestnut, *Castanospermum australe* (Hohenschutz *et al.* 1981). Although this large Australian tree is quite different in form from the herbaceous weeds *Astragalus*, *Oxytropis*, and *Swainsona*, all are members of the family Leguminosae. Moreover, the seeds are toxic to cattle (McKenzie *et al.* 1988). Castanospermine had not been tested for glycosidase inhibitory properties but since the tree had been introduced into Southern California as an ornamental, the authors and co-workers were able to re-isolate the

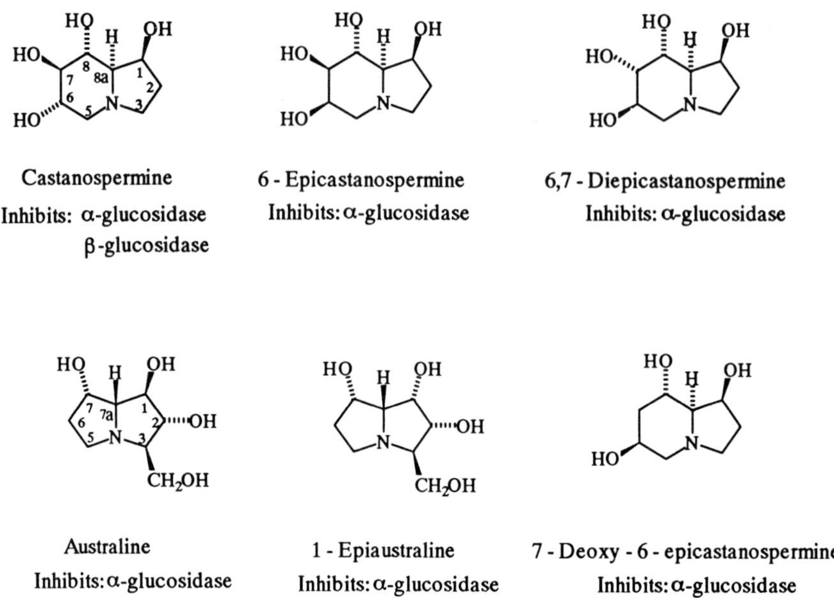

Fig. 9.4. Structures of polyhydroxy-indolizidine and -pyrrolizidine alkaloids isolated from *Castanospermum australe*.

alkaloid and determine its biochemical activity. It was found to be a potent inhibitor of α- and β-glucosidase, consistent with its structural similarity to glucose (Saul *et al.* 1983, 1984).

Thin-layer chromatographic examination of the alkaloid mixture remaining after crystallization of most of the castanospermine revealed a profusion of minor alkaloids, three of which were closely related to castanospermine, namely 6-*epi*castanospermine, 6,7-di*epi*castanospermine, and 7-deoxy-6-*epi*castanospermine (Fig. 9.4) (Molyneux *et al.* 1986, 1990, 1991*b*). All of these were found to be inhibitors of α-glucosidase although less potent than castanospermine. Surprisingly, 6-*epi*castanospermine, in which the stereochemistry of the hydroxyl groups is analogous to those of mannose, does not inhibit α- or β-mannosidase.

In addition to the above compounds a novel alkaloid was isolated which was named australine (Molyneux *et al.* 1988*a*). The compound was found to have a pyrrolizidine ring system in contrast to the more typical indolizidine skeleton (Fig. 9.4). Mass spectrometric and NMR analysis established the positions and relative stereochemistry of the hydroxyl groups, a conspicuous feature being the CH_2OH group, which resulted in loss of 31 a.m.u. in the mass spectrum and gave a characteristic signal at δ 65.6 in the ^{13}C NMR. The absolute stereochemistry was established by X-ray crystallography. In contrast to pyrrolizidine alkaloids which occur in many other plant genera, especially *Senecio*, australine has the $-CH_2OH$ group appended to the ring system at the 3- rather than the 1-position. Three additional australine epimers have subsequently been isolated from *C. australe* (Nash *et al.* 1988, 1990; Harris *et al.* 1989), namely 1-*epi*-, 3-*epi*-, and 7-*epi*-australine. These establish 3-hydroxymethylpyrrolizidines as a general class of alkaloids, which might be supposed to be more widely distributed in the Leguminosae. Australine is a potent inhibitor of α-glucosidase and the processing enzyme glucosidase I (Tropea *et al.* 1989).

Biosynthesis of castanospermine and australine

A very minor constituent of *C. australe* is the alkaloid *N*-hydroxymethyl-2-hydroxymethyl-3-hydroxypyrrolidine (Fig. 9.5) (Molyneux *et al.* 1991*b*),

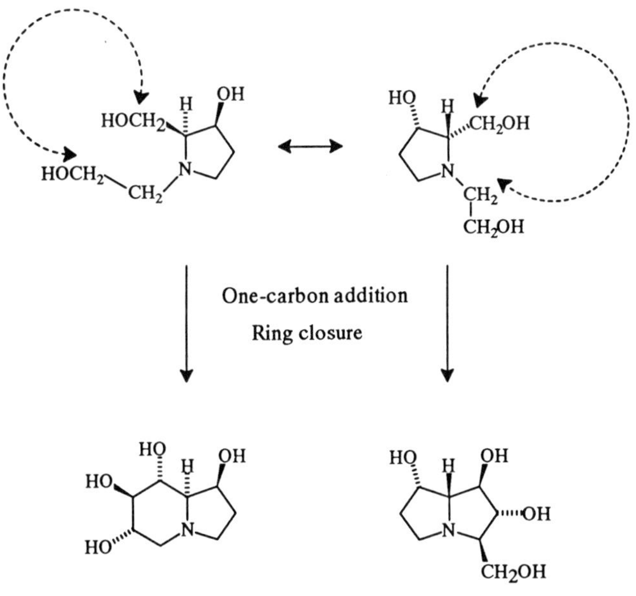

Fig. 9.5. Hypothetical biosynthetic route to castanospermine and australine, via a one-carbon addition and ring closure.

which is closely related to 2-hydroxymethyl-3-hydroxypyrrolidine isolated by Nash *et al.* (1985). The structure of the *N*-hydroxyethyl alkaloid suggests that it may be a common biosynthetic precursor of the castanospermines and australines. While swainsonine has been shown to be biosynthesized from pipecolic acid (Harris *et al.* 1988) the latter is not incorporated into castanospermine. A hypothetical alternative route to both the indolizidine and pyrrolizidine bicyclic systems from a pyrrolidine precursor is shown in Fig. 9.5. Addition of a single carbon moiety between the $-CH_2OH$ group and either the α or β carbon of the $-CH_2CH_2OH$ substituent of the pyrrolidine would yield either australine or castanospermine. Synthesis of the radio-labelled pyrrolidine and appropriate feeding experiments with *C. australe* would test this hypothesis.

Enzyme inhibitory properties and biological activity

All of the polyhydroxy-indolizidine and -pyrrolizidine alkaloids which have been isolated to date exhibit some degree of glycosidase inhibitory activity. The concentration of alkaloid required to produce 50 per cent inhibition of enzyme activity is shown in Table 9.3, for assays performed under comparable conditions. All of the alkaloids are competitive inhibitors. It is noteworthy that swainsonine, australine, and castanospermine are the most potent inhibitors. Epimerization of a single hydroxyl group brings about a significant reduction in activity, while loss of a hydroxyl group decreases the activity even more markedly. These results have been supported by the synthesis of many additional epimers of

Table 9.3 Concentration of polyhydroxy alkaloids to produce 50 per cent inhibition of specific glycosidases

Glycosidase	Alkaloid	Concentration (μM)
α-Glucosidase (amyloglucosidase)	Castanospermine	8
	6-*Epi*castanospermine	20
	6,7-Di*epi*castanospermine	84
	7-Deoxy-6-*epi*castanospermine	230
	Lentiginosine	32
	Australine	6
	1-*Epi*australine	26
β-Glucosidase	Castanospermine	21
α-Mannosidase	Swainsonine	2

castanospermine and swainsonine, none of which are as active as the natural products. However, a bicyclic ring system is not essential for inhibition since a number of polyhydroxy-pyrrolidine and -piperidine alkaloids show significant activity. It is apparent that prediction of the degree or specificity of activity is not easily correlated with structure. The structure–activity relationships of this group of alkaloids seem admirably suited for molecular modelling studies, preliminary attempts having already been made (Winkler and Holan 1989).

The glycosidase inhibitory properties of swainsonine and castanospermine have proven to be particularly significant in understanding glycoprotein processing, an essential function of all living cells. Glycoproteins are receiving increasing recognition as a very important group of biopolymers, especially in relation to cell–cell recognition. Interruption of the processing sequence, using the alkaloidal glycosidase inhibitors as very specific biochemical probes, can thus be used to elucidate the structural features of glycoproteins essential for proper functioning of various cell types.

This fundamental mode of action at the cellular level has stimulated the evaluation of swainsonine and castanospermine for their effects upon a wide variety of animal, plant, fungal, and viral systems. Some of the diverse activities which have been discovered are given in Table 4. Especially significant is the anti-viral, including anti-HIV, activity of castanospermine (Gruters *et al.* 1987; Tyms *et al.* 1987; Walker *et al.* 1987). A considerable effort is now being devoted to the synthesis of pharmacologically more effective, i.e. less water-soluble, derivatives (Margolin *et al.* 1990). Equally noteworthy is the anti-metastatic activity of swainsonine

Table 9.4 Biological activities and potential therapeutic applications of swainsonine and castanospermine

Alkaloid	Biological activity
Swainsonine	Animal model for human genetic mannosidosis
	Enhancement of immune response
	Inhibition of tumor growth
	Suppression of metastasis
Castanospermine	Animal model for Gaucher's disease in humans
	Suppression of hyperglycemic response (antidiabetic)
	Antifeedant to bean beetles and aphids
	Inhibitor of root elongation in plants
	Antiviral and antiretroviral activity
	Inhibition of infectivity and replication of HIV

towards malignant melanoma, and corresponding increase in survival rates, when it is administered to mice in the drinking water at very low dose rates (Humphries *et al.* 1990). Such findings will undoubtedly stimulate the evaluation of additional alkaloidal glycosidase inhibitors as potential therapeutic reagents.

Conclusions

Phytochemical examination and analysis is a *conditio sine qua non* for obtaining the information necessary to prevent losses of livestock due to consumption of the multitude of poisonous plants infesting rangelands in the Western United States. A rational research approach, well illustrated by the locoweed problem, can successfully prevent major economic losses due to toxic range plants. Moreover, a comprehensive understanding of the biochemical mode of action of the phytotoxic constituents can suggest structure–activity relationships leading to natural products with particular significance to human health and welfare.

Acknowledgements

The author wishes to acknowledge with deep gratitude the exceptional collaboration of his Agricultural Research Service colleagues at the Poisonous Plant Research Laboratory, Logan, Utah, led by Dr Lynn F. James; and his co-workers at the Western Regional Research Center, Albany, California. He is also indebted to the co-operation of the research groups led by Dr Alan D. Elbein (University of Texas at San Antonio and University of Arkansas Medical School), Dr Thomas M. Harris (Vanderbilt University), Dr George W. J. Fleet (Oxford University), and Dr Kenneth Olden (Howard University). Each of the these groups provided unique expertise which greatly expanded the scope of the research beyond the phytochemistry of toxic range plants.

References

Barneby, R. C. (1952). A revision of the North American Species of *Oxytropis* D.C. *Proc. California Academy of Sciences*, **27**, 177–312.

Barneby, R. C. (1964). *Atlas of North American Astragalus*. The New York Botanical Garden, NY.

Binns, W., Shupe, J. L., Keeler, R. F., and James L. F. (1965). Chronologic evaluation of teratogenicity in sheep fed *Veratrum californicum*. *J. Am. Vet. Med. Assoc.*, **147**, 839–42.

Cao, G. R., Li, S. J., Duan, D. X., Molyneux, R. J., James, L. F., Wang, K., and Tong, C. (1992). The toxic principle of Chinese locoweeds (*Oxytropis* and *Astragalus*): Toxicity in goats. In *Proceedings of the Third International Symposium on Poisonous Plants* (ed. L. F. James, R. F. Keeler, E. M. Bailey, P. R. Cheeke, and M. P. Hegarty), pp. 117–21. Iowa State University Press, Ames, Iowa.

Colegate, S. M., Dorling, P. R., and Huxtable, C. R. (1979). A spectroscopic investigation of swainsonine: An α-mannosidase inhibitor isolated from *Swainsona canescens*. *Aust. J. Chem.*, **32**, 2257–64.

Crawford, A. C. (1908). *Barium, a cause of the locoweed disease*. Bulletin 129. US Department of Agriculture, Bureau of Animal Industry, Washington, DC.

Dorling, P. R., Huxtable, C. R., and Vogel, P. (1978). Lysosomal storage in *Swainsona* spp. toxicosis: An induced mannosidosis. *Neuropath. Appl. Neurobiol.*, **4**, 285–95.

Dorling, P. R., Huxtable, C. R., and Colegate, S. M. (1980). Inhibition of lysosomal α-mannosidase by swainsonine, an indolizidine alkaloid isolated from *Swainsona canescens*. *Biochem. J.*, **191**, 649–51.

Dorling, P. R., Colegate, S. M., and Huxtable, C. R. (1989). Toxic species of the plant genus *Swainsona*. In *Swainsonine and related glycosidase inhibitors* (ed. L. F. James, A. D. Elbein, R. J. Molyneux, and C. D. Warren), pp. 14–22. Iowa State University Press, Ames, Iowa.

Gruters, R. A., Neefjes, J. J., Tersmette, M., de Goede, R. E. Y., Tulp, A., Huisman, H. G., *et al.* (1987). Interference with HIV-induced syncytium formation and viral infectivity by inhibitors of trimming glucosidase. *Nature*, **330**, 74–7.

Harris, C. M., Campbell, B. C., Molyneux, R. J., and Harris, T. M. (1988). Biosynthesis of swainsonine in the Diablo locoweed (*Astragalus oxyphysus*). *Tetrahedron Lett.*, **29**, 4815–18.

Harris, C. M., Harris, T. M., Molyneux, R. J., Tropea, J. E., and Elbein, A. D. (1989). 1-*Epi*australine, a new pyrrolizidine alkaloid from *Castanospermum australe*. *Tetrahedron Lett.*, **30**, 5685–8.

Hino, M., Nakayama, O., Tsurumi, Y., Adachi, K., Shibata, T., Terano, H., *et al.* (1985). Studies of an immunomodulator, swainsonine. I. Enhancement of immune response by swainsonine *in vitro*. *J. Antibiot.*, **38**, 926–35.

Hohenschutz, L. D., Bell, E. A., Jewess, P. J., Leworthy, D. P., Pryce, R. J., Arnold, E., and Clardy, J. (1981). Castanospermine, a 1,6,7,8-tetrahydroxyoctahydroindolizine alkaloid, from seeds of *Castanospermum australe*. *Phytochemistry*, **20**, 811–14.

Humphries, M. J., Matsumoto, K., White, S. L., Molyneux, R. J., and Olden, K. (1990). An Assessment of the effects of swainsonine on survival of mice injected with B16-F10 melanoma cells. *Clin. Exp. Metastasis*, **8**, 89–102.

Huxtable, C. R. and Gibson, A. (1970). Vacuolation of circulating lymphocytes in guinea pigs and cattle ingesting *Swainsona galegifolia*. *Aust. Vet. J.*, **46**, 446–8.

James, L. F. (1983). Neurotoxins and other toxins from *Astragalus* and related genera. In *Handbook of natural toxins*, Vol. 1., *Plant and fungal toxins* (ed. R. F. Keeler and A. T. Tu), pp. 445–62. Marcel Dekker, New York.

James, L. F., Keeler, R. F., Johnson, A. E., Williams, M. C., Cronin, E. H., and Olsen, J. D. (1980). *Plants poisonous to livestock in the Western States*. Agriculture Information Bulletin 415. US Department of Agriculture, Washington, DC.

James, L. F., Short, R. E., Panter, K. E., Molyneux, R. J., Stuart, L. D., and Bellows, R. A. (1989). Pine needle abortion in cattle: A review and report of 1973–1984 research. *Cornell Vet.*, **79**, 39–52.

James, L. F., Panter, K. E., Broquist, H. P., and Hartley, W. J. (1991). Swainsonine-induced high mountain disease in calves. *Vet. Hum. Toxicol.*, **33**, 217–9.

Keeler, R. F. and Binns, W. (1968). Teratogenic compounds of *Veratrum californicum* (Durand). V. Comparison of cyclopian effects of steroid alkaloids from the plant and structurally related compounds from other sources. *Teratology*, **1**, 5–10.

Maiden, J. H. (1901). *Plants reputed to be poisonous to stock in Australia*. Miscellaneous Publication 477. Department of Agriculture, New South Wales.

Margolin, A. L., Delinck, D. L., and Whalon, M. R. (1990). Enzyme-catalyzed regioselective acylation of castanospermine. *J. Am. Chem. Soc.*, **112**, 2849–54.

Marsh, C. D. (1909). *The locoweed disease of the plains*. Bulletin 112. US Department of Agriculture, Bureau of Animal Industry, Washington, DC.

Marsh, C. D. (1924). *Stock-poisoning plants of the range*. Bulletin 1245. US Department of Agriculture, Bureau of Animal Industry, Washington, DC.

McKenzie, R. A., Reichmann, K. G., Dimmock, C. K., Dunster, P. J., and Twist, J. O. (1988). The toxicity of *Castanospermum australe* seeds for cattle. *Aust. Vet. J.*, **65**, 165–7.

Molyneux, R. J. and Gómez-Sosa, E. (1991). Presencia del alcaloide indolizidinico swainsonine en *Astragalus pehuenches* (Leguminosae-Galegueae). *Bol. Soc. Arg. Bot.*, **27**, 59–64.

Molyneux, R. J. and James, L. F. (1982). Loco intoxication: Indolizidine alkaloids of spotted locoweed (*Astragalus lentiginosus*). *Science*, **216**, 190–1.

Molyneux, R. J., James, L. F., and Panter, K. E. (1985). Chemistry of toxic constituents of locoweed (*Astragalus* and *Oxytropis*) species. In *Plant toxicology* (ed. A. A. Seawright, M. P. Hegarty, L. F. James, and R. F. Keeler), pp. 266–78. Queensland Poisonous Plants Committee, Yeerongpilly.

Molyneux, R. J., Roitman, J. N., Dunnheim, G., Szumilo, T., and Elbein, A. D. (1986). 6-*Epi*castanospermine, a novel indolizidine alkaloid that inhibits α-glucosidase. *Arch. Biochem. Biophys.*, **251**, 450–7.

Molyneux, R. J., Benson, M., Wong, R. Y., Tropea, J. E., and Elbein, A. D. (1988*a*). Australine, a novel pyrrolizidine alkaloid from *Castanospermum australe*. *J. Nat. Prod.*, **51**, 1198–206.

Molyneux, R. J., Johnson, A. E., and Stuart, L. D. (1988*b*). Delayed manifestation of Senecio-induced pyrrolizidine alkaloidosis in cattle: Case reports. *Vet. Hum. Toxicol.*, **30**, 201–5.

Molyneux, R. J., Tropea, J. E., and Elbein, A. D. (1990). 7-Deoxy-6-*epi*-castanospermine, a trihydroxyindolizidine alkaloid glycosidase inhibitor from *Castanospermum australe*. *J. Nat. Prod.*, **53**, 609–14.

Molyneux, R. J., James, L. F., Panter, K. E., and Ralphs, M. H. (1991*a*). Analysis and distribution of swainsonine and related polyhydroxyindolizidine alkaloids by thin-layer chromatography. *Phytochem. Anal.*, **2**, 125–9.

Molyneux, R. J., Pan, Y. T., Tropea, J. E., Benson, M., Kaushal, G. P., and Elbein, A. D. (1991*b*). 6,7-Di*epi*castanospermine, a tetrahydroxyindolizidine alkaloid inhibitor of amyloglucosidase. *Biochemistry*, **30**, 9981–7.

Nash, R. J., Bell, E. A., Fleet, G. W. J., Jones, R. H., and Williams, J. M. (1985). The identification of a hydroxylated pyrrolidine derivative from *Castanospermum australe*. *J. Chem. Soc., Chem. Commun.*, 738–40.

Nash, R. J., Fellows, L. E., Plant, A. C., Fleet, G. W. J., Derome, A. E., Baird, P. D., *et al.* (1988). Isolation from *Castanospermum australe* and X-ray crystal structure of 3,8-di*epi*alexine, (1*R*,2*R*,3*S*,7*S*,8*R*)-3-hydroxymethyl-1,2,7-trihydroxypyrrolizidine [(2*S*,3*R*,4*R*,5*S*,6*R*)-2-hydroxymethyl-1-azabicyclo-[3.3.0]octan-3,4,6-triol]. *Tetrahedron*, **44**, 5959–64.

Nash, R. J., Fellows, L. E., Dring, J. V., Fleet, G. W. J., Girdhar, A., Ramsden, N. G., *et al.* (1990). Two alexines [3-hydroxymethyl-1,2,7-trihydroxypyrrolizidines] from *Castanospermum australe*. *Phytochemistry*, **29**, 111–14.

Nielsen, D. B., Rimbey, N. R., and James, L. F. (1988). Economic considerations of poisonous plants on livestock. In *The ecology and economic impact of poisonous plants on livestock production* (ed. L. F. James, M. H. Ralphs, and D. B. Nielsen), pp. 5–15. Westview Press, Boulder, Colorado.

Panter, K. E., Ralphs, M. H., Smart, R. A., and Duelke, B. (1987). Death camas poisoning in sheep: a case report. *Vet. Hum. Toxicol.*, **29**, 45–8.

Panter, K. E., Bunch, T. D., Keeler, R. F., Sisson, D. V., and Callan, R. J. (1990). Multiple congenital contractures (MCC) and cleft palate induced in goats by ingestion of piperidine-alkaloid containing plants: reduction in fetal movement as the probable cause. *J. Toxicol. Clin. Toxicol.*, **28**, 69–83.

Panter, K. E., Molyneux, R. J., Smart, R. S., Mitchell, L., and Hansen, S. (1993). English yew (*Taxus baccata*) poisoning in cattle: A case report. *J. Am. Vet. Med. Assoc.* (in press).

Pastuszak, I., Molyneux, R. J., James, L. F., and Elbein, A. D. (1990). Lentiginosine, a dihydroxyindolizidine alkaloid that inhibits amyloglucosidase. *Biochemistry*, **29**, 1886–91.

Saul, R., Chambers, J. P., Molyneux, R. J., and Elbein, A. D. (1983). Castanospermine, a tetrahydroxylated alkaloid that inhibits β-glucosidase and β-glucocerebrosidase. *Arch. Blochem. Biophys.*, **221, 593–7.**

Saul, R., Molyneux, R. J., and Elbein, A. D. (1984). Studies on the mechanism of castanospermine inhibition of α- and β-glucosidases. *Arch. Biochem. Biophys.*, **230**, 668–75.

Schneider, M. J., Ungemach, F. S., Broquist, H. P., and Harris, T. M. (1983). (1*S*,2*R*,8*R*,8a*R*)-1,2,8-Trihydroxyoctahydroindolizine (swainsonine), an α-mannosidase inhibitor from *Rhizoctonia leguminicola*. *Tetrahedron*, **39**, 29–32.

Tropea, J. E., Molyneux, R. J., Kaushal, G. P., Pan, Y. T., Mitchell, M., and Elbein, A. D. (1989). Australine, a pyrrolizidine alkaloid that inhibits amyloglucosidase and glycoprotein processing. *Biochemistry*, **28**, 2027–34.

Tyms, A. S., Berrie, E. M., Ryder, T. A., Nash, R. J., Hegarty, M. P., Mobberley, M. A., *et al.* (1987). Castanospermine and other plant alkaloid inhibitors of glucosidase activity block the growth of HIV. *Lancet*, 1025–6.

Walker, B. D., Kowalski, M., Goh, W. C., Kozarsky, K., Krieger, M., Rosen, C., *et al.* (1987). Inhibition of human immunodeficiency virus syncytium formation and virus replication by castanospermine. *Proc. Natl. Acad. Sci. USA*, **84**, 8120–4.

Williams, M. C. and Barneby, R. C. (1977). The occurrence of nitro-toxins in North American *Astragalus* (Fabaceae). *Brittonia*, **29**, 310–26.

Winkler, D. A. and Holan, G. (1989). Design of potential anti-HIV agents. 1. Mannosidase inhibitors. *J. Med. Chem.*, **32**, 2084–9.

10. Production and application of phytochemicals from an agricultural perspective

MICHAEL WINK

Universität Heidelberg, Institut für Pharmazeutische Biologie, Im Neuenheimer Feld 364, 6920 Heidelberg, Germany

Introduction

Animals and microorganisms (except the chemo- or photo-autotrophic bacteria) directly or indirectly rely on plants as a major and ultimate source of food. We can safely assume that plants have developed strategies against herbivorous animals or phytopathogenic microorganisms during evolution. Furthermore, plants compete with other plants (of the same or of different species) for light, water, and nutrients.

How do plants defend and protect themselves against microorganisms, which include bacteria, fungi, and viruses; herbivores (insects, mammals, molluscs, or nematodes); and other plants? We are well aware of the defensive strategies of animals against microbes and predators (Edmunds 1974; Blum 1981; Harborne 1988): the complex immune system with its cellular and humoral components is a well-studied area in the context of vertebrate—microbe interactions (Alberts *et al.* 1989). Against predating animals, weapons, armour, crypsis, thanatosis, deimatic behaviour, aposematism, flight or defence chemicals (usually called 'poisons') have evolved (Edmunds 1974).

It is evident that most of these possibilities are not available for plants with their sessile and 'passive' life style. Instead, we can distinguish the following defence mechanisms (Levin 1976; Deverall 1977; Swain 1977; Harborne 1988; Wink 1988, 1992*a,b*, 1993*a–c*), which are not independent and which may interact cooperatively and synergistically. We should be aware that many species have additionally evolved specialized traits and strategies this context. Strategies include:

1. If plants are wounded or if parts of them are eaten, this is usually not as fatal as the similar situation in animals, since plants can easily replace a lost leaf or branch (so-called 'open growth').

2. Mechanical protection provided by hooks, thorns, spikes, trichomes, glandular hairs, and stinging hairs (which are often supported by defence chemicals, such as histamine, or acetylcholine).
3. The formation of a thick bark in roots and stems can be considered as a sort of armour, the presence of hydrophobic cuticular layers as a penetration barrier directed against microbes which, at the same time, prevents the uncontrolled evaporation of water.
4. Laticifers and resin ducts full of latex or resin, respectively, which glue the mandibles of a chewing insect.
5. Cell walls are biochemically rather inert with reduced digestibility to many organisms because of their complex cellulose, pectin, suberin, and lignin molecules. Callose, suberin, and lignin are often accumulated at the site of infection or wounding and form a penetration barrier.
6. The most important strategy however, is the production and storage of defence chemicals, which are abundant and a typical trait of all plants (Table 10.1):
   (a) Plant surfaces are usually covered by a hydrophobic layer consisting of cuticular waxes which often incorporate other antibiotic and deterrant/repellent allelochemicals such as flavonoids.

Table 10.1 Number of known allelochemicals

Nitrogen-containing allelochemicals	
Alkaloids	> 10 000
Amines	> 100
Non-protein amino acids	> 400
Cyanogenic glycosides	> 50
Glucosinolates	> 100
Lectins	> 100
Protease inhibitors	> 50
Other allelochemicals (glycosides, terpenes)	
Monoterpenes	> 1 000
Sesquiterpenes	> 1 500
Diterpenes	> 1 000
Triterpenes/steroids	> 800
Tetraterpenes	> 350
Polyketides	> 700
Polyacetylenes	> 750
Flavonoids	> 1 200
Phenylpropanoids	> 500

(b) Synthesis of inhibitory or toxic proteins (e.g. lectins, protease inhibitors, toxalbumins) or enzymes (e.g. chitinase, β-1,3-glucanase, hydrolases, nucleases) which degrade microbial cell walls or other microbial constituents, or peroxidase and phenolase, which help to inactivate phytotoxins produced by many bacteria and fungi. These proteins are either stored in the vacuole or are secreted as exoenzymes into the cell wall or the extra-cellular space (review: Chrispeels 1991; Wink 1993*c*). In addition, storage proteins (of cereals and legumes) are often deficient in particular essential amino acids, such as lysine or methionine. If animals feed on these seeds exclusively, their growth is substantially retarded.

(c) As a widely distributed and important trait, secondary metabolites with deterrent/repellent or toxic properties against microorganisms, viruses, and/or herbivores may be produced. These compounds can be constitutively expressed, they may be activated by wounding (e.g. cyanogenic glycosides, glucosinolates, coumaryl glycosides, alliin, ranunculin, etc.), or their *de-novo* synthesis may be induced by elicitors (so-called 'phytoalexins'), infection, or herbivory.

These defence products are often synthesized and stored at strategically important sites (epidermal tissues or in cells adjacent to an infection), or in plant parts that are especially important for reproduction and survival (flowers, fruits, seeds, bark, roots).

In animals, we can observe the analogous situation in that many insects and other invertebrates (especially those, which are sessile and unprotected by armour, i.e. many marine species (Fenical 1986)), but also some vertebrates store protective secondary metabolites which are often similar in structure to plant allelochemicals. In many instances, the animals have obtained these toxins from their host-plants (Rosenthal and Janzen 1979; Duffey 1980; Blum 1981; Harborne 1988). Hardly any zoologist or ecologist doubts that the principal function of these secondary metabolites (which are often termed 'toxins' in this context) in animals is that of protection against predators or microorganisms.

More than 30 000 'natural products' or 'secondary metabolites' have been reported from plants so far (Table 10.1) (Swain 1977; Harborne 1988; Wink 1988). Since only 5–10 per cent of all higher plants which consist of >300 000 species have been analysed phytochemically in some detail, the overall real number of secondary products is certainly very large and exceeds 100 000 compounds.

Although the biological function of many plant-derived secondary metabolites has not been studied experimentally, it is now generally assumed that these compounds are important for the survival and fitness of a plant and that they are not useless waste products, as was suggested earlier in this century (Paech 1950; Mothes 1976). In each instance, it has to be determined whether a particular compound is active against microorganisms (viruses, bacteria, fungi), against herbivores (molluscs, arthropods, vertebrates), or against competing plants (so-called allelopathy).

Further functions are the attraction of pollinating or seed-dispersing animals, e.g. by coloured compounds, such as betalains (within the Centrospermae), anthocyanins, carotenoids, flavonoids, or fragrances, such as terpenes, amines, and aldehydes. Also physiological roles, such as UV protection (by flavonoids or coumarins), intra- or inter-specific signalling, or N-transport or N-storage are plausible.

Allelochemicals are often not directed against a single organism, but against a variety of potential enemies, or they may combine the roles of both deterrents and attractants; for example, anthocyanins and many essential oils are attractants in flowers, but are at the same time also insecticidal and antimicrobial. Thus, many natural products have *multiple functions* (for further examples see below).

It might be argued that the defence hypothesis cannot be valid since most plants, even those with extremely poisonous metabolites (from the human point of view), are nevertheless attacked by pathogens and herbivores. However, we have to understand and accept that chemical defence is not an absolute process. Rather, it constitutes a general barrier which will be effective in most circumstances, i.e. most potential enemies are repelled or deterred. Plants with allelochemicals at the same time represent an ecological niche for potential pathogens and herbivores. During evolution a few organisms have generally been successful in specializing towards that niche, i.e. on a particular toxic plant in that they found a way to sequester the toxins or become immune to them (Bernays 1982; Bernays and Chapman 1987; Bernays and Graham 1988; Harborne 1988). This is especially apparent in the largest class of animals, the insects (probably with several million species), which are often highly host-plant specific. The number of these 'specialists' is exceedingly small for a given plant species as compared to the number of potential enemies that are present in the ecosystem. We can compare this situation with our immune system: It works against the majority of microorganisms, but fails towards a few viruses, bacteria, fungi, and protozoa, which have overcome this defence barrier by clever strategies. Nobody would call the immune system and the antibodies useless because of these few adapted specialists! We should adopt the same argumentation when we consider the plants' defenses by secondary metabolites.

Exploitation and production of phytochemicals

Since many secondary metabolites have evolved in Nature as biologically active compounds with particular effects on other organisms, some of them are useful to mankind as pharmaceuticals, fragrances, flavours, colours, stimulants, or pesticides. In addition, many allelochemicals provide interesting molecular probes for the biochemist and lead structures for the organic medicinal chemist to be developed into new and more active compounds.

Pharmaceuticals

The world market for plant-derived chemicals in the fields of pharmaceuticals, fragrances, and flavours alone exceeds several billion $ per year and thus the agriculture of the respective plants is of economic importance (Balandrin *et al.* 1985). Whereas the whole plant is used as a crude drug in many instances (Deutsches Arzneibuch, DAB 9; Europäisches Arzneibuch, PhEu), several species are processed after harvest and the phytochemicals of interest are extracted (Table 10.2) and either used as extracts, mixtures, or as pure compounds.

The compounds listed in Table 10.2 have different economic impacts: Whereas some have retail markets that can exceed $50 million per year (e.g. cardiac glycosides, tropane alkaloids, *Papaver* alkaloids, quinine, *Catharanthus* alkaloids, essential oils), other markets are small and in the range of several million dollars per year or even smaller (e.g. sparteine, ricin, piperine, lipids, etc.).

Production of pharmaceuticals by intact plants

Since most natural products have a complex stereochemistry, their chemical synthesis is usually difficult and often hardly economic, especially if the total market is small. Therefore, plants still serve as a source for most of them (Table 10.2). Whereas a few plant species, that are used in pharmacy, are still collected from specimens in the wild, most species are regularly cultivated in many countries around the world, where some farmers have specialized in 'speciality crops'. Thus, growing medicinal plants is an interesting alternative to the production of commonly cultivated crop-plant species, which often suffer from a surplus production, at least in many countries of the Western world. The plant material is harvested and shipped to processing companies which are often based in Europe (especially Germany, Switzerland, Great Britain), North America, or Japan.

Table 10.2 Examples of postharvest processing (extraction, purification) of phytochemicals which are used pharmaceutically

Species	Compounds of interest
Atropa belladonna	Hyoscyamine
Hyoscyamus niger	Hyoscyamine
Datura stramonium	Scopolamine
Duboisia myoporoides	Scopolamine, hyoscyamine
Scopolia carniolica	Hyoscyamine
Nicotiana spp.	Nicotine, anabasine
Cephaelis ipecacuanha	Emetine
Cinchona spp.	Quinine, quinidine
Rauvolfia serpentina	Reserpine, ajmaline
Pausinystalia yohimbe	Yohimbine
Catharanthus roseus	Vinblastine, vincristine
Vinca minor	Vincamine
Chondodendron tomentosum	Tubocurarine
Strychnos toxifera	Toxiferine I
Strychnos nux-vomica	Strychnine, brucine
Coffea spp.	Caffeine
Camellia sinensis	Theophylline
Theobroma cacao	Theobromine, fat
Paullinia cupana	Caffeine
Ephedra spp.	L-Ephedrine
Physostigma venenosum	Physostigmine
Pilocarpus jaborandi	Pilocarpine
Lobelia inflata	Lobeline
Veratrum album	Protoveratrine
Colchicum autumnale	Colchicine, demecolcine
Taxus baccata	Taxol
Sanguinaria canadensis	Sanguinarine
Maclaeya cordata	Sanguinarine
Cytisus scoparius	Sparteine
Papaver somniferum	Morphine, thebaine, noscapine
Erythroxylum novogratense	Cocaine
Capsicum frutescens	Capsaicine
Piper nigrum	Piperine
Digitalis purpurea	Cardiac glycosides
Digitalis lanata	Cardiac glycosides
Urginea maritima	Proscillaridin
Strophantus gratus	g-Strophantin (ouabain)
Aesculus hippocastanum	Aescin
Glycyrrhiza glabra	Glycyrrhizin

Table 10.2 (*continued*)

Species	Compounds of interest
Aloe spp.	Aloin A,B
Podophyllum peltatum	Podophyllotoxin
Silybum marianum	Silymarin
Fagopyrum esculentum	Rutin
Sophora japonica	Rutin
Ammi visnaga	Khellin, visnadin
Quercus infectoria	Tannin
Eucalyptus globulus	1,8-Cineol, essential oil
Mentha x piperita	Menthol, essential oil
Zingiber officinale	Essential oil, oleoresin
Melissa officinalis	Essential oil
Salvia officinalis	Essential oil
Lavandula angustifolia	Essential oil
Thymus vulgaris	Thymol, essential oil
Rosmarinus officinalis	Essential oil
Cinnamomum camphora	Camphor
Citrus aurantium	Essential oil
Citrus limon	Essential oil
Pinus spp.	Essential oil
Juniperus communis	Essential oil
Matricaria recutita	Essential oil
Syzygium aromaticum	Eugenol, essential oil
Carum carvi	Essential oil
Pimpinella anisum	Essential oil
Foeniculum vulgare	Essential oil
Valeriana officinalis	Essential oil, valepotriates
Commiphora molmol	Essential oil
Ginkgo biloba	Ginkgolides A,B,C, bilobalide
Curcuma spp.	Curcumin
Artemisia absinthum	Sesquiterpene lactones
Humulus lupulus	Iso-humulon, lupulon
Olea europaea	Oil
Prunus dulcis	Oil
Arachis hypogaea	Oil
Brassica napus	Oil
Sesamum indicum	Oil
Glycine max	Oil
Linum usitatissimum	Oil
Cocos nucifera	Oil
Ricinus communis	Oil, ricin

Table 10.2 (*continued*)

Species	Compounds of interest
Oenothera biennis	Oil
Hydnocarpus kurzii	Oil
Copernicia prunifera	Waxes
Simmondsia chinensis	Waxes
Viscum album	Viscotoxin
Abrus precatorius	Abrine
Canavalia ensiformis	Concanavalin A
Echinacea purpurea	Polysaccharides
Allium sativum	Alliin, essential oil

A number of medicinal plants have been improved through intensive plant breeding with respect to yields or resistance towards pathogens, e.g. plants producing tropane alkaloids, cardiac glycosides, or essential oils. It is likely that this is an area which could be profitably improved in the future. Genetic engineering may have an impact in this field, too, by either increasing yields or by altering the existing patterns of phytochemicals. An interesting approach regarding the latter aspect, was reported recently by Hashimoto and Yamada (1992). They transformed plants of *Atropa belladonna* with the hyoscyamine-6β-hydroxylase gene, thus converting hyoscyamine via 6-hydroxyhyoscyamine into scopolamine, which is in higher demand on the pharmaceutical market than atropine. The resulting plants accumulate only scopolamine in their aerial parts.

Production by in-vitro *cell and tissue culture*

During the last three decades considerable R&D efforts were directed towards the biotechnological production of plant-based pharmaceuticals in cell and tissue culture, which theoretically could provide a more economic and convenient source of phytochemicals than intact plants (Zenk 1982; Berlin 1984). Although a few cell culture systems were established which produced considerable amounts of a secondary product, such as shikonin, berberine, rosmarinic acid, anthraquinones, betalaines, and several phenolics (Morris *et al.* 1986; Constabel and Vasil 1987; Kurz 1989; Charlwood and Rhodes 1990), the *in-vitro* production of most of the compounds which are valuable for pharmacy (Table 10.2) was usually a failure. These compounds still cannot be produced in callus or cell suspension cultures in reasonable quantities or economic ways. It became more and more clear that some pathways are turned off in undifferentiated cell culture systems or that necessary storage spaces are missing, leading to the rapid degradation of the wanted product (Morris *et al.* 1986; Constabel and Vasil 1987; Kurz 1989; Charlwood and Rhodes 1990).

In those situations, in which a secondary product is being made in roots, such as tropane alkaloids, nicotine, berberine, and others, root cultures (either normal or hairy roots, induced by *Agrobacterium rhizogenes*) proved to be interesting culture systems and often resulted in comparable or even higher yields than differentiated plants (review: Rhodes *et al.* 1990). But root cultures and even more so shoot cultures, grow slowly and are much more difficult to handle in fermentor systems than microbial or suspension-cultured cells.

Perspectives

As long as phytochemicals are not in short supply from plants cultivated in the field (e.g. tropane alkaloids, cardiac glycosides), it is probably not economic to use root or shoot cultures instead, since the overhead costs for the biotechnological production plants are orders of magnitude higher than for field cultivation and the costs for down-stream processing are almost identical in both approaches.

For rare and expensive phytochemicals (e.g. vincristine) the biotechnological approach is still interesting, but here the *in-vitro* process either fails or is still disappointing. We have to concede that this failure comes from our lack of knowledge concerning the underlying biochemical and genetical control mechanisms. However, using the now available methods of molecular biology it should be possible to switch on the genes that are necessary for the biosynthesis and storage of a particular compound in those undifferentiated cells, where their expression is normally turned off. Alternatively, the genes encoding the enzymes of a secondary product pathway could be expressed in bacteria or yeasts in order to produce the pharmaceutical in question (Wink 1989). But much more work on the enzymology and accumulation of allelochemicals and the corresponding genes is necessary, a development which certainly needs many years of future research.

Applications of phytochemicals in agriculture

This section will not address those plant-derived pharmaceuticals which are used in husbandry (most of the relevant points were discussed above), but will focus on the problem of using phytochemicals as natural pesticides (herbicides, fungicides, and insecticides) in plant cultivation. Special emphasis will be on natural insecticides (see also Chapters 3 and 4 in this volume) which seems to be a most interesting field.

Modern agriculture can only exist if chemical means are employed to protect crops against microorganisms, especially fungi (fungicides), competing plants (herbicides) and insects (insecticides), and sometimes nematodes (nematicides) and snail and slugs (molluscicides). The reasons why cultivated crops are so much more susceptible to their enemies than their wild counterparts, are manyfold:

1. Mass cultivation of any plant provides an attractive ecological niche for potential enemies and thus the selection of them is favoured by modern large-scale agriculture.
2. Many crop plants have lost their natural resistance that was mediated through secondary metabolites. These traits were neglected during plant breeding for high yields or were abolished on purpose, as outlined later for lupins.
3. Synthetic pesticides usually contain one active ingredient, whereas natural resistance is often mediated by a cocktail of poisons which makes adaptations by herbivores much more difficult.
4. The equilibrium between pests and their natural enemies is often disturbed in modern agriculture (see Chapter 6 on tritrophic interactions in this volume).

Because of the increasing resistance of pests against synthetic pesticides, new plant protectives are always in demand. Furthermore, synthetic pesticides are often very persistent in soil and water (especially halogenated hydrocarbons) and can accumulate in the food chain. Since many of them are also toxic, safe alternatives are always in demand.

Farming in third-world countries is confronted with even more severe problems. Most of these countries have a fast growing population which needs to be fed. On the other hand, the economy of many of these countries often does not allow the purchase of safe and efficient pesticides, which are in use in Europe or the USA. Instead, cheap alternatives (such as DDT) are used which have been banned in developed countries because of the inherent environmental or health problems. Here, a cheap and safe strategy is of urgent importance to promote food self-sufficiency, which can be achieved better by 'low-input agriculture' than by the 'high-tech agriculture' of the Western world.

The question is therefore, whether plant-derived pesticides could be an alternative. As illustrated in the following, these compounds are also active pesticides and usually do not persist in the environment for long, since they are rapidly degraded by microorganisms or they are unstable under sunlight. Furthermore, if these compounds can be obtained from plants grown locally, this would help third-world countries to become food self-sufficient and less dependent on second- and first-world countries and petrochemical industries. However, since these allelochemicals evolved in nature as defence compounds, they cannot be considered as harmless and a thorough investigation of their safety and toxicology will be necessary before exploiting them in the field (see 'Outlook').

In order to appreciate the problems and chances of applying natural insecticides, a small number of case studies (Frear 1948; Heal *et al.* 1950; Jacobson and Crosby 1971; Jacobson 1975, 1988; Schmutterer *et al.* 1981; Schmutterer and Ascher 1984, 1987; Stoll 1986) will be described in the following sections. The compounds mentioned do not necessarily represent the natural products which are best suited for their purpose, but at least some experience exists regarding their exploitation as natural insecticides.

Nicotine

Nicotine (**1**) is the main alkaloid of many *Nicotiana* species, such as *N. tabacum*. In *N. rusticana* leaves it reaches concentrations of 2–8 per cent. Nicotine is a slightly lipophilic alkaloid, that can easily penetrate skin and biomembranes by simple diffusion. Thus, it can be quickly resorbed by target organisms. Nicotine is a strong agonist of the acetylcholine (ACH) receptor in insects and vertebrates, and acts as a deadly poison even at low concentrations (lethal dose (LD_{50}) in mice after intravenous application (i.v.) is 0.3 mg/kg). It modulates signal transfer in neuronal synapses and at the neuromuscular plate.

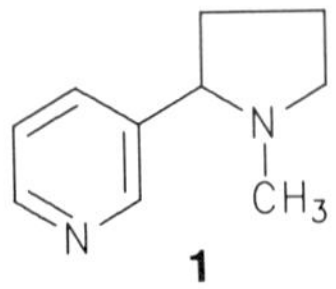

N
CH_3
N
1

It has been long known, that extracts from *Nicotiana* species can be effectively used to kill ectoparasites and agricultural pests (insects of various classes). Nicotine, which is produced as a byproduct of the tobacco industry, was used as a 40 per cent solution of nicotine sulphate in the US ('Black leaf 40'). As a contact poison it is most effective as soap, i.e. as the laurate, oleate, or naphthenate. As a stomach poison a combination with bentonite (a vulcanic mineral) has come into use (Merck Index 1989).

Since nicotine is highly toxic, many cases of intoxication were reported from farm workers, which resulted in nausea, vomiting, evacuation of bowel and bladder, mental confusion, twitching, and convulsions. Whereas nicotine was commonly used until the Second World War, it was widely abandoned, because of its toxicity, in the US and Europe, when synthetic insecticides became more available. Also derivatives of nicotine, such as nornicotine, anabasine, anatabine, and nicotyrine have insecticidal properties.

Nicotine is lethal for most insects, except the tobacco hornworm, *Manduca sexta*. *M. sexta* is a specialist and can be fed on daily nicotine doses which are fatal for man (Heidrich and Wink, unpublished). It is suspected that the insensitivity of this insect is achieved by a mutation of the ACH receptor; through target site modification nicotine no longer binds to the receptor and thus cannot perform its toxic activity. It is hoped to test this hypothesis at the molecular level in the author's laboratory.

Since nicotine is such a potent insecticide, its use might still be interesting even today provided that safe formulations and applications can be developed. For example, in cases, when pests become resistant to synthetic compounds or in third-world countries, where nicotine could be produced at relatively low costs.

Rotenone

The isoflavanone rotenone (**2**) is produced by plants of the genus *Derris* (Fabaceae), such as *D. elliptica*, a small scrub originating from tropical rain forests of the Malayian archipelago. Roots of *Derris* species are especially rich in rotenone (up to 6–8 per cent). Also roots of *Tephrosia* and *Lonchocarpus* store rotenone derivatives.

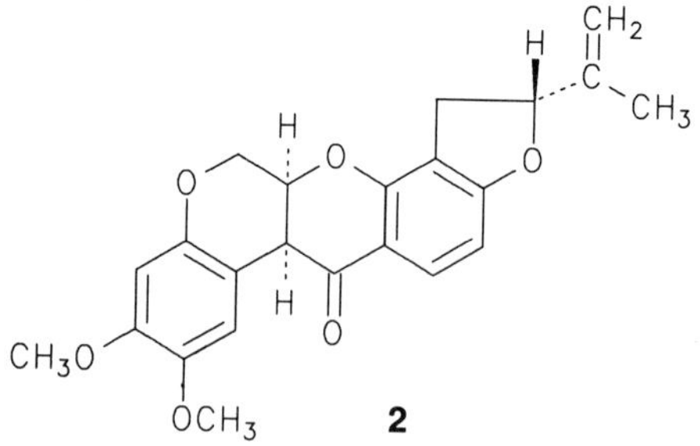

2

Rotenone is a powerful inhibitor of the electron chain in mitochondria and thus a potent toxin for all animals, including pests. Natives have used extracts of *Derris* roots as an insecticide and as a fish poison for some time. After 1848 until the advent of DDT, rotenone became one of the most widely used insecticides worldwide (Atwal 1976). The USA imported substantial amounts of *Derris* roots from Java, Sumatra, the Philippines, and Malaysia; for example, import figured 3954 t in 1939, 6727 t in 1947, 2356 t in 1948, and 1500 t in 1963 (Atwal 1976). Within Europe (e.g. in Austria and Switzerland), *Derris* extracts were usually combined with pyrethrins (see below).

For its use as an insecticide, root powders of *Derris* are either mixed with neutral soap (1 kg powder + 1.5 kg soap in 100 litres water), paraffin oils, or with talcum or clay dust (1 kg *Derris* + 5 kg dust). About 30 kg/ha of the dust mixture are needed for an effective control of pests (Neumann *et al.* 1986; Stoll 1986). *Derris* preparations are effective in the control of various aphids, caterpillars, thrips, diptera, beetles, nematodes, and even fungi, and no resistance to this pesticide has been reported.

Rotenone is hardly phytotoxic, but does show some human toxicity (Merck Index 1989): inhalation or ingestion of large doses may cause numbness of oral mucosa, nausea and vomiting, muscle tremors, and tachypnea. Lethal effects are caused by respiratory paralysis. Chronic poisoning may produce fatty acid changes in liver and kidney. Direct contact sometimes causes a mild irritation of skin or mucosae.

Quassin

Quassia amara and *Picrasma excelsa* (Simaroubaceae) contain bitter-tasting seco-triterpenes, such as quassin (**3**), neoquassin, and 18-hydroxyquassin, which were used as a bitter tonic, a hop imitate, an anthelmintic, and as an insecticide. *Q. amara* is a small tree (2–6 m height) which originates from Brazil and Guiana and is cultivated in Colombia, Panama, and

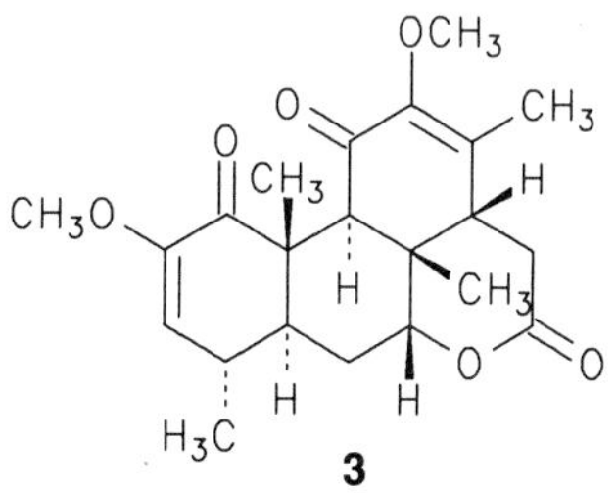

3

the West Indies. *Picrasma excelsa* is a larger tree (up to 20 m height) and inhabits Jamaica and the Caribbean islands.

Quassia extracts, prepared from the wood of both trees, were widely used against ectoparasites (lice) and flies. Extracts were prepared by cooking 500 g of wood chips in 10 litres water. Cooled extracts are mixed with 2 kg soap and dissolved in 5 litres water. The solutions are filled up to make 100 litres spray solution (Neumann *et al.* 1986; Stoll 1986).

Quassin acts as a contact, stomach, and systemic poison in insects (ectoparasites, lepidopteran larvae, beetles, and flies) and nematodes. *Quassia* extracts are also used for fly poison on flypaper. Quassin seems to be a selective pesticide with no recorded toxic effects on vertebrates. Applied in high concentrations, irritation of stomach and vomiting can occur in man. Since the triterpene is very bitter it should not be sprayed on edible crops prior to consumption.

Ryanodine

Ryania speciosa (Flacourtiaceae) is a tropical American (Amazonian basin, Caribbean) small tree and produces the insecticidal ryanodine (**4**). Dried leaves, stems and roots of *Ryania* are milled and mixed with inert talcum or clay dust in a ratio of 4:6. Since ryanodine breaks down in aqueous solution, application as a dust is indicated, although this form of application is less convenient than spraying.

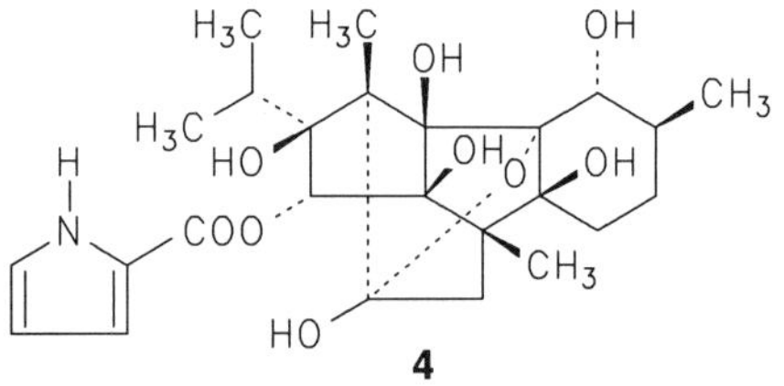

H_3C H_3C OH OH CH_3 H_3C HO OH O OH H N COO CH_3 HO

4

Ryanodine is a stomach and contact poison and constitutes a selective pesticide against cockroaches, *Tribolium*, other beetles, and lepidopteran larvae. It shows relatively low vertebrate toxicity (LD_{50}, per oral application (p.o.) in rats, mice, rabbits, guinea pigs was 1200, 650, 650, and 2500 mg/kg, respectively) (Merck Index 1989).

Ryania powder is well-suited to protect stored grain, such as maize and wheat. For example, 400 g *Ryania* powder can protect 1 t of maize or wheat for 2 years against *Tribolium castaneum*. Ryanodine was patented in 1946 by Merck & Co ('Ryanex, Ryanicide'), but its use is limited due to the shortage of raw material.

Pyrethrins

Flowers of the perennenial *Chrysanthemum cinerariifolium* (syn. *Pyrethrum cinerariifolium*) (Asteraceae) which originate from Dalmatia and Montenegro, produce a series of terpene esters (pyrethrin (**5**), cinerin, pyrethrosin, pyretol, chrysanthemine, etc.) with insecticidal properties. Also other *Chrysanthemum* species, such as *C. coccineum* and *C. marschallii* are exploited for the production of pyrethrins. At present, *Chrysanthemum* species are cultivated throughout the world, (e.g. in Japan, Kenya, Tanzania, Rwanda, Ecuador, Yugoslavia, California, Mexico, Chile, Brasil, and Russia). The *Chrysanthemum* material produced in developing countries is usually exported to Europe and USA.

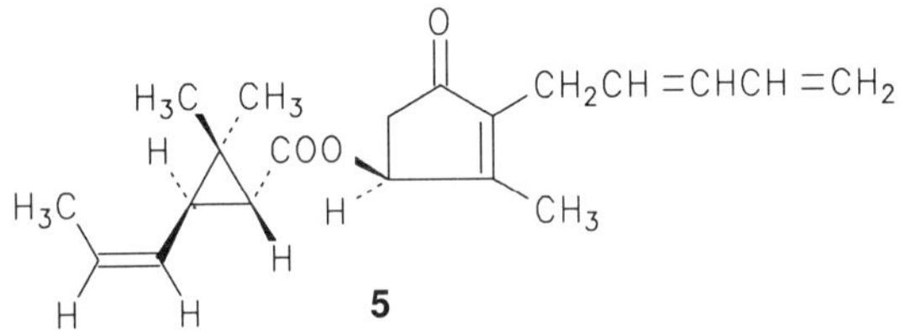

5

The insecticidal compounds include pyrethrin I and II and cinerine I and II and are especially abundant in carpels of young flowers, when at least 3/4 of the disc florets have opened. Crop yields may be as high as 200–1000 kg/ha. The active principles can be extracted by organic solvents, and since they are not soluble in water, they are applied as aerosols or as dusts which contain 20 per cent pyrethrin I and II and 5 per cent cinerins. Pyrethrins are quickly inactivated by light and oxygen but few resistant pests have developed as compared to synthetic pesticides.

Pyrethrins (see also Chapter 3) are active as a nervous and contact poison in insects, but also act as a deterrent. Pyrethrins have been found to be effective against many insects (beetles, lepidopteran larvae, aphids, flies, cochroaches, ants, mosquitos, locusts, thrips, etc.). The insecticidal activity can be enhanced by the use of synergists, like sesamine from sesame oil or piperonylbutoxide (which inhibits mixed function cytochrome oxidases *p*450), by antioxidants, such as hydroquinone and tannin, or by activators such as the ethyleneglycolether of pinene.

Pyrethrins are slightly toxic to warm-blooded vertebrates (LD_{50} in rats, p.o., 1.2 g/kg). They can cause severe allergetic dermatitis and systemic allergic reactions. Higher doses may cause nausea, vomiting, headache, and other CNS disturbances in man (which are also common side effects of synthetic insecticides).

Synthetic derivates, the so-called 'pyrethroids', have been synthesized and are much more stable and active (up to a factor of 1000) than the natural products. In Germany, it has been discussed whether pyrethroids should be banned because of their side effects (according to a note in the German Apothekerzeitung No. 14 of 30 March, 1992).

Azadirachtins

The neem tree *Azadirachta indica* (Syn. *Melia azadirachta*) and the chinaberry tree (*Melia azederach*) (Meliaceae) produce seeds which are rich in insect deterrent and insecticidal tetranortriterpenoids, such as 'azadirachtin' (**6**) (see Chapter 2). This plant and its constituents have been developed and tested intensively during the last two decades for use as a natural insecticide (Schmutterer *et al.* 1981; Schmutterer and Ascher 1984, 1987; Schmutterer 1990). Neem products were studied mainly at the laboratory level until 1984, from then on, practical aspects of plant cultivation, seed collection and storage, extraction procedures, and application obtained more attention (Dreyer and Hellpap 1991). This plant is already widely used in Asian countries and will have a wide impact in other countries of the third world (Schmutterer 1990) as a cheap and efficient pesticide, especially in the small-scale cultivation of vegetables.

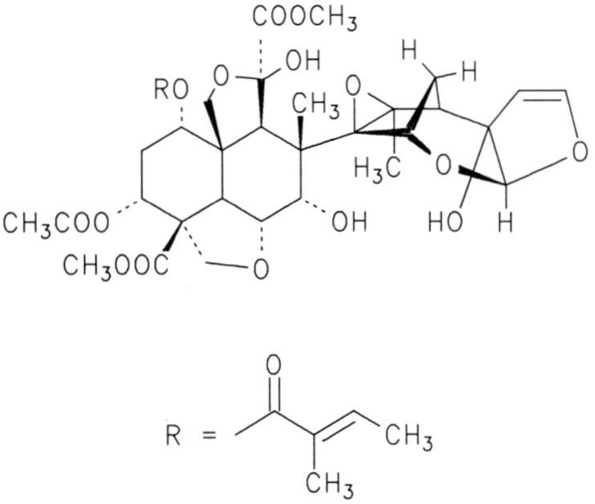

COOCH3
OH
RO
O
CH3
O
H
H
O
H3C
O
CH3COO
OH
HO
H
CH3OOC
O

O
R =
CH3
CH3
6

Extracts of the neem tree have a number of advantages:

1. They are effective in the control of a wide range ($n > 200$ species) of insects as contact and stomach poisons, because azadirachtin interferes with the metamorphosis (especially with the action of the moulting hormone ecdysone) of insects, a process absent in vertebrates. They are thus non-toxic to warm-blooded animals and non-mutagenic (Jacobson 1986). In addition, azadirachtin can act as a feeding deterrent. Azadirachtin is a comparatively active product: in the laboratory, 1 mg/kg shows clear effects.
2. The neem tree can be cultivated easily and can exist on dry, unfertile soil, and does not compete for space and nutrients with other crops. At present, neem trees are being grown all over Asia, Africa, and recently also in Tropical America (Dreyer and Hellpap 1991). In India, about 150 000 tons of neem oil are produced yearly already. The neem tree produces many seeds, which can be easily picked up from the ground.
3. Extracts can be prepared from milled seeds and leaves by ethanol or methanol. Aqueous extracts are not as effective because azadirachtin is only incompletely extracted by water. Nevertheless for practical applications in third-world countries, aqueous extracts are used. They are prepared from 50–100 g of milled seeds per litre and 400–600 litre/ha are applied (Dreyer and Hellpap 1991). In the long run, small processing plants which can apply solvent extractions are probably necessary since 3 to 7 g azadirachtin can be produced from 1 kg seeds.
4. Resistance of pests against azadirachtin and other neem products have not been recorded so far.

A commercial product, Margosan-O has been developed in the USA, but has been registered only for non-food plants and forestry (P. Düweke in *Die Zeit* 1991).

Quinolizidine alkaloids (QA)

Whereas the above mentioned plants and ingredients have already been introduced into the agricultural praxis, a number of potential candidates have reached a preparatory stage (Jacobson 1958, 1975, 1982). One of them is quinolizidine alkaloids **(7)**–**(11)**, which the author and his co-workers have studied intensively during the last 15 years (reviews in: Kinghorn and Balandrin 1984; Wink 1985*a*; 1987*a*,*c*, 1988, 1992*a*,*b*, 1993*a*–*c*).

QA constitute the main secondary products of many Leguminosae, especially in the genera *Lupinus*, *Genista*, *Cytisus*, *Baptisia*, *Thermopsis*, *Sophora*, and *Ormosia* (Wink 1992*a*,*b*, 1993*a*–*c*). The main structural

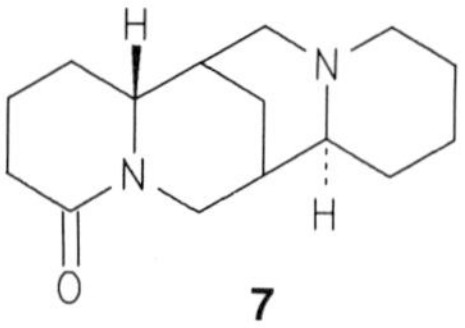

7

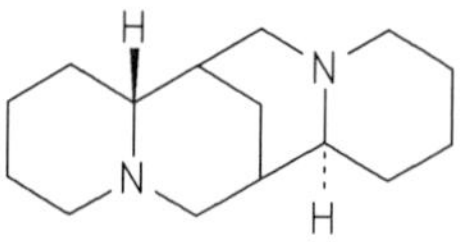

8

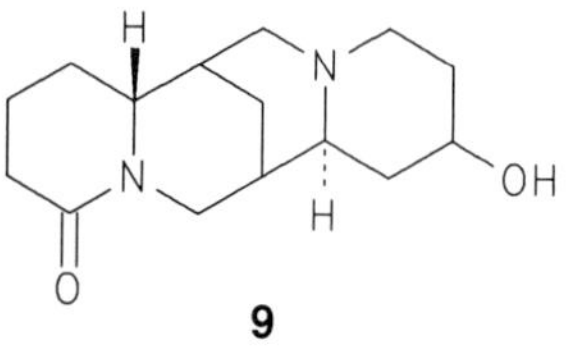

9

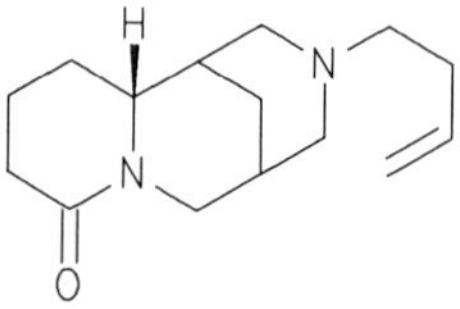

10

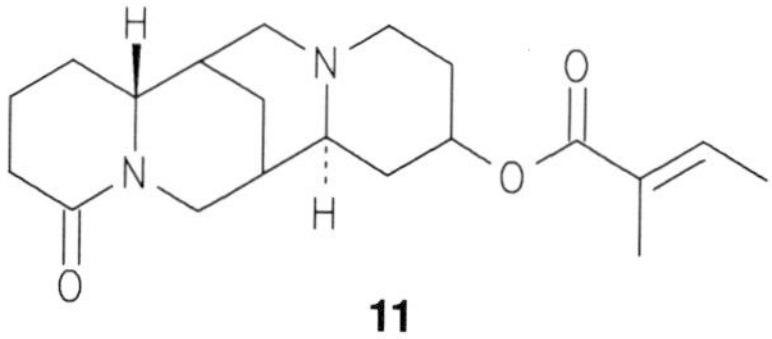

11

types belong to lupanine (7)/sparteine (8), multiflorine, 13α-hydroxylupanine (9), tetrahydrorhombifoline (10), 13α-tigloyloxylupanine (11), aphylline, anagyrine/cytisine, lupinine, matrine, and respective derivatives (Kinghorn and Balandrin 1984; Wink 1992*b*,*c*).

Table 10.3 presents an overview of the main biochemical and physiological events which are relevant for the biosynthesis, transport, storage, and degradation of QA in lupins and broom (Wink and Hartmann 1981, 1982, 1985; Wink 1983*b*, 1985*a*,*b*, 1986, 1987*b*, 1988, 1990, 1991, 1992*a*,*b*, 1993*a*–*c*; Wink and Witte 1984, 1985, 1991; Wink *et al.* 1984; Mende and Wink 1987).

Insecticidal and deterrent properties of QA We investigated the functions of QA in lupins and found that their main function is that of chemical defence besides minor functions in N transport in the phloem and N storage in the seed (Wink and Witte 1984).

QA were found to be feeding deterrents for a number of oligo- and poly-phagous insects, including aphids, moth and butterfly larvae, beetles, grasshoppers, flies, bees, and ants, i.e. QA are deterrents over a wide range of insect orders (Table 10.4) (Wink 1987*a*, 1993*a*,*b*).

The few electrophysiological experiments performed with QA so far indicate that sparteine and other QA elicit responses at chemical

Table 10.3 Overview of the biochemistry of quinolizidine alkaloids (after Wink 1984*a*,*b*; 1985*a*,*b*; 1987*c*, 1991, 1992*a*,*b*, 1993*a–c*)

Occurrence:	Fabaceae: *Lupinus*, *Cytisus*, *Genista*, *Laburnum*, *Thermopsis*, *Baptisia*, *Sophora*, etc.
Biosynthesis:	Sequence: Lysine ⟶ cadaverine ⟶ lupanine Enzymes involved: Lysine decarboxylase, oxosparteine synthase Alkaloid synthesis only in green tissue Localization of alkaloid biosynthesis in leaf chloroplast Regulation by light (pH, thioredoxin, precursor availability) Diurnal fluctuation of enzymes and alkaloids
Accumulation:	All parts of a plant accumulate alkaloids Vacuole intracellular storage compartment Epidermal/subepidermal tissues main site of alkaloid storage in leaves and stems Storage in seeds
Transport:	Long distance transport in phloem Passage across tonoplast with aid of an alkaloid transporter; Mg-ATP and K^+ necessary for uptake Also carrier-mediated transport across plasmalemma
Degradation:	Diurnal degradation in all organs During germination and growth of the seedling (mobilization of alkaloidal nitrogen)

sensilla of *Entomoscelis*, *Phormia*, and *Pieris* which lead to food rejection (Schoonhoven 1972; Mitchell and Sutcliffe 1984; Blades and Mitchell 1986). In addition to deterrence, QA seem to be toxic to many insects (Table 10.4). The mechanisms underlying the toxic effects observed in insects have not been elucidated. However, it was shown that QA can interfere with protein biosynthesis (Korcz *et al.* 1987; Wink 1992*a*, 1993*a*; Wink and Twardowski 1992). Since protein biosynthesis is an important and vulnerable process in most organisms, it is not surprising that QA have toxic properties for a wide range of organisms (Table 10.5). Furthermore, cytisine and other QA bind to acetylcholine (ACH) receptors with high affinity (K_d = 0.4 nM) and act as strong agonists similar in action to nicotine. Other targets could be K^+ channels, which are influenced by sparteine in vertebrates, or glutamate receptors, which can be modulated by matrine in cray fish neurons (Table 10.5; Kinghorn and Balandrin 1984; Wink 1992*a*,*b*, 1993*a–c*). The list is certainly far from complete.

Table 10.4 Deterrence or toxic effects of quinolizidine alkaloids in insects (after Wink 1992*a*, 1993*a*)

Insect species	Alkaloid	Effect		
		Insecticidal	Deterrent	Growth retardant
Aphidae				
Acyrthosiphon pisum	Sparteine		+	
	Lupinine		+	
	Cytisine		+	
Trialeurodes brassicae	QA extract	+		
Lepidoptera				
Spodoptera eridania	Sparteine			+
	Lupanine			+
Choristoneura fumiferana	13-Tigloyloxylupanine		+	
	13-*trans*-Cinnamoyl-oxylupanine		+	
Syntomis mogadorensis	Sparteine		+	
	Lupanine		+	
	Cytisine		+	
	17-Oxosparteine		+	
Manduca sexta	Sparteine		+	
Plutella maculipennis	Lupanine	+		
	Sparteine	+		
	13-Tigloyloxylupanine	+		
Pieris brassicae	Sparteine		+	
Homoptera				
Dysdercus spp.	Lupanine	+		
	Sparteine	+		
	13-Tigloyloxylupanine	+		

Table 10.4 (*continued*)

Insect species	Alkaloid	Effect		
		Insecticidal	Deterrent	Growth retardant
Orthoptera				
Melanoplus bivitatus	Lupinine		+	
Coleoptera				
Entomoscelis americana	Sparteine		+	
Callosobruchus maculatus	Sparteine	+		
Leptinotarsa decemlineata	QA extract	+		+
Phaedon spp.	Lupanine	+		
	Sparteine	+		
	13-Tigloyloxylupanine	+		
Diptera				
Phormia regina	Sparteine		+	
Ceratitis capitata	Lupanine	+		
	Sparteine	+		
	13-Tigloyloxylupanine	+		
Hymenoptera				
Apis melifera	Sparteine	+	+	
Formica rufa	Sparteine		+	
	Lupanine		+	
	Cytisine		+	

Table 10.5 Biological activities of quinolizidine alkaloids (QA) (for insect related activities see Table 10.4) (after Wink 1992*a*, 1993*a*)

Organism/Target	Alkaloid	Effect
1. Allelopathy		
Poa spp.	QA mixture	Inhibition of seed germination
Lactuca sativa	QA mixture	Inhibition of seed germination; reduction of radicle length
	Sparteine	Inhibition of seed germination
		Inhibition of radicle growth
	Lupanine	Inhibition of seed germination
	Cytisine	Inhibition of seed germination
	13-Tigloyloxylupanine	Inhibition of seed germination
Lepidum sativum	Cytisine	Inhibition of radicle growth
	Sparteine	Inhibition of radicle growth
	Lupanine	Inhibition of radicle growth
Raphanus sativum	Sparteine	Inhibition of radicle growth
Sinapis alba	Sparteine	Inhibition of radicle growth
2. Antiviral activity		
Potato X-virus in *Nicotiana*	Sparteine	Inhibition of viral multiplication
3. Antibacterial activity		
Streptococcus viridis (+)	Sparteine	Growth inhibition
Micrococcus luteus (+)	Sparteine	Growth inhibition
Mycobacterium phlei (+)	Sparteine	Growth inhibition
Bacillus megaterium (+)	Sparteine	Growth inhibition
Serratia marcescens (−)	Sparteine	Growth inhibition

Table 10.5 (*continued*)

Organism/Target	Alkaloid	Effect
3. Antibacterial activity (*continued*)		
Staphylococcus aureus (+)	13-Hydroxylupanine	Growth inhibition
	Lupanine	Growth inhibition
	Sparteine	Growth inhibition
	Angustifoline	Growth inhibition
Bacillus subtilis (+)	13-Hydroxylupanine	Growth inhibition
	Lupanine	Growth inhibition
	Sparteine	Growth inhibition
	Angustifoline	Growth inhibition
Bacillus thuringiensis (+)	13-Hydroxylupanine	Growth inhibition
	Lupanine	Growth inhibition
	Sparteine	Growth inhibition
	Angustifoline	Growth inhibition
Escherichia coli (−)	13-Hydroxylupanine	Growth inhibition
	Lupanine	Growth inhibition
	Sparteine	Growth inhibition
	Angustifoline	Growth inhibition
Pseudomonas aeruginosa (−)	13-Hydroxylupanine	Growth inhibition
	Lupanine	Growth inhibition
	Sparteine	Growth inhibition
	Angustifoline	Growth inhibition

Table 10.5 (*continued*)

Organism/Target	Alkaloid	Effect
4. Antifungal activity		
Aspergillus oryzae	Sparteine	Growth inhibition
Alternaria porri	Sparteine	Growth inhibition
Piricularia orycae	Sparteine	Growth inhibition
Helminthosporium carbonum	Sparteine	Growth inhibition
Rhizoctonia solani	Sparteine	Growth inhibition
Fusarium oxysporum	Sparteine	Growth inhibition
Erysiphe graminis f.sp.*hordei*	Lupanine	Inhibition of conidia germination
	Sparteine	Inhibition of conidia germination
	13-Tigloyloxylupanine	Inhibition of mildew development
5. Antinematode and antihelminth activity		
Bursaphelenchus xylophilus	*N*-Methylcytisine	Nematicidal
	Anagyrine	Nematicidal
	Cytisine	Nematicidal
Angiostrongylus cantonensis	*N*-Methylcytisine	Reduction of motility (spastical)
	Matrine	Reduction of motility (paralytical)
Dipylidium caninum	*N*-Methylcytisine	Reduction of motility (spastical)
	Matrine	Reduction of motility (paralytical)
Fasciola hepatica	*N*-Methylcytisine	Reduction of motility (spastical)
	Matrine	Reduction of motility (paralytical)

Table 10.5 (*continued*)

Organism/Target	Alkaloid	Effect
6. Mollusc deterrence		
Helix pomatia	Sparteine	Feeding deterrent
	Lupanine	Feeding deterrent
	Cytisine	Feeding deterrent
Biomphalaria glabrata	2,3-Dehydro-*O*-(2-pyrrolycarbonyl) virgiline	Molluscicidal
7. Vertebrate toxicity		
(a) Molecular/cellular targets:		
Acetylcholine receptor	Cytisine and other QA	Binding to ACH receptor, agonist
K^+-channels	Sparteine	Inhibition of K^+ channels
Nerve cells	Matrine	Inhibition of glutamate action
	13-Hydroxylupanine	Inhibition of Phe-*t*RNA binding and elongation
		Inhibition of *in-vitro* translation
	Lupanine	Inhibition of Phe-*t*RNA binding and elongation
		Inhibition of *in-vitro* translation
	Sparteine	Inhibition of Phe-*t*RNA binding and elongation
		Inhibition of *in-vitro* translation
	Angustifoline	Inhibition of Phe-*t*RNA binding and elongation
	13-Tigloyloxylupanine	Inhibition of Phe-*t*RNA binding
		Inhibition of *in-vitro* translation
	17-Oxosparteine	Inhibition of Phe-*t*RNA binding
		Inhibition of *in-vitro* translation

Table 10.5 (*continued*)

Organism/Target	Alkaloid	Effect
7. Vertebrate toxicity (*continued*)		
	Cytisine	Inhibition of phe-*t*RNA binding
		Inhibition of *in-vitro* translation
Mutagenicity	Anagyrine	Congenital malformations in calves
	Cytisine	Teratogenic in chicks and rabbits
(b) Toxicity:		
Cytotoxicity	Matrine	Antitumour activity in Ehrlich ascites cells
		Antitumour activity in mouse sarcoma
	Matrine-*N*-oxide	Antitumour activity in mouse sarcoma
General toxicity $LD_{50/100}$	Sparteine	LD_{100} i.p. Guinea pig 23–30 mg/kg
		LD_{50} i.p. rat 42–44 mg/kg
		s.c. rat, 68–75 mg/kg
		LD_{50} i.p. mouse, 55(m)–67(f) mg/kg
		i.v. mouse, 17(m)–20(f) mg/kg
		p.o. mouse, 350(m)–510(f) mg/kg
		LD_{100} p.o. rabbit, 450 mg/kg
		Lethal dose, i.v. rabbit, 20–30 mg/kg
		Lethal dose i.v. dog, 50–70 mg/kg
		Lethal dose i.v. pigeon, 40–50 mg/kg

Table 10.5 (*continued*)

Organism/Target	Alkaloid	Effect
General toxicity $LD_{50/100}$ (*continued*)	Lupanine	LD_{100}, i.p. Guinea pig 22–25 mg/kg
		LD_{50}, i.p. mouse, 80 mg/kg
		i.p. rat, 180–192 mg/kg
		i.p. Guinea pig, 210 mg/kg
		LD_{50} i.p. mouse, 175 mg/kg
		p.o. mouse, 410 mg/kg
		LD_{50} i.p rat, 177 mg/kg
		p.o. rat, 1464 mg/kg
	Lupinine	LD_{100}, i.p. Guinea pig, 28–30 mg/kg
	Epilupinine	LD_{50}, i.p. rat, 200–400 mg/kg
	13-Hydroxylupanine	LD_{100}, i.p. Guinea pig, 228 mg/kg
		s.c. Guinea pig, 456 mg/kg
		LD_{50}, i.p. rat, 199 mg/kg
		LD_{50}, i.p. mouse, 172 mg/kg
	17-Oxolupanine	LD_{50}, i.p. mouse 690 mg/kg
	Cytisine	LD_{100}, s.c. cat, 3 mg/kg
		s.c. dog, 4 mg/kg
		s.c. goat, 109 mg/kg
	Matrine	LD_{50}, i.p. mouse 150 mg/kg
	Matrine-*N*-oxide	LD_{50}, i.p. mouse 750 mg/kg
		i.v. mouse, 150 mg/kg

(+), (−): gram+ or gram− bacterium.

Further biological activities of QA

1. *Allelopathy*: Some evidence suggests activity against other competing plants (Wink 1983*a*, 1992*a*, 1993*a,b*) (Table 10.5). During germination QA are exported from the roots into the soil. The main alkaloids detected were 13-tigloyloxylupanine (11) and other esters. These alkaloids displayed the strongest effects on germination and seedling growth *in vitro*.
2. *Antiviral activity*: QA such as sparteine (8), inhibited the multiplication of potato-X virus in laboratory experiments (Wink 1987*a*). But under field conditions alkaloid-rich lupins were infected by a number of viruses which are transmitted by aphids that do not live on alkaloid-rich (so-called 'bitter') lupins but on other legumes or other crops. These aphids nevertheless probe bitter lupins and transmit the virus.
3. *Antimicrobial activity*: The growth of gram-negative and gram-positive bacteria and of certain fungi (Wink 1984*a*; Wippich and Wink 1985; Tyski *et al.* 1988) is significantly reduced by some QA *in vitro* (Table 10.5). However, 'bitter' lupins suffer from a number of fungal pathogens, such as *Rhizoctonia*, *Fusarium*, *Phoma*, *Ascochyta*, *Colletotrichum*, *Uromyces*, and *Erysiphe*, which are obviously not very sensitive towards QA. In case of the fungus *Phomopsis leptostromiformis* it is tempting to speculate that lupin–fungus interactions are not parasitic but actually symbiotic: lupins are known to produce a disease in sheep and other herbivores, the so-called 'lupinosis'. Originally, alkaloids were assumed to be the causative agents. Recently, a group of fungal toxins, which are hexapeptides called phomopsins, has been found to be responsible. Phomopsin levels up to 1.5 mg/kg seeds or foliage have been detected. Phomopsins are strong poisons with LD_{50} values (p.o.) in rats and sheep of 1–2 mg/kg. Their molecular target is tubulin, and the inhibition is as strong as that observed for the known tubulin poison, vincristine. As a consequence, mitosis is blocked and the longterm consequences are hepatocarcinomas (Culvenor and Petterson 1986). Since the fungus does not kill the host-plant, the production of a potent phytotoxin could be helpful for the plant in the defence against herbivores in addition to chemical defence by QA.
4. *Toxicity to non-insect herbivores*: QA deter or repel the feeding of a number of non-insect herbivores (and other animals), e.g. nematodes, snails, rabbits, cows, or are directly toxic or mutagenic to them (Table 10.5). For example, anagyrine causes malformations, the so-called 'crooked calf disease' in young sheep and calfs, when their mothers feed on lupins or broom containing anagyrine (ingested at 7–11 mg/kg per day during day 40 and 75 of gestation) (Keeler 1976).

Alkaloid extracts of *Lupinus angustifolius* which contain lupanine, angustifoline, and 13-hydroxylupanine did not show mutagenic activity in a *Salmonella* and Chinese hamster cell system (Culvenor and Petterson 1986). This suggests that the mutagenic effect is limited to anagyrine and perhaps other α-pyridone alkaloids, but not caused by the lupanine-type QA.

QA are obviously bitter to man and other mammals: Mean detection levels in man are 0.00085 per cent for sparteine, 0.0021 per cent for lupanine, and 0.017 per cent for 13-hydroxylupanine. In animals, the sensitivity to bitter lupins increases sheep > rabbit > Guinea pig > mouse and pig (Wink 1992*a*, 1993*a*,*b*). This indicates that mammalian herbivores are able to detect and to avoid harmful levels of dietary QA. Interestingly, young and adult geese (*Anser anser*, *A. indicus*, and *Branta canadensis*) do not detect the bitter sparteine and feed deliberately on an artificial diet with up to 1 per cent sparteine (Wink *et al.* 1993).

Vertebrate toxicity was assessed for those QA that are abundant and frequently found in plants (Table 10.5): alkaloids of the sparteine-lupanine-type are relatively toxic when injected (i.v or i.p.) but less so when given orally, an important fact, if we consider using these alkaloids as natural pesticides. The α-pyridone alkaloids, such as cytisine are approximately one order of magnitude more toxic.

As discussed above, the toxic effects may be due to inhibition of K^+ ion channels, to activation of the acetylcholine receptor, to inhibition of protein synthesis, and to other mechanisms which have not been elucidated (Wink 1992*a*,*b*, 1993*a*–*c*).

Ecological relevance of QA QA concentrations are sufficiently high in lupins and brooms to guarantee their inhibitory effects observed *in vitro* (Wink 1992*a*,*b*, 1993*a*–*c*). In addition, QA contents can be increased by wounding: this short-term effect was highest under greenhouse conditions, but also measurable in the field (Wink 1983*b*; Johnson *et al.* 1989). In this context QA localization in epidermal tissues (Wink *et al.* 1984) can be interpreted as a strategically important adaptation, since this tissue has to ward off small herbivores and pathogens in the first instance.

Plants which produce QA such as *Cytisus*, *Genista*, *Chamaecytisus*, *Laburnum*, *Lupinus*, *Baptisia*, and *Thermopsis*, are relatively abundant in temperate climate zones of the Old and New World and are often dominant members of plant associations, present in pastures, meadows, heath, and ruderal habitats. QA plants seldom suffer severely from the attack of insects and other herbivores and comparably few adapted insects have been described. Therefore, it seems likely that QA are definitely important for the fitness of lupins, an assumption which could be tested experimentally.

Lupins have relatively large seeds which contain up to 40–50 per cent protein, up to 20 per cent lipids, and 2–8 per cent alkaloids. In order to use lupin seed for animal or human nutrition, *Homo sapiens*, for several thousand years, used to cook the seeds and leach out the alkaloids in running water. This habit has been reported for the Egyptians and Greeks in the Old World, and for the Indians and Incas of the New World (Bleitgen *et al.* 1979). The resulting seeds taste sweet, in contrast to the alkaloid-rich ones which are very bitter.

At the turn of this century, German plant breeders initiated a programme to grow alkaloid-free lupins, the so-called 'sweet lupins'. Although extremely rare in Nature (less than 1 plant in 100 000 specimens), the efforts were largely successful, and at present, 'sweet' varieties with an alkaloid content lower than 0.01 per cent exist for *Lupinus albus*, *L. mutabilis*, *L. luteus*, *L. angustifolius*, and *L. polyphyllus*. As far as we know, the sweet varieties only differ from their original bitter wild forms, by their degree of alkaloid accumulation. This offers the chance to test experimentally, whether bitter lupins have a higher fitness than sweet ones with regard to microorganisms and herbivores (Table 10.6).

The results of these experiments were clearcut (Table 10.6): In the greenhouse, where plants are protected from herbivores or pathogens, no clear advantage was seen. But when planted in the field, without being fenced in and without man-made chemical protection, a dramatic effect was regularly encountered, especially with regard to herbivores.

Rabbits (*Cuniculus europaeus*) and hares (*Lepus europaeus*) clearly prefer the sweet plants and leave the bitter plants almost untouched, at least as long as there was an alternative food source (Waller and Nowacki 1978; Wink 1985*a*, 1987*a*, 1988). A similar picture was seen for a number of insect species, such as aphids, beetles, thrips, and leaf-mining flies (Table 10.6), i.e. the sweet forms were attacked, whereas the alkaloid-rich ones were largely protected. The list of insects attracted by sweet lupins is even longer and includes *Agrotis* spp. (Lep.), *Phorbia platura* (Dipt.), *Tipula* spp. (Dipt.), *Agriotes* spp. (Col.), *Smithurus viridis* (Collembola), *Ciampa arietaria* (Lep.), *Heliothis punctigera*, *H. armigera* (Lep.), *Phytomyza hortcola* (Dip.), *Sitona griseus* (Coll.), *Acyrthosiphon pisi* (Hom.), *A. kondoi* (Hom.), *A. craccivora* (Hom.), *Myzus persicae* (Hom.), *Calocoris norvegicus* (Het.), *Adelphocoris lineolatus* (Het.), and *Lygus rugulipennis* (Het.). Plant breeders have observed that also bacterial, fungal, and viral diseases are more abundant in the sweet forms, but this effect has not been documented in sufficient detail.

These data clearly support the importance of alkaloid production for chemical defence against herbivores. Since other allelochemicals (such as flavonoids, isoflavones, anthocyanins, phenolics, saponins, protease inhibitors, stachyose and verbascose, erucic acid, phytic acid, etc.) are present

Table 10.6 Importance of lupin alkaloids for the resistance of lupins against herbivores (after Wink 1988, 1992*a*)

Herbivore species	Effect Alkaloid content of lupins high ('bitter')	 low ('sweet')
Non-adapted herbivores:		
1. Vertebrates		
Sheep	−	+
Hares (*Lepus europaeus*)	−	+
Rabbits (*Oryctolagus europaeus*)	−	+
2. Insects		
Leaf miner (Agromyzidae)	−	+
Sitona lineatus (Coleoptera)	−	+
Acyrthosiphon pisum (Aphidae)	−	+
Aphis fabae (Aphidae)	−	+
Myzus spp. (Aphidae)	−	+
Thrips (*Frankliniella* spp.)	−	+
Adapted herbivores:		
3. Insects		
Macrosiphum albifrons (Aphidae)	+	−

−, no or very low infestation/herbivory; +, high or complete infestation/herbivory.

Bitter lupins have alkaloid levels in their leaves of 1–4 mg/g f.w., sweet lupins of 0.01–0.05 mg/g f.w.

in both forms, some residual resistance is still maintained even in the absence of QA. This is especially true for roots, which are a rich source for antifeedant and antifungal isoflavones.

The lupin example also tells us about the standard philosophy and problems of plant breeding. With our present knowledge of the ecological importance of QA for the fitness of lupins, it seems doubtful whether the selection of sweet lupins was a wise decision. In order to grow them we have to build fences and worse, to employ man-made chemical pesticides, which have a number of well-documented disadvantages. It can be assumed that similar strategies, i.e. to breed away unwanted chemical traits, have been chosen with other agricultural crops (such as cabbage, turnip, rape seed, tomato, potato, cassava, or barley), with the consequence, that the overall fitness was much reduced (Wink 1988). We can easily observe their reduced fitness by trying to leave crop species to themselves in the wild: they will quickly disappear and not colonize new habitats. If we do the same with wild species, usually the opposite is

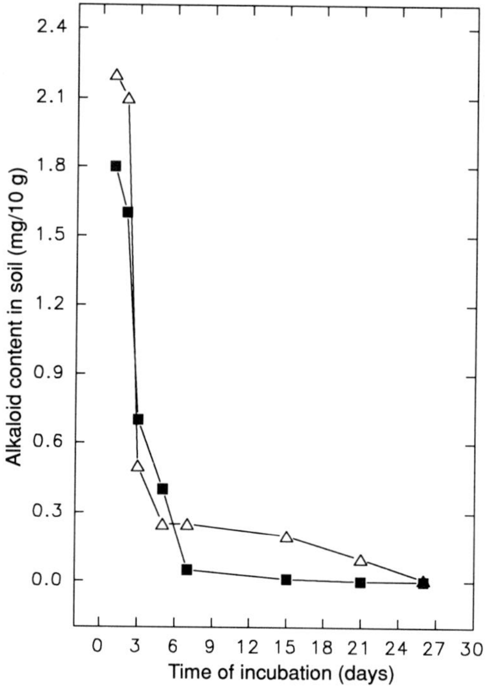

Fig. 10.1. Degradation of quinolizidine alkaloids in soil. Soil samples from two locations, on which lupins had been cultivated, were incubated with pure sparteine (2 mg per 10 g soil) under natural climatic conditions. Samples were taken at the time intervals given and evaluated by capillary GLC (after Gross and Wink 1986).

observed and these 'newcomers' are often very successful in establishing new territories.

For lupins, we have proposed two alternatives (Wink 1991):

1. To select for lupin mutants which do not translocate the alkaloids to the seeds, since seeds do not produce QA but store them. In this case, the plant would retain its chemical resistance but would provide the valuable alkaloid-free seeds at the same time.

2. To grow alkaloid-rich plants but to process the seeds, which are important because of their high levels of proteins and oil (in the case of *L. mutabilis*). We could devise large-scale technological procedures to remove alkaloids from the seeds after harvest (similarly to sugar raffination from sugar beets).

At present a few companies are actively exploring these possibilities, such as the Mittex Anlagenbau in Ravensburg, Germany. One idea is to produce simultaneously pure protein, lipids, amino acids, and dietary fibres from bitter seeds. A spin-off product would be alkaloids, which could be either used in medicine (in Germany sparteine is exploited as a drug to treat heart antiarrhythmia) or in agriculture as a natural plant protective, i.e. insecticide. A pilot lupin refinery will be built soon by the Mittex company with the help of the EC (W. Jaeggle 1992).

Considering the biological activities of QA (insecticidal in solutions with 0.1–0.5 per cent alkaloid), it seems promising to exploit the alkaloids as natural plant protectants, similar to the use of other allelochemicals, such as pyrethroids or the constituents of the Neem tree. QA have the advantage that they do not accumulate in the soil, but are degraded within 10 days (Fig. 10.1) (Gross and Wink 1986). Experiments with cell cultures indicate that QA can be degraded rapidly by plants (Wink 1985*b*; Wink and Witte 1985), thus an accumulation in crop plants is not very likely. Since QA will be produced as a byproduct of a normal crop, which is grown for proteins, lipids, and fibres, QA can be a comparatively cheap insecticide. Since both, proteins and insecticides, are in short supply in many countries around the world (especially in the third world), lupin cultivation and processing might be a valuable alternative, although support must be given from developed countries to built such plants. Since lupins have nitrogen fixing *Rhizobia*, lupin cultivation improves the soil and results in higher yield in the next crop, such as wheat. However, it might be argued that QA are toxic in higher animals. As compared to nicotine or rotenone and to many other synthetic insecticides, QA are of moderate toxicity (Table 10.5) and can be tolerated if necessary precautions are taken. Thus, lupin alkaloids are a promising newcomer in the field of natural pesticides.

On a long-term perspective it might be interesting to transfer the genes which are responsible for production of lupin alkaloids from one species to another in order to improve the fitness of the receiving species (Hanke and Wink 1992). But much more basic knowledge is necessary in order to consider all the relevant factors (e.g. biosynthesis, transport, and degradation) involved.

Other allelochemicals with activities against herbivores and microorganisms

A recent survey analysed the existing literature for reports on the biological activity of the largest class of allelochemicals, the alkaloids (Wink 1993*a*), i.e. whether they are antibiotic, allelopathic, or deterrent/toxic against animals (Table 10.7). Although more than 10 000 alkaloids

Table 10.7 Evaluation of allelochemical activities of alkaloids. Based on a computer search of the literature (Chemical abstracts (CA) online, Biosis) until 1990 (after Wink 1993*a*)

Activity	Number of active alkaloids
Against microorganisms:	
antibacterial	185
antifungal	124
antiviral	43
Against other plants:	
allelopathy	52
Against animals:	
insect deterrence/toxicity	148
vertebrate toxicity	> 300
Cellular/molecular targets:	
cytotoxicity	152
interaction with proteins	166
interaction with DNA/RNA	37
mutagenicity/teratogenicity	48
interactions with biomembranes	7

are known, few (approximately 2–5 per cent) have been analysed for their biochemical properties, and even fewer for their ecophysiological roles. Furthermore, the corresponding studies were usually designed to find useful medicinal or sometimes agricultural applications of alkaloids. This has to be kept in mind, since if an alkaloid is termed inactive in the literature this usually means less active than a standard compound already established as a medicinal compound (such as penicillins in antimicrobial screenings). In many medicinal experiments relatively low doses are applied because of the toxic properties of many alkaloids. If the same compound had been tested at relevant concentrations (which normally means elevated concentrations) that are present in the plant, an ecologically relevant activity might have been detected. Another restriction is that the activities of alkaloids have been tested with organisms that are sometimes irrelevant for plants but medicinally important. However, if a compound is active against *Escherichia coli* it is likely that is is also active against other gram-negative, but plant-relevant bacteria. Nevertheless, most of the data obtained in these studies provide important information and imply a potential exploitation either in medicine or in agriculture.

Although many alkaloids have antimicrobial and allelopathic properties, their main activity seems to be against herbivores (Table 10.7) (both insects and vertebrates), where they modulate the activity of essential molecular targets, such as receptors of neurotransmitters, ion channels and transporters, or processes such as replication, transcription, protein biosynthesis, electron chains, or membrane stability (details in Wink 1993*a*).

This analysis (Table 10.7) does not imply that bioactive compounds are not found in the groups of mono-, sesqui-, di-, and tri-terpenes, anthraquinones, glucosinolates, cyanogenic glycosides, non-protein amino acids, saponins, cardiac glycosides, quinones, phenolics, tannins, and catechins. The opposite can be seen from the data given in Jacobson (1958, 1975, 1982), Jacobson and Crosby (1971), Swain (1977), Levin (1976), Rosenthal and Janzen (1979), Harborne (1988), and Rosenthal and Berenbaum (1991), but a detailed analysis, comparable to that of alkaloids is not yet available.

These data show that Nature provides a large set of active compounds which could be useful in the future as natural plant protectants. We only need to explore and exploit them!

Outlook

There is an urgent demand for effective and ecologically tolerable plant protectants in developing countries, but this is also true in the Western world. As a result of stricter legislation, an increase of resistant and new pests, escalating costs for R&D in synthetic pesticides, and a growing ecological awareness of the public, 'natural pesticides' have become an interesting alternative.

Many natural products are active insecticides, fungicides, or herbicides (Table 10.7). Since plants usually produce a complex mixture of defence chemicals attacking different targets in pests, it is more difficult for a herbivore or pathogen to develop resistance to them, compared to the situation with synthetic regulators that consist of a single compound only. If these natural pesticides could be produced easily and at low costs from local plants, both subsistence and cash crop farming may benefit. As a result, developing countries could maintain or gain independence from petrochemical industries and build up food self-sufficiency.

A good natural pesticide should have the following qualities:

(1) effectivity, such as insecticidal activity or insect deterrence;

(2) low toxicity in vertebrates and beneficial insects;

(3) more than one mode of action, to reduce resistance of pests;

(4) low persistence in plants, soil, and water;
(5) extraction and formulation should be simple, cheap, and easy;
(6) producing plants should be grown locally;
(7) producing plants should grow easily and should allow a multipurpose utilization;
(8) plants should not compete for good soils, fertilizer, and water.

Although not all criteria can be fulfilled in most cases, some of the examples given above come rather close to these demands, e.g. neem and lupins.

Since natural products evolved as defence compounds, some adverse effects towards animals and man should be expected. Therefore, natural phytochemicals must not be applied without proper evaluation of short- and long-term toxicity and environmental compatibility.

Pesticide legislation and registration differ between countries. Whereas they are often lacking completely in developing countries, they are very restrictive and highly expensive in developed countries. In the USA, EPA can grant permits for 'minor uses of pesticides' and 'biorational pesticides (compounds without direct toxicity)', which can apply to some natural pesticides. In comparison, legislation in EC-countries is more strict and does not distinguish between conventional and natural pesticides. In Germany, the pesticide law accepts so-called 'pflanzenstärkende Mittel, plant restoration substances', which require less testing. Usually, only nutrients are considered in this category, and not insecticides.

In general, a full registration requires the following tests for any single compound in a prescription:

(1) acute toxicity tests (LD_{50});
(2) 90 days oral tests (performed with rats with different amounts of a pesticide);
(3) mutagenicity tests;
(4) residue tests (if residues appear in the plant or water, they have to be tested for carcinogenicity and mutagenicity);
(5) environmental side-effects (fate of pesticides in water, soil, and air).

These studies are costly and demand approximately DM 300 000–500 000 for a single substance. At present, natural mixtures are not dealt with separately and each single compound which is present needs to be evaluated. Since the resulting markets are usually restricted, most companies are reluctant to invest money in such an enterprise. One reason is the present patent laws of Western countries: As soon as the structure or the bioactivity of a natural product has been published, this compound and the respective application can no longer be patented. And many of the known active allelochemicals, such as those discussed above are published substances. For a company, the incentive, to pay for the necessary R&D and registration of a non-patentable substance is therefore usually small, because any other company could exploit the compound without any additional costs, once it had been registered at the licensing agencies. For big companies, there is another obstacle, they evaluate their overhead costs in addition and if these figures are taken into account, the R&D costs for a single insecticide immediately jump to around DM 20 million and higher. Since there are always rival synthetic compounds around, the decision is always towards the synthetic approach.

As a consequence, natural pesticides have not been developed at all or not in the same way as synthetic pesticides. Therefore, the preconceived view is often heard, that natural products are either not effective or too toxic to be used. I am certain, that after investing the necessary R&D work in these compounds, natural pesticides applicable in both developing and developed countries will be produced.

If we want to develop those non-patented natural products, which have demonstrated potentials as natural pesticides for the agriculture of third world but also of Western countries, the initiative for R&D in the field of product evaluation, formulation, application, and registration must come from political agencies and governments, since private investors will be reluctant in most cases to invest large amounts of money. Since developing countries are often poor and lack qualified personnel and research facilities, support must come from developed countries, where the necessary infrastructure and expertise are available, such as USA, Europe, or Japan.

Besides developing new pesticides, we should not forget to improve our present agricultural praxis, to avoid large-scale monocultures, to develop biological strategies against herbivores and pathogens, as done in the integrated pest control, and to breed for more resistant crops. Or in the long run, to use genetic engineering to establish transgenic plants which are more resistant towards insects, because they have obtained the genes for the biosynthesis of insecticidal allelochemicals, such as lectins, protease inhibitors, or even alkaloids. First steps in this direction, that have been performed already include the functional expression of several genes of alkaloid biosynthetic pathways, such as lysine decarboxylase (Herminghaus *et al.* 1991), ornithine decarboxylase (Hamill *et al.* 1990), tryptophan decarboxylase (DeLuca *et al.* 1989), hyoscyamine hydroxylase (Hashimoto and Yamada 1992), or strictosidine synthase (McKnight *et al.* 1991) in transgenic plants, which often only resulted in the accumulation of an intermediate of the respective pathway. We can anticipate, that full pathways will be cloned in the future which will hopefully lead to plants with improved defence qualities.

Acknowledgement

The author's research was supported by grants of the Deutsche Forschungsgemeinschaft. The author would like to thank Dr P. Marechal for making available an unpublished manuscript on plant-derived pesticides.

References

Alberts, B., Bray, D., Lewis, J., Raff, M., Roberts, K., and Watson, J. D. (1989). *Molecular biology of the cell* (2nd. edn). Garland, New York

Atwal, A. S. (1976). *Agricultural pests of India and South-East Asia.* Kumar, Kalyani.

Balandrin, M. F., Klocke, J. A., Wurtele, E. S., and Bollinger, W. H. (1985). Natural plant chemicals: sources of industrial and medicinal materials. *Science*, **228**, 1154–60.

Berlin, J. (1984). Plant cell cultures—A future source for natural products? *Endeavour*, **8**, 5–8.

Bernays, E. (1982). The insect on a plant—a closer look. In *Insect–plant relationships* (ed. J. H. Visser and A. K. Minks), pp. 3–17. Pudoc, Wageningen.

Bernays, E. A. and Chapman, R. (1987). The evolution of deterrent responses in plant-feeding insects. In *Perspectives in chemoreception and behaviour* (ed. R. Bernays, M. Chapman, and J. Stoffolano). Springer, New York, Heidelberg.

Bernays, E. A. and Graham, M. (1988). On the evolution of host specificity in phytophagous arthropods. *Ecology*, **69**, 886–92.

Blades, D. and Mitchell, B. K. (1986). Effects of alkaloids on feeding by *Phormia regina. Entomol. Exp. Appl.*, **41**, 299–304.

Bleitgen, R., Gross, R., and Gross, U. (1979). Die Lupine, ein Beitrag zur Nahrungsversorgung in den Anden. 5. Einige Beobachtungen zur traditionellen Entbitterung von Lupinen in Wasser. *Z. Ernährungswiss.*, **18**, 104–11.

Blum, M. S. (1981). *Chemical defenses of arthropods.* Academic press, New York

Charlwood, B. V. and Rhodes, M. J. C. (1990). *Secondary products from plant tissue culture.* Clarendon Press, Oxford.

Chrispeels, M. J. (1991). Sorting of proteins in the secretory system. *Annu. Rev. Plant Physiol. Plant Mol. Biol.*, **42**, 21–53.

Constabel, F. and Vasil, I. K. (1987). *Cell culture in phytochemistry.* Academic Press, New York.

Culvenor, C. C. J. and Petterson, D. S. (1986) Lupin toxins-alkaloids and phomopsins. *Proc. 4th Intl. Lupin conference, Geralton*, pp. 188–98. International Lupin Association, South Perth.

DeLuca, V., Marineau, C., and Brisson, N. (1989). Molecular cloning and analysis of cDNA encoding a plant tryptophan decarboxylase: Comparison with animal dopa decarboxylase. *Proc. Natl. Acad. Sci. USA*, **86**, 2582–6.

Deverall, B. J. (1977). *Defense mechanisms in plants*. Cambridge University Press, Cambridge, London.

Dreyer, M. and Hellpap, C. (1991). Niem-ein vielversprechendes natürliches Insektizid für klein-flächige Gemüseproduktion in tropischen und subtropischen Ländern. *Z. Pflanzenkrankh. u. Pflanzensch.*, **98**, 428–37.

Duffey, J. (1980). Sequestration of plant natural products by insects. *Annu. Rev. Entomol.*, **25**, 447–77.

Edmunds, M. (1974). *Defense in animals*. Longman, Harlow.

Fenical, W. (1986). In *Alkaloids: Chemical and biological perspectives* (ed. S. W. Pelletier), Vol. 4, pp. 276–330. Wiley, New York.

Frear, D. E. H. (1948). *A catalogue of insecticides and fungicides*. II. *Chemical fungicides and plant insecticides*. Chronica Botanica Comp., Waltham, MA, USA.

Gross, R. and Wink, M. (1986) Degradation of sparteine in soil. *Lupin Newsletter*, **9**, 15–18.

Hamill, J. D., Robins, R. J., Parr, A. J., Evans, D. M., Furze, J. M., and Rhodes, M. J. C. (1990). Overexpressing a yeast ornithine decarboxylase gene in transgeneic roots of *Nicotiana rustica* can lead to enhanced nicotine production. *Plant Mol. Biol.*, **15**, 27–38.

Hanke, M. and Wink, M. (1992). Die Suche nach den Genen der Lupinenalkaloid-Biosynthese. In *Lupinen 1991-Forschung, Anbau und Verwertung* (ed. M. Wink), pp. 157–65. Universität Heidelberg.

Harborne, J. B. (1988). *Introduction to ecological biochemistry* (3rd edn). Academic Press, New York.

Hashimoto T. and Yamada, Y. (1992). Biosynthesis of scopolamine and an application for genetic engineering of medicinal plants. In *Plant tissue culture and gene manipulation for breeding and formation of phytochemicals* (ed. K. Oono, T. Hirabayashi, S. Kikuchi, H. Handa, and K. Kajiwara), pp. 255–9. National Institute of Agrobiological Research, Tsukuba.

Heal, R. E., Rodgers, E. F., Wallace, R. T., and Starness, O. (1950). A survey of plants of insecticidal activity. *Lloydia*, **13**, 89–162.

Herminghaus, S., Schreier, P. H., McCarthy, J. E. G., Landsmann, J., Bottermann, J., and Berlin, J. (1991). Expression of a bacterial lysine decarboxylase gene and transport of protein into chloroplasts of transgenic tobacco. *Plant Mol. Biol.*, **17**, 475–86.

Jacobson, M. (1958). *Insecticides from plants. A review of the literature 1941–1953*. USDA, Agric. Handbook 154.

Jacobson, M. (1975). *Insecticides from plants. A review of the literature 1954–1971*. USDA, Agric. Handbook 461. US Government Printing Office, Washington, DC.

Jacobson, M. (1982). Plants, insects and man: Their interrelationships. *Econ. Bot.*, **36**, 346–54.

Jacobson, M. (1986). The neem tree: natural resistance par excellence. In *Natural resistance of plants to pests* (ed. M. B. Green and P. A. Hedin). ACS Symp. Ser. 296. American Chemical Society, Washington, DC.

Jacobson, M. (1988). *Focus on phytochemical pesticides. I: Neem tree*. CRC Press, Boca Raton, Florida.

Jacobson, M. and Crosby, D. G. (1971). *Naturally occurring insecticides*. Marcel Dekker, New York.

Jaeggle, W. (1992). Agricultural refining of billes lupins into derivates with added value. In *Lupinus mutabilis*: its adaptation and production under European climatic conditions. *EUR* 14102, 127–46 (Brussels).

Johnson, N. D., Rigney, L., and Bentley, B. L. (1989). Short-term induction of alkaloid production in lupines. Differences between N_2-fixing and nitrogen-limited plants. *J. Chem. Ecol.*, **15**, 2425–34.

Keeler, R. F. (1976). Lupin alkaloids from teratogenic and nonteratogenic lupins: III. Identification of anagyrine as the probable teratogen by feeding trials. *J. Toxicol. Environ. Health*, **1**, 887–98.

Kinghorn, A. D. and Balandrin, M. F. (1984). Quinolizidine alkaloids of the Leguminosae: Structural types, analysis, chemotaxonomy, and biological activities. In *Alkaloids: Chemical and biological perspectives* (ed. W. S. Pelletier), Vol. 2, pp. 105–48. Wiley, New York.

Korcz, A., Markiewicz, M., Pulikowska, J., and Twardowski, T. (1987). Species-specific inhibitory effects of lupine alkaloids on translation in plants. *J. Plant Physiol.*, **128**, 433–42.

Kurz, W. (ed.) (1989). *Primary and secondary metabolism in cell cultures*. Springer, Heidelberg.

Levin, D. A. (1976). The chemical defences of plants to pathogens and herbivores. *Annu. Rev. Ecol. Syst.*, **7**, 121–59.

McKnight, T. D., Bergey, D. R., Burnett, R. J., and Nessler, C. L. (1991). Expression of enzymatically active and correctly targeted strictosidine synthase in transgenic tobacco plants. *Planta*, **185**, 148–52.

Mende, P. and Wink, M. (1987). Uptake of the quinolizidine alkaloid lupanine by protoplasts and vacuoles of *Lupinus polyphyllus* cell suspension cultures. *J. Plant Physiol.*, **129**, 229–42.

Merck Index (1989). (11th edn). Merck & Co, Rahway.

Mitchell, B. K. and Sutcliffe, J. F. (1984). Sensory inhibition as a mechanism of feeding deterrence: effects of three alkaloids on leaf beetle feeding. *Physiol. Entomol.*, **9**, 57–64.

Morris, P., Scragg, A. H., Stafford, A., and Fowler, M. W. (1986). *Secondary metabolism inplant cell cultures*. Cambridge University Press.

Mothes, K. (1976). Secondary plant substances as materials for chemical high quality breeding in higher plants. In *Biochemical interaction between plants and insects* (ed. J. W. Wallace and R. L. Mansell). Plenum Press, London, New York.

Neumann, T., Plumhoff, D., Gebauer, M., and Roos, W. (1986). *Dörfliche Gewinnung und Nutzung von Pflanzenzubereitungen für den Pflanzen- und Vorratschutz*. GTZ/GATE, Eschborn.

Paech, K. (1950). *Biochemie und Physiologie der sekundären Pflanzenstoffe*. Springer, Berlin, Heidelberg.

Rhodes, M. J. C., Robins, R. J., Hamill, J. D., Parr, A. J., Hilton, M. G., and Walton, N. J. (1990). Properties of transformed root cultures. In *Secondary products from plant tissue culture* (ed. B. V. Charlwood and M. J. C. Rhodes), pp. 201–25. Clarendon Press, Oxford.

Rosenthal, J. and Berenbaum, M. R. (1991). *Herbivores. Their interaction with secondary plant metabolites* (2nd edn). Academic Press, New York.
Rosenthal, J. and Janzen, D. (1979). *Herbivores: Their interactions with plant secondary metabolites*. Academic press, New York.
Schmutterer, H. (1990). Properties and potential of natural pesticides from the neem tree. *Annu. Rev. Entomol.*, **35,** 271–97.
Schmutterer, H. and Ascher, K. R. S. (1984). *Proc. 2nd Int. Neem Conf., Rauischholzhausen, 1983*. GTZ, Eschborn.
Schmutterer, H. and Ascher, K. R. S. (1987). *Proc. 3rd Int. Neem Conf., Nairobi, 1986*. GTZ, Eschborn.
Schmutterer, H., Ascher, K. R. S., and Rembold, H. (1981). *Proc. 1st Int. Neem Conf., Rottach-Egern, 1980*. GTZ, Eschborn.
Schoonhoven, L. M. (1972). Secondary plant substances and insects. In *Structural and functional aspects of phytochemistry* (ed. C. V. Runekless and T. C. Tso). *Rec. Adv. Phytochem.*, **5**, 197–224.
Stoll, G. (1986). *Natural crop protection based on local farm resources in the tropics and subtropics*. J. Markgraf, Aichtal.
Swain, T. (1977) Secondary compounds as protective agents. *Annu. Rev. Plant Physiol.*, **28,** 479–501
Tyski, S., Markiewicz, M., Gulewicz, K., and Twardowski, T. (1988). The effect of lupin alkaloids and ethanol extracts from seeds of *Lupinus angustifolius* on selected bacterial strains. *J. Plant Physiol.*, **133,** 240–2.
Waller, G. R. (1987). *Allelochemicals: role in agriculture and forestry*, ACS Symp. Ser. 330. American Chemical Society, Washington, DC.
Waller, G. R. and Nowacki, E. (1978). Alkaloid biology and metabolism in plants. Plenum Press, New York, London.
Wink, M. (1983*a*). Inhibition of seed germination by quinolizidine alkaloids. Aspects of allelopathy in *Lupinus albus* L. *Planta*, **158,** 365–8.
Wink, M. (1983*b*). Wounding-induced increase of quinolizidine alkaloid accumulation in lupin leaves. *Z. Naturforsch.*, **38c,** 905–9.
Wink, M. (1984*a*). Chemical defense of Leguminosae. Are quinolizidine alkaloids part of the antimicrobial defense system of lupins? *Z. Naturforsch.*, **39c,** 548–52.
Wink, M. (1984*b*). Chemical defense of lupins. Mollusc-repellent properties of quinolizidine alkaloids. *Z. Naturforsch.*, **39c,** 553–8.
Wink, M. (1985*a*). Chemische Verteidigung der Lupinen:Zur biologischen Bedeutung der Chinolizidinalkaloide. *Plant Syst. Evol.*, **150,** 65–81.
Wink, M. (1985*b*). Metabolism of quinolizidine alkaloids in plants and cell suspension cultures. In *Primary and secondary metabolism of plant cell cultures* (ed. K. H. Neumann, W. Barz, and E. Reinhard), pp. 107–16. Springer, Heidelberg.
Wink, M. (1986). Storage of quinolizidine alkaloids in epidermal tissues. *Z. Naturforsch.*, **41c,** 375–80.
Wink, M. (1987*a*). *Chemical ecology of quinolizidine alkaloids*. ACS Symp. Ser. 330, pp. 524–33. American Chemical Society, Washington, DC.
Wink, M. (1987*b*). Site of lupanine and sparteine biosynthesis in intact plants and *in vitro* organ cultures. *Z. Naturforsch.*, **42c,** 868–72.

Wink, M. (1987*c*). Quinolizidine alkaloids: Biochemistry, metabolism, and function in plants and cell suspension cultures. *Planta Med.*, **53**, 509–14.
Wink, M. (1988). Plant breeding: Importance of plant secondary metabolites for protection against pathogens and herbivores. *Theor. Appl. Genet.*, **75**, 225–33.
Wink, M. (1989). Genes of secondary metabolism: Differential expression in plants and *in vitro* cultures and functional expression in genetically transformed microorganisms. In *Primary and secondary metabolism in cell cultures* (ed. W. Kurz). Springer, Heidelberg.
Wink, M. (1990). Physiology of secondary product formation in plants. In *Secondary products from plant tissue culture* (ed. B. V. Charlwood and M. J. C. Rhodes), pp. 23–41. Clarendon Press, Oxford.
Wink, M. (1991). Plant breeding: High or low alkaloid levels? In *Proc. 6th Intl. Lupin Conf., Temuco*, pp. 326–34. International Lupin Association, Temuco, Chile.
Wink, M. (1992*a*). Role of quinolizidine alkaloids in plant–insect interactions. In *Focus and insect plant interactions*, Vol. 4, (ed. E. A. Bernays), pp. 131–66. CRC Press, Boca Raton.
Wink, M. (1992*b*). Die chemische Verteidigung der Pflanzen und die Anpassungen der Pflanzen-fresser. In *Lupinen 1991-Forschung, Anbau und Verwertung* (ed. M. Wink), pp. 130–56. Universität Heidelberg.
Wink, M. (1993*a*). Allelochemical properties and the raison d'être of alkaloids. In *The alkaloids* (ed. G. Cordell), Vol. 43, pp. 1–118. Academic press, London, New York.
Wink, M. (1993*b*). Quinolizidine alkaloids. In *Methods in plant biochemistry*, Vol. 8, pp. 197–239. Academic press, London, New York.
Wink, M. (1993*c*). The plant vacuole: A multifunctional compartment. *J. Exp. Bot.*, **44**, Suppl. 231–46.
Wink, M. and Hartmann, T. (1981). Sites of enzymatic synthesis of quinolizidine alkaloids and their accumulation in *Lupinus polyphyllus. Z. Pflanzenphysiol.*, **102**, 337–44.
Wink, M. and Hartmann, T. (1982). Localization of the enzymes of quinolizidine alkaloid biosynthesis in leaf chloroplast of *Lupinus polyphyllus. Plant Physiol.*, **70**, 74–7.
Wink, M. and Hartmann, T. (1985). Enzymology of quinolizidine alkaloid biosynthesis. In *Natural products chemistry 1984* (ed. R. I. Zalewski and J. J. Skolik), pp. 511–20. Elsevier, Amsterdam.
Wink, M. and Twardowski, T. (1992). Allelochemical properties of alkaloids. Effects on plants, bacteria and protein biosynthesis. In *Allelopathy: Basic and applied aspects* (ed. S. J. H. Rizvi and V. Rizvi), pp. 129–150. Chapman & Hall, London.
Wink, M. and Witte, L. (1984). Turnover and transport of quinolizidine alkaloids: Diurnal variation of lupanine in the phloem sap, leaves and fruits of *Lupinus albus* L. *Planta*, **161**, 519–24.
Wink, M. and Witte, L. (1985). Quinolizidine alkaloids as nitrogen source for lupin seedlings and cell suspension cultures. *Z. Naturforsch.*, **40c**, 767–75.

Wink, M. and Witte, L. (1991). Storage of quinolizidine alkaloids in *Macrosiphum albifrons* and *Aphis genistae* (Homoptera: Aphididae). *Entomol. Gener.*, **15**, 237–54.

Wink, M., Heinen, H. J., Vogt, H., and Schiebel, H. M. (1984). Cellular localization of quinolizidine alkaloids by laser desorption mass spectrometry (LAMMA 1000). *Plant Cell Rep.*, **3**, 230–3.

Wink, M., Hofer, A., Bilfinger, M., Martin, M., and Schneider, D. (1993). Geese and plant dietary allelochemicals—Food palatability and geophagy. *Chemoecology*, (in press).

Wippich, C. and Wink, M. (1985). Biological properties of alkaloids. Influence of quinolizidine alkaloids and gramine on the germination and development of powdery mildew, *Erysiphe graminis* f.sp. *hordei*. *Experientia*, **41**, 1477–8.

Zenk, M. H. (1982). Pflanzliche Zellkulturen in der Arzneimittelforschung. *Naturwissenschaften*, **69**, 534–6.

11. Polyphenols — phytochemical chameleons

EDWIN HASLAM

Department of Chemistry, University of Sheffield, Sheffield S3 7HF, UK

Introduction

Natural polyphenols (*syn.* vegetable tannins) are complex higher plant secondary metabolites. According to earlier definitions they are water soluble, possess relative molecular masses in the range 500 to 3000, and besides giving the usual phenolic reactions, they have the ability to precipitate some alkaloids, gelatin, and other proteins, from solution. These complexation reactions are not only of intrinsic scientific interest as studies in molecular recognition but they have important and wide-ranging practical applications—in the manufacture of fashion leathers, in foodstuffs and beverages, in herbal medicines, and in chemical defence and pigmentation in plants. Polyphenols are *phytochemical chameleons.* Their interactions with other molecular species is, on occasion, an important facet of the chemistry of plants and it also underlies man's use of these materials in a number of practical spheres. In other situations these same interactions may well be undesirable if not totally unacceptable.

Polyphenol structure

Emil Fischer at the turn of the century made some characteristically brilliant and perceptive contributions to the study of the constitution of the gallotannins from Chinese and Aleppo (Turkish) galls. His work and that of Paul Karrer and Karl Freudenberg stimulated a great deal of interest amongst chemists. These initial enthusiasms waned, however, as the great complexity of many plant extracts (often natural, but frequently induced by the many and varied post-mortal processes required to derive them) was realized. By the 1950s the topic had become one of the dark impenetrable areas of Organic Chemistry. Its renaissance coincided with the advent of new methods of analysis and separation in the 1950s and the 1960s. Today the composition of many plant extracts can usually be adequately defined in terms of their polyphenolic (tannin) and simple phenolic constituents. There is thus, for the first time, a firm base from which to embark upon studies of the biological properties of plant polyphenols and in particular their complexation reactions (Haslam 1989). If there are exceptions to this generalization then these relate to the polyphenolic metabolites obtained from the wood and bark of trees. Here post-mortal events multiply the complexities of normal metabolism. As a result the nature and total composition of many of these commercially important polyphenolic extracts remain uncertain.

It is now possible to describe in broad terms the nature of plant polyphenols. They are secondary metabolites widely distributed in various sectors of the higher plant kingdom. They are distinguished by the following general features:

1. *Water solubility*. Although, when pure, some plant polyphenols may be sparingly soluble in water, in the natural state polyphenol–polyphenol interactions usually ensure some minimal solubility in aqueous media.
2. *Molecular weights*. Natural polyphenols encompass a substantial molecular weight range from 500 to 3–4000. Suggestions that polyphenolic metabolites occur which retain the ability to act as tannins but possess molecular weights up to 20 000 must be doubtful in view of the solubility proviso.
3. *Structure and polyphenolic character*. Polyphenols, per 1000 relative molecular mass, possess some 12–16 phenolic groups and 5–7 aromatic rings. They are based upon two broad structural patterns: condensed proanthocyanidins and galloyl and hexahydroxydiphenoyl esters and their derivatives.

Condensed proanthocyanidins

The fundamental structural unit in this group is the phenolic flavan-3-ol (‘*catechin*’) nucleus (**1**). Condensed proanthocyanidins exist as oligomers (soluble) containing two to five or six ‘*catechin*’ units and polymers (insoluble). The flavan-3-ol units are linked principally through the 4 and the 8 positions. In most plant tissues the polymers are of greatest quantitative significance but there is also usually found a range of soluble molecular species—monomers, dimers, trimers, etc.

Oligomeric condensed proanthocyanidins have been held (Bate-Smith and Metcalfe 1957; Bate-Smith 1962) to be most commonly responsible for the many distinctive properties of plants typically attributed to ‘condensed tannins’. One correlation to which Bate-Smith repeatedly drew attention was that between woody and non-woody plants. The distinction, he suggested, was sufficiently pronounced for it to be possible to speak

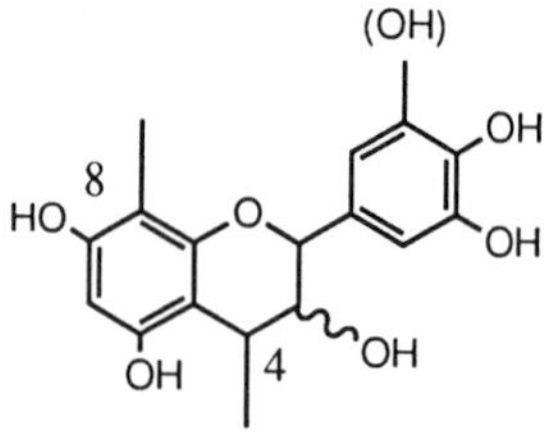

1 Flavan-3-ol oligomer unit

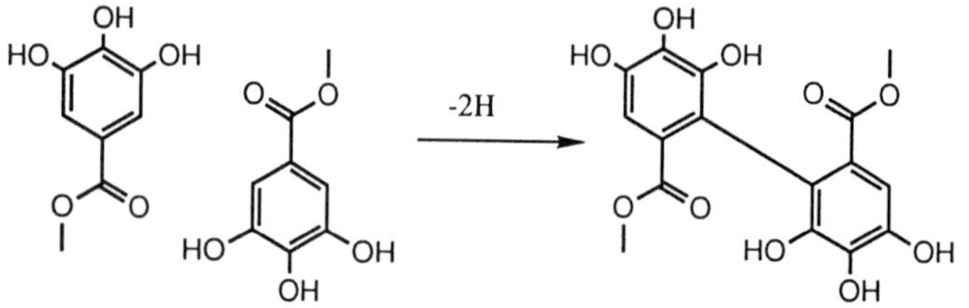

2 galloyl ester

-2H

3 hexahydroxydiphenoyl ester

of a typically 'woody' pattern of phenolic constituents in the leaves of certain plants. The same 'woody' phenolic constituents—apparent in a wide range of monocotyledons, gymnosperms, ferns dicotyledons—are, however, not found in mosses, algae, fungi, and lichens. Bate-Smith noted the structural analogies and similarities amongst these 'woody phenolics' to lignin and he postulated that they might well be concerned, in some way, with the presence of a vascular system in those plants which contain them. According to Bate-Smith three classes of phenolic constituent overwhelmingly predominate in the leaves of vascular plants. One of these is that group of compounds then referred to as leuco-anthocyanins but now known as condensed proanthocyanidins. On the basis of solubility differences Sir Robert and Lady Robinson (Robinson and Robinson 1931; Robinson 1939) subdivided the leucoanthocyanins (condensed proanthocyanidins) into three classes:

(i) those that are insoluble in water and the usual organic solvents or give only colloidal solutions;

(ii) those readily soluble in water but not readily extracted therefrom by means of ethyl acetate;

(iii) those capable of extraction from aqueous solution by ethyl acetate.

In so far as the total complement of condensed proanthocyanidins (procyanidins and prodelphinidins) found in plant tissues is concerned the soluble oligomeric forms (monomers, dimers, trimers ...) are in metabolic terms but the 'tip of the iceberg'. According to the Robinson's classification they represent category (iii) above. For the generality of plants it is quite clear that condensed proanthocyanidins which fall within the two other categories (i and ii) invariably strongly predominate over the more freely soluble forms. They are, metaphorically speaking, the base of the 'metabolic iceberg'. Indeed in the tissues of some plants such as ferns and fruit, e.g. the persimmon (*Diospyros kaki*), there is an overwhelming preponderance of these forms. They are also of frequent occurrence in plant gums and exudates. Shen *et al.* (1986) put forward a novel explanation for the occurrence and structural composition of those condensed proanthocyanidins 'insoluble in water and the usual organic solvents' (category (i) above). They suggested that these forms of condensed proanthocyanidins are covalently bound to a carbohydrate (or other polymer) matrix within the plant cell.

Galloyl and hexahydroxydiphenoyl esters (2 and 3) and their derivatives

These metabolites are almost invariably found as multiple esters with D-glucose and a great many can be envisaged as derived from the key biosynthetic intermediate β-1,2,3,4,6-pentagalloyl-D-glucose. Derivatives of hexahydroxydiphenic acid are assumed to be formed by oxidative coupling of vicinal galloyl ester groups in a galloyl D-glucose ester.

Gallic acid is most frequently encountered in plants in ester form. These may be classified into several broad categories:

(1) simple esters;

(2) depside metabolites (*syn* gallotannins);

(3) hexahydroxydiphenoyl and dehydrohexahydroxydiphenoyl esters (*syn* ellagitannins) based upon:
   (a) $^{4}C_{1}$ conformation of D-glucose;
   (b) $^{1}C_{4}$ conformation of D-glucose;
   (c) 'open-chain' derivatives of D-glucose.

Simple esters

Various simple esters of gallic acid have been described from plant sources. These have been catalogued (Haslam 1989). They range in composition from esters with sugars, cyclitols, quinic and shikimic acids, to those with phenols and phenolic glycosides and in many respects resemble the numerous esters of the hydroxycinnamic acids which are found in plants.

Depside metabolites

The ability to metabolize depside derivatives of gallic acid may be used as a guide to inter-relationships in particular plant families and there is, on present evidence, a close association of this form of metabolism with the Rhoideae tribe in the Anacardiaceae. Many products of this form of metabolism were often grouped together in the earlier literature under the generic term 'gallotannin'. The most familiar example is Chinese gallotannin or tannic acid (**4**) (galls, *Rhus semialata*) which has the overall

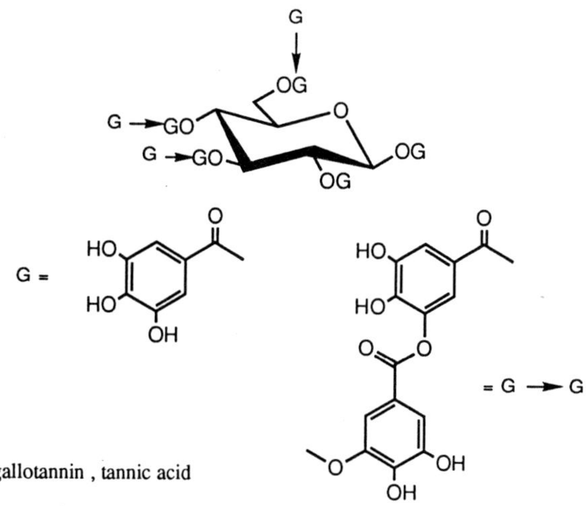

G
OG
G → GO
G → GO
O
OG
OG
G =
HO
HO
OH
O
HO
HO
O
O
O
OH
OH
O
= G → G

4 Chinese gallotannin , tannic acid

composition of a hepta-octagalloyl-β-D-glucose in which, on average, two to three additional galloyl groups are esterified to a pre-existing β-1,2,3,4,6-pentagalloyl-D-glucose core. As it is extracted the gallotannin is heterogeneous and contains a series of closely related molecules of varying overall composition. These range from, β-1,2,3,4,6-pentagalloyl-D-glucose itself to compounds with up to five or six additional galloyl ester groups linked as *m*-depsides to this core. The proportion of each type of molecular species present determines the final overall composition of the gallotannin. Using ^{13}C NMR the additional galloyl depside residues have been determined to be predominantly, but not exclusively, attached to the galloyl groups at C-3, C-4, and C-6. This polygalloyl-

D-glucose derivative is the most widely distributed gallotannin in plants although others have also been described (Haslam 1989).

Hexahydroxydiphenoyl esters and related metabolites

In the thirty year period from ~ 1950–1980 the Heidelberg school of Otto Schmidt and Walter Mayer made distinguished and seminal contributions to the study of naturally occurring hexahydroxydiphenoyl esters. They also provided the basis for the subsequent development of an intellectually satisfying biosynthetic rationale for the derivation of this group of natural polyphenolic esters (*syn* ellagitannins) from galloyl esters. This work has underpinned all the developments which have followed in this field—particularly the notable studies of the Japanese schools of Nishioka and Okuda. The number of discrete compounds now identified in this class is around 400!

Strong circumstantial evidence exists to suggest that the vast majority of these substances are derived biosynthetically from β-1,2,3,4,6-pentagalloyl-D-glucose by processes of dehydrogenation and rearrangement (Fig. 11.1). The principal modes of transformation are:

(**a**) Oxidative C–C coupling of galloyl ester groups (4–6 and 2–3) in the thermodynamically most stable $^4C_1$ conformation of β-1,2,3,4,6-pentagalloyl-D-glucose, followed by oxidative oligomerization (C–O coupling).

(**b**) Oxidative C–C coupling of galloyl ester groups (3–6, 1–6, and 2–4) in the thermodynamically least stable $^1C_4$ conformation of β-1,2,3,4,6-pentagalloyl-D-glucose. Metabolites of this class are also characterized by the presence of the dehydrohexahydroxydiphenoyl ester group and its derivatives. Examples have also recently been reported where these metabolites have undergone further oxidative (C–O coupling) oligomerization.

(**c**) Oxidative transposition of β-1,2,3,4,6-pentagalloyl-D-glucose to give 'open-chain' derivatives of D-glucose.

Polyphenol complexation

General considerations

The ability to associate with other metabolites, although it manifests itself most strikingly in the case of polyphenols, is nevertheless a property *per se* of the phenolic nucleus itself. There seems little doubt that the

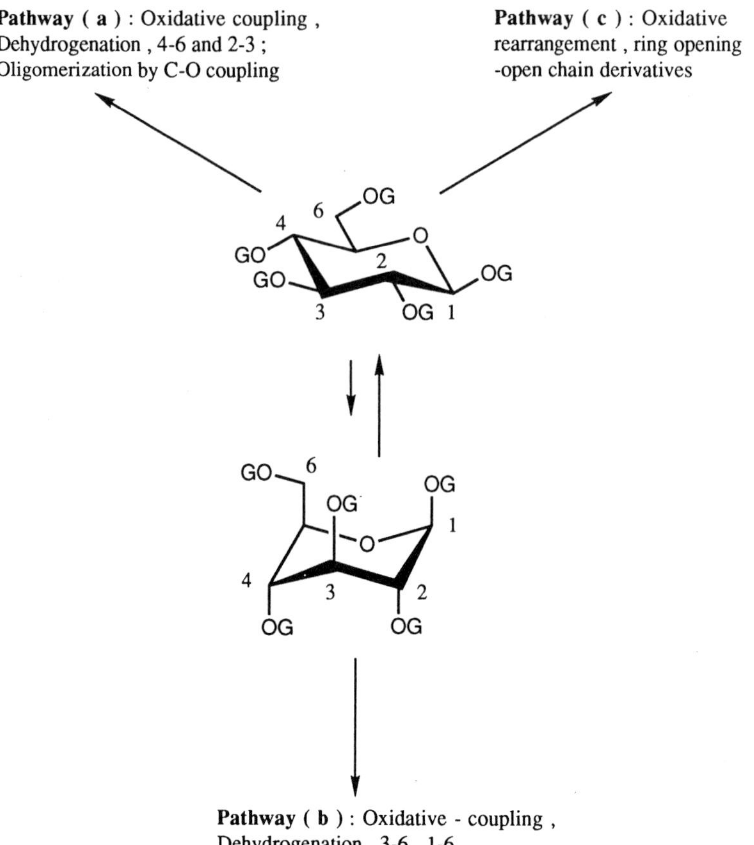

Fig. 11.1. Oxidative transformations of β-1,2,3,4,6-pentagalloyl-D-glucose (G = Galloyl) in higher plants. Formation of ellagitannins.

efficacy of polyphenols as complexing agents derives principally from their relative molecular size and from the possession within the same molecule of a multiplicity of phenolic groups and associated aromatic nuclei. These give rise not only to a number of important physical and chemical properties but also to the ability to act as multidentate ligands in processes of complexation.

The processes of complexation may be *reversible* or *irreversible* and may involve co-substrates—proteins, polysaccharides, alkaloids, anthocyanins, etc.—from within the same organism or extrinsic to that organism.

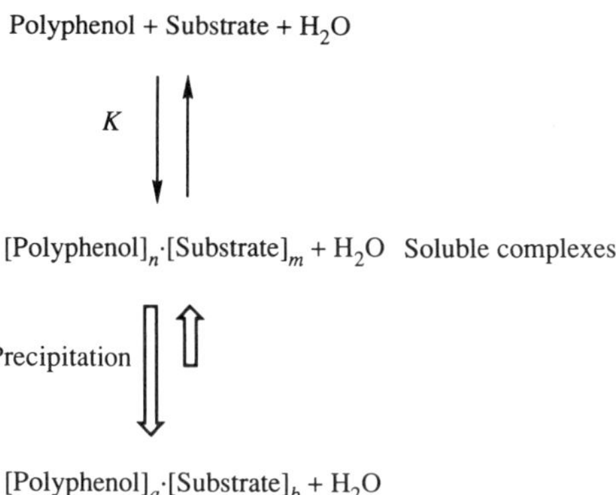

Fig. 11.2. Reversible polyphenol association.

Thus leather is formed as a result of the interaction of plant polyphenols with the protein collagen of hides and animal skins; the astringency of many fruits and beverages derives from the interaction of the salivary proteins in the mouth with polyphenolics in the fruit or beverage; the hard exoskeleton (cuticle) of many insects is formed by the oxidative copolymerization of phenolics and proteins in the freshly excreted cuticle. Reversible complexation of polyphenols may be considered as a two stage process in the first of which the polyphenol and the co-substrate, by the deployment of various non-covalent forces, are in equilibrium with the soluble complexes (Fig. 11.2). As the position of this equilibrium changes then, as a second stage, these soluble complexes may well aggregate and precipitate from solution. The whole process is however usually reversible and under suitable conditions the precipitated complexes may be redissolved.

The close juxtaposition of polyphenol and co-substrate which complexation brings about and the influence of external agencies (e.g. oxygen, metal ions, acid) may well promote secondary *irreversible* processes in which covalent bond formation takes place between the polyphenol* and the co-substrate*, or in which entirely new types of product are formed (Fig. 11.3). The process of complexation then becomes an irreversible one. Polyphenols are prone to oxidation, under the influence of enzymes or metal ion, or autocatalytically in mildly basic media, to give *ortho*-quinones which are extremely reactive intermediates. Likewise polyphenolic proanthocyanidins are susceptible to acid-catalysed rupture of the interflavan bond in media of pH 4.0 and less. This leads to the formation of highly electrophilic carbocation species which, like the *ortho*-quinones,

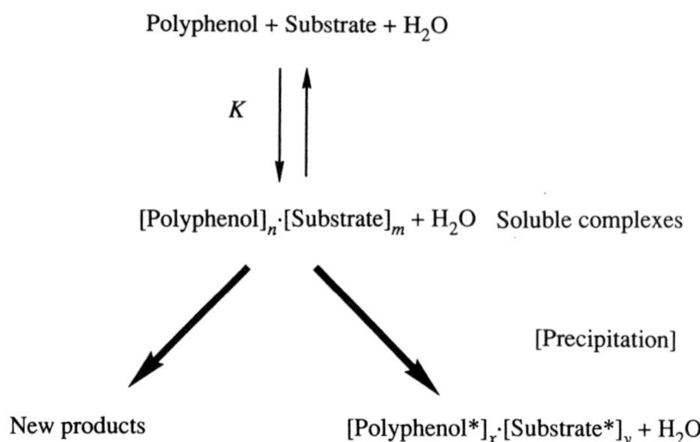

Fig. 11.3. Irreversible polyphenol association.

are able to react rapidly with other substrates. In the case of proteins reaction most probably occurs with nucleophilic groups (NH_2, SH) leading to covalently bound polyphenol–protein complexes. Irreversible polyphenol complexation reactions are very widely encountered. Some typical examples are:

(1) enzymic and non-enzymic browning of fruits and fruit juices, and in the fermentation processes leading to the preparation of teas and coffees;

(2) permanent haze formation in beers and lagers;

(3) ageing of red wines which leads to changes in pigmentation and astringency;

(4) humic acid formation as organic matter is degraded in the generation of humus in soils;

(5) necrosis—the protection of plant tissues against infective agents, parasites, and predators;

(6) sclerotization—the development of the hard exoskeleton in insects;

(7) 'fixed tannage' in leather manufacture.

Molecular aspects

The following factors have been identified as those which most strongly influence intermolecular association with polyphenols:

(a) Hydrophobic effects; '$\pi-\pi$ stacking' with aromatic substrates.

(b) Solvation and desolvation in aqueous media.

(c) Hydrogen bonding.

(d) The presence in co-substrates of tertiary amide groups, e.g. caffeine and prolyl peptides.

(e) Inorganic salts and metal ions; these may manifest themselves as general salt effects or in the form of specific metal ion complexation with the substrates.

(f) Temperature and pH.

(g) In the case of macromolecular substrates, such as proteins, conformational flexibility in both polyphenol and the macromolecule co-substrate assists complexation. Association is both time dependent and dynamic (as a 'hand in glove') and gives rise to multidentate binding of the polyphenol to the macromolecule.

Various observations point to the crucial importance of solubility and hydrophobic effects (a and b above) in polyphenol complexation, and hence to the significance of solvation and desolvation processes in the aqueous environment. Intermolecular recognition depends on the differences between interactions of the isolated 'host' and 'guest' molecules and their environments—for polyphenols this is almost invariably an aqueous one—and the interactions developed in the 'host–guest' complex. 'Hydrophobic effects' result from the differences in the energetics of interaction between hydrophobic groups on the 'host' and 'guest' and the water medium and their interactions with each other, which thereby reduces the total surface area of such groups exposed to water (Fig. 11.4).

The importance of 'hydrophobic effects' and solubility in an aqueous medium can be most readily seen by reference to β-1,2,3,4,6-pentagalloyl-D-glucose and two polyphenols believed to be derived biosynthetically from it. β-1,2,3,4,6-Pentagalloyl-D-glucose is amorphous, has a limited solubility in water ($\sim$1.0 mM at 20 °C), is readily extracted from aqueous media by ethyl acetate, and has a distribution coefficient K [octan-1-ol/H_2O] = 32. Both *in vivo* and *in vitro* β-1,2,3,4,6-pentagalloyl-D-glucose is a highly effective 'tannin'.

Vescalagin and castalagin (**6**) are unique 'open-chain' [cf. category (**c**) above, galloyl and hexahydroxydiphenoyl esters] diastereoisomeric polyphenols from *Quercus* and *Castanea* species. They are formally six hydrogen atoms less than their presumed biosynthetic precursor β-1,2,3,4,6-pentagalloyl-D-glucose (**5**). However both are nicely crystalline, highly soluble in water, not extracted therefrom by ethyl acetate, and show a

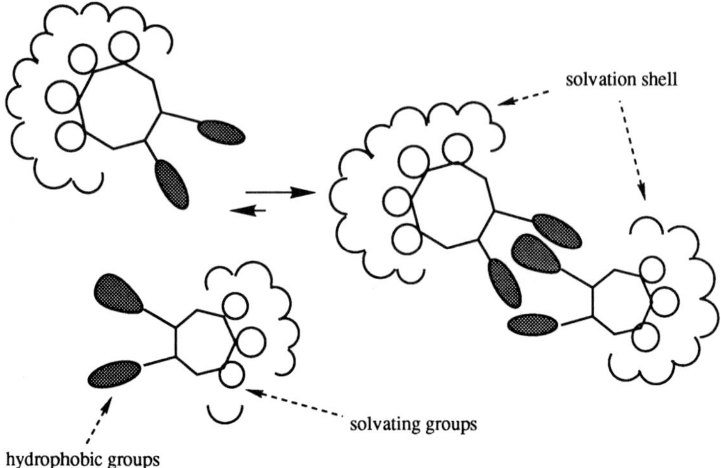

Fig. 11.4. 'Hydrophobic effects'.

- 6H

5 β-1,2,3,4,6-pentagalloyl-D-glucose
G = galloyl

6a castalagin , C-1 , α - OH
6b vescalagin , C-1 , β - OH

K [octan-1-ol/H_2O] = 0.1. Both, in the conventional *in-vitro* tests for 'tannins', respond very poorly indeed. Vescalagin and castalagin possess 15 phenolic hydroxyl groups and five aromatic nuclei, identical to the number of such groups in β-1,2,3,4,6-pentagalloyl-D-glucose (it should, however, also be noted that vescalagin and castalagin are conformationally rigid and inflexible whilst β-1,2,3,4,6-pentagalloyl-D-glucose is conformationally mobile).

Early work on the questions surrounding polyphenol complexation was almost exclusively concerned with attempts to elucidate the identity of the 'active sites' on protein molecules and the nature of the interaction between these sites and the polyphenolic substrates which led to strong association. These investigations, and particularly those of two of the leading workers Grassman (Grassman and Stadler 1961) and Gustavson (1956), favoured the peptide bonds as the principal 'active sites' and this led to the view that complexation was mediated by the development of intermolecular hydrogen bonds between the polyphenol, with its multiplicity of phenolic groups, and the polypeptide.

More recent studies have however highlighted the inherent complexity of molecular recognition processes and have underlined the importance of developing a global view of the problem. Whilst not denying the possibility of there being specific sites to which the polyphenol may finally become anchored by hydrogen bonding, the emphasis is now more clearly on the whole process beginning with the initially solvated species. Increasingly the association of polyphenols with other species—proteins, polysaccharides, caffeine, cyclodextrins, anthocyanins, anthracyclinones —whatever their size and complexity is seen to be driven by *hydrophobic effects*. The multiplicity of phenolic groups in the polyphenol cannot, however, be ignored. Their presence points, intuitively, to some role for hydrogen bonding in the associative processes.

In an aqueous medium the development of hydrogen bonding as a means of molecular recognition requires, as a precondition, that hydrogen bonds involving each of the substrates to water are themselves first broken. The energetics of interaction thus depend on the stability of the hydrogen bonds both *made* and *broken*. Although isolated monofunctional hydrogen bonds may therefore be expected to contribute minimally to molecular recognition processes, polyfunctional hydrogen bonding networks, formed cooperatively and radiating from the polyphenol substrate, probably contribute significantly as part of a second phase of polyphenol complexation. This concept seems particularly apposite when seeking to rationalize the very strong binding of polyphenols to the loose, flexible structures of proline rich proteins (PRPs), first noted by Hagerman and Butler (1981). There is thus ample evidence, particularly in aprotic systems, that the carbonyl function in tertiary amides (e.g. prolyl peptides, caffeine) is a very good hydrogen bond acceptor.

Molecular recognition of macromolecules such as proteins by polyphenols is therefore best described by the 'hand in glove metaphor'. In the initial stages the process is driven by 'hydrophobic effects'. There is however not a static matching of binding groups in the 'host' and 'guest', rather the association is, in its second phase, time dependent and dynamic (cf. the importance of conformational flexibility noted earlier). The required segmental mobilities of 'host' and 'guest' to bring, say, appropriate tertiary amide and phenolic groups into close juxtaposition to form hydrogen bonds are probably small, entropically favourable, and therefore not energetically prohibitive. The binding energy in this type of model derives from the summation of a relatively large number of contacts developed in a time dependent and dynamic manner.

Polyphenol complexation—some practical case studies

The interaction of polyphenols with other molecular species is an important facet of the chemistry of plants and underlies man's use of these materials in a number of practical spheres. Although the most commonly recognized and often most significant examples involve proteins and polysaccharides the association of polyphenols with small molecules —caffeine and anthocyanins—is also very important. Particular attention is drawn below to this aspect of polyphenol complexation and to situations in which the primary driving force towards association is '$\pi-\pi$' stacking of aromatic nuclei in the polyphenol and the co-substrate.

Leather

One of man's earliest uses of plant materials, such as plant galls, rich in polyphenolic metabolites, was in the conversion of raw animal hides and skins into leather. It is generally agreed that the essence of vegetable tannage is to protect the molecular form of the collagen fibres in the animal skin. Present views suggest that this is achieved by packing the amorphous regions of the fibrils with plant polyphenols, since the close packed crystalline regions do not, it is argued, require protection. Evidence suggests that this is complete when the collagen has absorbed about half its own weight of polyphenols. It is interesting to note, in the light of the earlier discussion, that after tannage a certain proportion of the polyphenols may be removed by solvent extraction from the finished leather (presumably that portion held in association with the collagen by non-covalent complexation; Fig. 11.2). Other polyphenolic material however remains irreversibly bound to the fibre and this, it is presumed, is brought about over the longer term by covalent bond formation between the polyphenol and the protein under the influence of external agencies such as acid, air, and light (Fig. 11.3).

Herbal medicines

The discovery that certain plants and herbs had curative or palliative effects in the treatment of illness and disease was of paramount importance in the promotion of the scientific study of plants. In addition to their extensive early use in tanning and dyeing and in the manufacture of inks infusions of plant galls have traditionally also provided powerful astringents useful in the treatment of cases of dysentery, diarrhoea, cholera, and as an injection in gonorrhea. A number of oriental medicinal plants and crude drugs are rich in polyphenolics based upon gallic acid and its derivatives. Many of these traditional herbal remedies survive and find their place in modern pharmacopoeias. In recent years they have been the subject of intense scrutiny in Japan in the schools of Nishioka and Okuda (Okuda 1981).

Astringency of fruits and beverages

Polyphenolic substances contribute significantly if not uniquely to the astringency of wines, fruits and fruit juices, teas, and other beverages. Polyphenols thus have a harsh astringent taste and produce in the palate a feeling of roughness and 'puckeriness', constriction, and dryness. As a sensation of taste that of astringency is not confined to a particular region of the mouth or tongue but is experienced as a diffuse stimulus. Moreover it may take a significant time to develop. A mucous membrane covers all the exposed surfaces of the mouth which are moistened by the secretions of the salivary glands. According to Bate-Smith (1954*a*,*b*) the primary reaction whereby astringency develops is via precipitation of proteins and mucopolysaccharides in the mucous secretions. It is interesting to note in this context that Butler and his collaborators (1987 have shown that mammalian herbivores produce unique proline rich (up to 45 per cent) salivary proteins (PRPs) which have a very high affinity for polyphenols. In humans these PRPs appear to be constitutive and are present in amounts which reflect the approximate levels of polyphenols and related phenolics in the normal diet. It has been suggested that these proline rich salivary proteins (PRPs) constitute the first line of defence against polyphenols in the digestive tract.

Loss of astringency is one of the major changes which takes place during the ripening of many edible fruits. It is generally agreed that this property (astringency) is due to the presence of polyphenols, but although some astringent fruit show a reduction in polyphenols upon ripening others do not. The biochemistry underlying these changes and their inter-relationship is one that remains to be fully explored. It has been suggested (Haslam and Lilley 1988) that in certain cases during the ripening process the cellular structure of the fruit changes and soluble pectin fractions are released. These it has been postulated modify and disrupt the ability of polyphenols to bind to glycoproteins in the mouth and hence reduce the observed astringency.

Sorghum and nutrition

Considerable attention has focused on various aspects of *Sorghum bicolor* (L) Moench, the traditional cereal crop of much of Africa, but particularly on the agronomic benefits and the dietary and antinutritional effects of sorghum polyphenols. More than other common cereals sorghum relies on intrinsic chemical defence—in the form of polyphenolic metabolites ('tannins')—to repel herbivores, to resist pathogens, and inhibit the growth of potential competitors. The presence of relatively high levels of condensed proanthocyanidins in sorghum cultivars is so frequently associated with unpalatability and resistance to bird damage that the terms 'bird resistant' and 'high tannin' have come to be synonymous. Feeding trials have also consistently pointed to the antinutritional effects of 'high tannin' sorghums in animal diets, although the mechanisms involved are still something of a matter for debate.

Black teas

The green tender and immature, rapidly growing shoots of the tea plant (*Camellia sinensis*) provide the raw material for the manufacture of both black and green teas. Since most of the distinctive sensory characteristics of teas—such as colour, taste, and aroma—are associated directly or indirectly with the oxidative transformations of phenolic flavan-3-ols of the green tea leaf during fermentation these substances have attracted much attention over the past 40 years. They occur in the cytoplasmic vacuoles of leaf cells and analysis shows that (−)-epigallocatechin-3-*O*-gallate always predomimates (up to 50 per cent of the phenolic flavan-3-ols) accompanied by lesser quantities of (−)-epigallocatechin, (−)-epicatechin-3-*O*-gallate, and (−)-epicatechin. It is pertinent to note that the tea plant displays a significant variation on the normal patterns of phenolic flavan-3-ol metabolism, see above. *Camellia sinensis* biosynthesizes flavan-3-ols as their gallate esters to the virtual exclusion of the more normally encountered oligomeric and polymeric condensed proanthocyanidins (after many exhaustive studies Nonaka *et al.* (1983) reported the isolation of four proanthocyanidin gallate esters (~0.07 per cent fresh weight) from green tea). All these compounds generally lack the properties of the so-called vegetable tannins but they form weak complexes with caffeine which may partially precipitate from solution. The fermentation of green tea leaf oxidatively transforms these phenolic flavan-3-ols into products (theaflavins and thearubigins, Roberts 1958, 1962; Sanderson 1972; Fig. 11.5) which display many of the typical properties of vegetable tannins and lend to manufactured tea many of its familiar characteristics as a beverage.

Theaflavin (M_R = 564) and its associated monogallate esters (M_R = 716) and digallate ester (M_R = 870) are all benztropolone derivatives formed by the mixed coupled oxidation of the principal flavan-3-ol precursors (Roberts 1958; Sanderson 1972; Collier *et al.* 1973). The thearubigins are responsible for much of the colour of a black tea infusion and they contribute significantly to the qualities of 'strength and mouth feel'. Much speculation has surrounded the mechanism of their formation and their structure, and they appear to be characteristic of the complex ill-defined polymers which result when fresh plant tissues are damaged and phenolic substrates are transformed by *in-vivo* enzyme-catalysed oxidation to *ortho*-quinones which subsequently polymerize in a random and promiscuous manner.

Incidentally it should be clearly noted that tea *does not*, as various reports have suggested from time to time, contain tannic acid (Chinese gallotannin). The biosynthesis of tannic acid is closely associated with plants of the Anacardiaceae (Haslam 1989). It is not metabolized by the tea plant (*Camellia sinensis*) nor is it formed in the fermentation of green tea leaf. Suggestions made periodically that tannic acid in tea may have various biological and physiological effects, e.g. the recent suggestion (Cramer 1990; Wilcox and Weinberg 1991) that it may impair women's fertility, are therefore erroneous. The term tannin is of course scientifically imprecise but it is not synonymous with tannic acid. Experimental work in these areas must surely concentrate on the phenolic metabolites originally present in the plant and which subsequently contribute to the beverages (e.g. tea, cocoa, coffee, wine) derived therefrom.

The formation of the caffeine–polyphenol complex ('cream') in a cup of tea is thought to be a reversible physico-chemical process in neutral or acidic media. It occurs via the initial formation of soluble caffeine–polyphenol complexes which subsequently aggregate and precipitate (Fig. 11.2). Structurally caffeine (**7**) has features which are, in a sense,

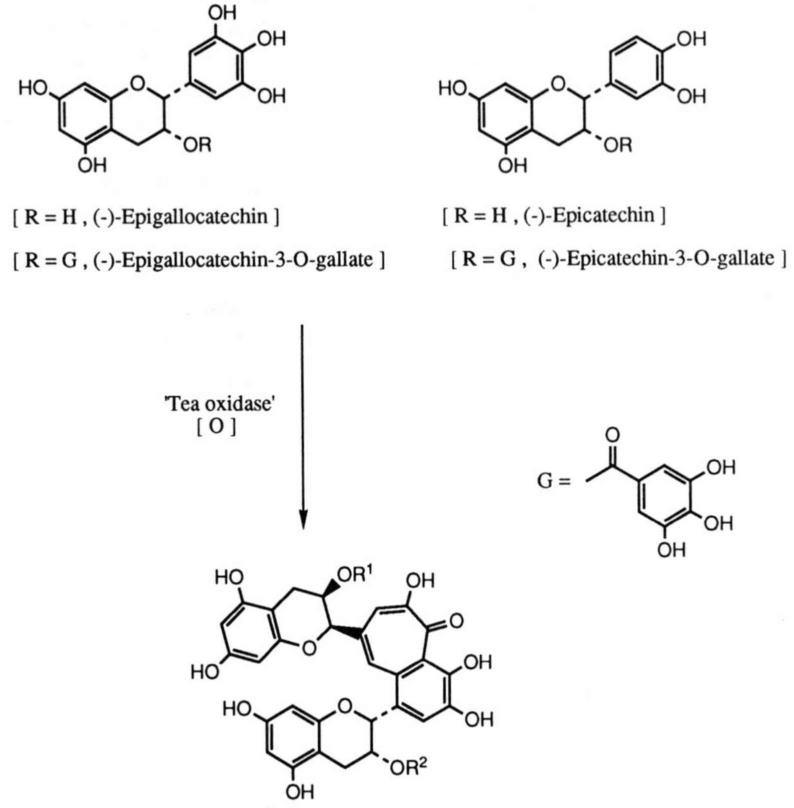

[R = H , (-)-Epigallocatechin]

[R = G , (-)-Epigallocatechin-3-O-gallate]

[R = H , (-)-Epicatechin]

[R = G , (-)-Epicatechin-3-O-gallate]

'Tea oxidase'
[O]

G =

[$R^1 = R^2$ = H , Theaflavin , M_R = 564]
[R^1 = G , R^2 = H ; R^2 = G , R^1 = H ; Theaflavin monogallates , M_R = 716]
[$R^1 = R^2$ = G , Theaflavin digallate , M_R = 870]
λ_{max} ~ 370 and 460 nm .

Fig. 11.5. Black tea polyphenols—oxidative generation of theaflavins.

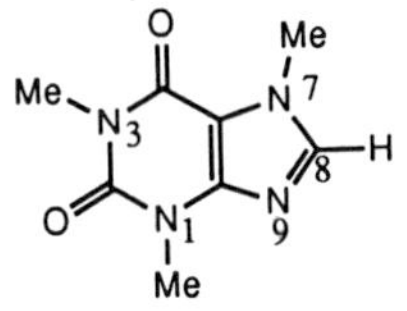

7 Caffeine

reminiscent of peptides derived from proline (viz. the two tertiary amide groups, —CO—NMe—) and studies of the form of its molecular complexation with polyphenols are also therefore of relevance to the role of proline in the promotion of protein–polyphenol association. Whilst it is not possible to directly extrapolate to behaviour in solution, X-ray data from various caffeine–phenol complexes (Martin *et al.* 1986, 1987) point to the importance of the following primary intermolecular forces in complexation:

(1) apolar 'hydrophobic interactions', 'π–π' stacking;

(2) hydrogen bonding;

(3) coordination around a central metal ion.

The crystal structure of the 1:1 methyl gallate–caffeine complex thus typically shows a layer lattice structure. In this array caffeine and methyl gallate molecules are arranged in alternating layers, approximately parallel, with an interplanar separation of 3.3 to 3.4 Å (Fig. 11.6). This 'π–π' stacking is complemented by an extensive in-plane system of hydrogen

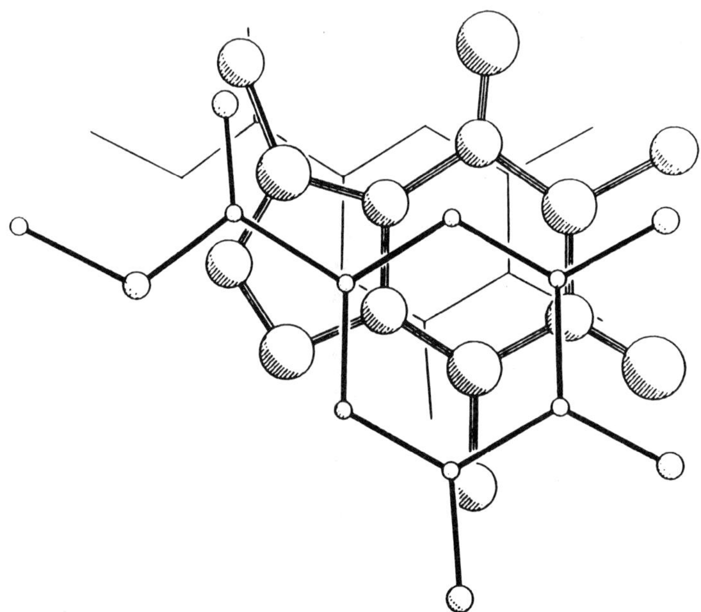

Fig. 11.6. Caffeine–methyl gallate complex: X-ray analysis—vertical 'π-π' stacking.

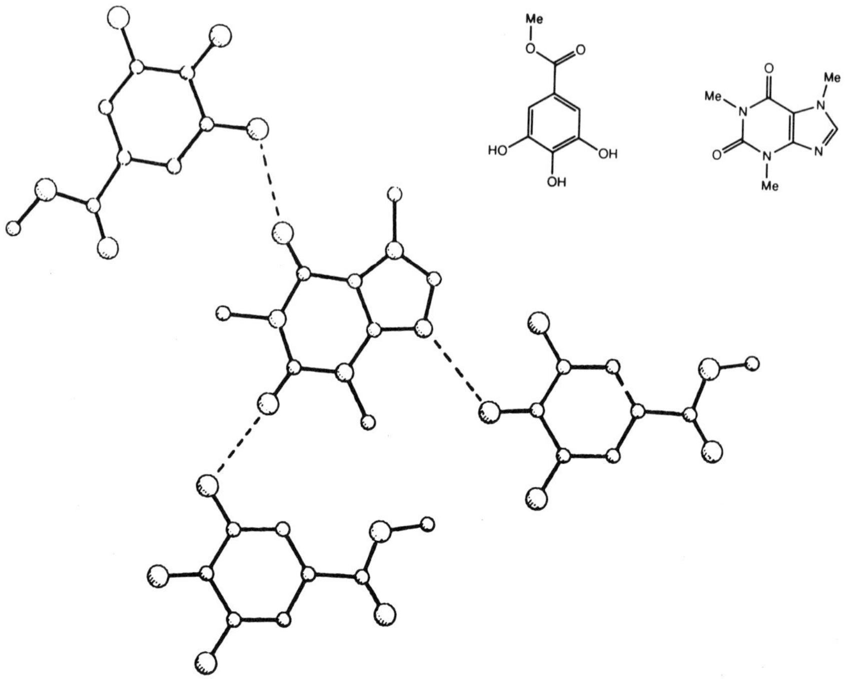

Fig. 11.7. Caffeine–methyl gallate complex: X-ray analysis—in-plane hydrogen bonding.

bonds between the three phenolic groups of methyl gallate (as proton donors) and, the two ketoamide groups and the basic N-9 of caffeine (as proton acceptors) (Fig. 11.7).

This tendency of caffeine to form apolar '$\pi-\pi$' stacking arrangements with phenolic substrates is a typical 'hydrophobic effect'. It also occurs in solution and may be observed by means of ^{1}H NMR spectral analysis as the anisotropic deshielding of the aromatic protons of the phenol and the three (—N—Me) groups and H-8 of the caffeine molecule. This property may be exploited to determine the association constants for the formation of 1:1 complexes between caffeine and various polyphenols (Table 11.1).

These data suggest that theaflavin and its gallate esters have a strong affinity for caffeine, comparable in the case of theaflavin digallate, to that of the archetypal tannin β-1,2,3,4,6-pentagalloyl-D-glucose. The question of how the ability of these polyphenols to form soluble complexes with caffeine correlates with their ability to precipitate caffeine

Table 11.1 Association constants (K, $dm^3\ mol^{-1}$) for the formation of 1:1 complexes between caffeine and polyphenols in deuterium oxide. Determined by ^1H NMR spectroscopy using the chemical shift changes observed for H-8 of caffeine

Polyphenol	M_R	K (45 °C)	K (60 °C)	K^* (45 °C)
(−)-Epicatechin	290	34.5	—	—
(−)-Epigallocatechin	306	35.6	—	—
(−)-Epigallocatechin-3-*O*-gallate	458	52.8	—	—
(+)-Catechin	290	26.1	—	—
(+)-Catechin-3-*O*-gallate	442	38.2	—	—
β-1,2,3,4,6-Pentagalloyl-D-glucose	940	130.2	81.6	58.0
Theaflavin	564	—	—	35.0
Theaflavin monogallate	716	—	—	42.3
Theaflavin digallate	870	—	—	53.7

* D_2O solutions containing 10 per cent D_4 methanol.

from aqueous media (Fig. 11.2) and hence how these substances participate individually in the formation of the 'tea cream' is an important one. The two processes of (i) association and (ii) aggregation are physically and chemically quite different. However in competitive experiments it may be seen (Table 11.2) that the extent of precipitation of a polyphenol from a mixture by caffeine is broadly, but not directly, related to the association constants (K, Table 11.1).

Clearly caffeine possesses a number of features which optimize its effectiveness as a small molecule for complexation with phenolic substrates. The phenolic groups of polyphenols are good proton donors in hydrogen bonding systems. Likewise the tertiary amide carbonyl groups of caffeine are good proton acceptors. It thus seems eminently reasonable that hydrogen bonding may ultimately make specific contributions to the stability of caffeine–polyphenol complexes through the entropy gain as bound water molecules of solvation are released. The 'hydrophobic effect' in which the caffeine and polyphenol molecules engage in '$\pi-\pi$' stacking arrangements is nevertheless probably the dominant influence on the association phenomenon.

Anthocyanin co-pigmentation

A surprisingly small number of pigments is responsible for the very wide range of fruit and floral colours observed in nature. The *betacyanins*

Table 11.2 Selective precipitation of polyphenols by caffeine:

(a) in D_2O

Polyphenol	Polyphenol precipitated (per cent) i	ii
*β-1,2,3,4,6-Pentagalloyl-D-glucose ($K = 130.0$)†	83	92
*(+)-Catechin ($K = 26.1$)†	30	46
*(−)-Epigallocatechin-3-*O*-gallate ($K = 52.8$)†	17	31

Caffeine concentrations: i, 4.5 mmol dm^{-3}; ii, 9.0 mmol dm^{-3}.
*All initial concentrations = 1.5×10^{-3} mol dm^{-3}.
†K (dm^3 mol^{-1}) for the formation of a 1:1 complex with caffeine in D_2O at 45 °C.

(b) in D_2O

Polyphenol	Polyphenol precipitated (per cent) iii
**β-1,2,3,4,6-Pentagalloyl-D-glucose	87
**Theaflavin	50
**(−)-Epigallocatechin-3-*O*-gallate	21

Caffeine concentration: iii, 6.0 mmol dm^{-3}.
**All initial concentrations = 1.0×10^{-3} mol dm^{-3}.

(c) in D_2O containing 10 per cent D_4 MeOH

Polyphenol	Polyphenol precipitated (per cent) iv
***Theaflavin ($K = 35$)‡	16
***Theaflavin digallate ($K = 54$)‡	21
***(−)-Epigallocatechin-3-*O*-gallate	9

Caffeine concentration: iv, 4.5 mmol dm^{-3} in D_2O containing 10 per cent D_4 MeOH.
***All initial concentrations = 1.5×10^{-3} mol dm^{-3} in D_2O containing 10 per cent D_4 MeOH.
‡K (dm^3 mol^{-1}) for the formation of a 1:1 complex with caffeine in D_2O containing 10 per cent D_4 MeOH at 45 °C.

are restricted to the Centrospermae and whilst *carotenoids* contribute extensively to yellow, orange, and red pigmentation, *flavonoids* and *anthocyanins* are almost universal in their distribution. The characteristic colour of red wines, cherry, whortleberry, elderberry, red and black currants, and many other fruit besides derive from anthocyanins. These same pigments also give the distinctive colours to the petals of innumerable flowering plants from salmon and pink, through scarlet, magenta, and violet, to the deep blue of the gentian and the delphinium.

Two great figures, both Nobel prize winners in Chemistry, are inextricably linked with the story of the chemistry of the anthocyanins. The work of Richard Willstatter and Robert Robinson spanned the years before and after the 1914–1918 war. During the 1930s, with J.B.S. Haldane, Sir Robert and Lady Robinson also engaged upon the first biochemical studies of anthocyanins and the chemical and genetic basis of flower colour variation. These pioneering researches drew attention to the phenomenon of co-pigmentation. In contrast to the autonomous carotenoids and chlorophylls, whose colours remain without the need for any accompanying substances, anthocyanins are generally not independent and are often in need of auxilliary molecules (co-pigments) for the full expression of colour.

Some of the principal factors important in the development of the ultimate colour of fruits and flowers are listed below (see also Chapter 15):

(1) variation in pigment class (e.g. chlorophylls, carotenoids, betacyanins, flavonoids, and anthocyanins);

(2) concentration of pigments;

(3) pigment mixtures;

(4) genetic variations within a particular pigment class;

(5) anthocyanins—colour variation dependent upon pH, metal ions, and co-pigments.

The great structural diversity present in the carotenoids contrasts with the surprisingly small number of anthocyanins. Six aglycones dominate anthocyanin structures (**8**). Sugars are attached most frequently at O-3 and O-5 and in many instances these sugars are acylated by hydroxycinnamic or benzoic acids. The degree of 'blueness' of individual anthocyanins is influenced by the hydroxylation/methoxylation status of the anthocyanin 'B' ring, [cf. (**4**) above]. However a free hydroxyl group at one or more of the positions 4′,5,7 is essential in order to generate *in vivo* all of the colours responsible for fruit and flower pigmentation from anthocyanins. The question of how these relatively few anthocyanin

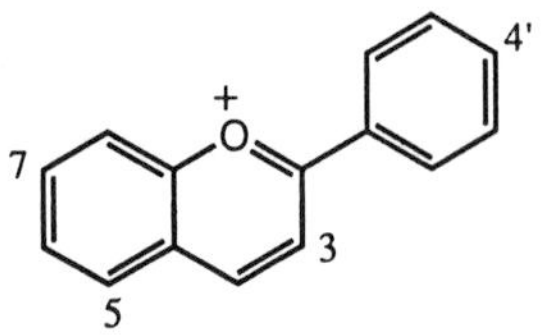

8 2-phenylbenzopyrilium nucleus

structures give rise to such a wonderful variety of floral colours has been one of unending fascination. In an early paper Willstatter attributed this property (and in particular the origin of the blue floral pigments) to variations in the pH of the flower cell sap. Shibata, a Japanese botanist, and later Hayashi proposed an alternative metal ion complex theory according to which the blue colours of anthocyanins derive from complexation with metal ions such as Mg^{2+} and Al^{3+}. Although the importance of metal ion complexation in anthocyanin pigmentation remains unresolved the pH of the cell vacuole, the concentration of the anthocyanin, and the phenomenon of co-pigmentation have emerged as three factors of utmost importance in the determination of the depth and tone of colour due to anthocyanins in fruit and flowers.

In an elegant series of papers Asen *et al.* (1971, 1972, 1975) made several pertinent observations related to the distribution and analysis of anthocyanins *in vivo* in plant tissues. Microscopic examination showed a compartmentalized and sharply delimited location of pigments. In the petals of most flowers colour due to flavonoids is enriched in the epidermal cells, and the adjacent sub-epidermal cells are colourless. With bi-coloured roses carotenoids are invariably concentrated in the outer and anthocyanins (concentrations of the order 10^{-2} M) in the inner face of the petal. The majority of pigmented cell saps have pH values in the region of 4.5–5.5, although extremes of 2.8 (in a begonia cultivar) and 7.5 (in morning glory cv Heavenly Blue) were also recorded. Interestingly, in the context of earlier observations of Sir Robert and Lady Robinson, the pH of epidermal cells from various parts of the same flower were often quite different. Thus in the fuschia (cv Black Knight) the pH of the pink calyx was 3.8 and that of the deep purple corolla was 5.4. Likewise increase in pH is a major factor in the colour changes (usually blueing) of flowers as they age. The optical path length of pigmented cells varied from 20 to 50 μm and Asen *et al.* concluded that it was improbable that anthocyanins alone could contribute to the colour of these tissues whose pH is within the compass 4–6, since in this pH range, in the concentrations specified, and with optical path lengths as determined,

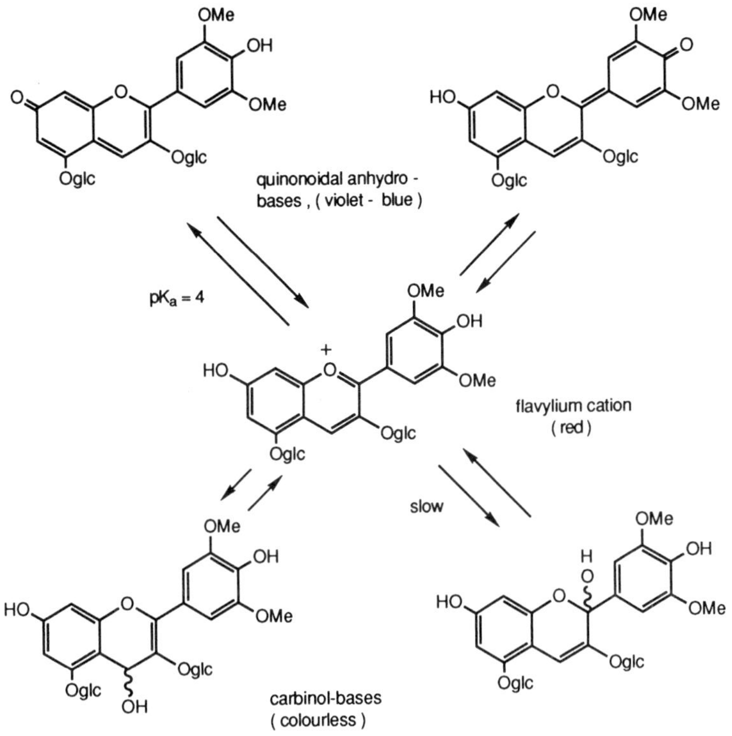

Fig. 11.8. Principal anthocyanin equilibria in aqueous media.

most anthocyanins are virtually colourless. They concluded that the phenomenon of co-pigmentation offers the most logical explanation of the infinite variations of flower colour possible within a given pH range.

Brouillard (1988; Brouillard *et al.* 1990) has affirmed that this ability of anthocyanins to exist as several stable colourless forms is an essential prerequisite to a full expression of floral pigmentation via a two stage decolourization and colour stabilizing mechanism. Under very weakly acidic conditions four anthocyanin structural types exist in equilibrium (demonstrated for the case of malvin; Fig. 11.8): the flavylium cation, the quinonoidal anhydro-bases, the colourless carbinol bases, and (not shown in the figure) the pale yellow reversed chalcones. Equilibration between the quinonoidal bases and the carbinol bases occurs exclusively via the flavylium cation and Brouillard and Dubois (1977) quote *K* [carbinol base/quinonoidal base] as 1.6×10^2 at 4 °C. The p*K* values which govern all these structural changes fall within the 2.5–7.5 pH range. Around neutrality there is a mixture of both the neutral and ionized (anionic) forms of the quinonoidal bases. In the more acidic media the flavylium ion predominates. In the intermediate pH range two situations appertain: (i) pH 3.5–4.5 a mixture of the flavylium ion and the neutral quinonoidal bases is found and (ii) pH 4.5–6.0 when the concentration of the flavylium ion becomes vanishingly small and the quinonoidal bases increasingly predominate.

Concentration is an important factor influencing the colour of anthocyanin solutions—a 100-fold increase in cyanin concentration (10^{-4} to 10^{-2}M) at pH 3.16 leads to a 300-fold increase in absorbance and a change in λ_{max}. Asen and his colleagues explained these deviations from the Beer–Lambert law by invoking the idea of anthocyanin stabilization by self-association. The anthocyanin molecules are believed to associate ('$\pi-\pi$' stacking) in chiral (helical) aggregates driven by hydrophobic effects, entirely analogously to the case of caffeine and phenolic metabolites discussed above. This type of interaction protects the chromophore against attack by nucleophilic reagents, particularly water (the principal constituent of the cell vacuole), thus preventing the formation of the colourless carbinol base forms of the anthocyanin.

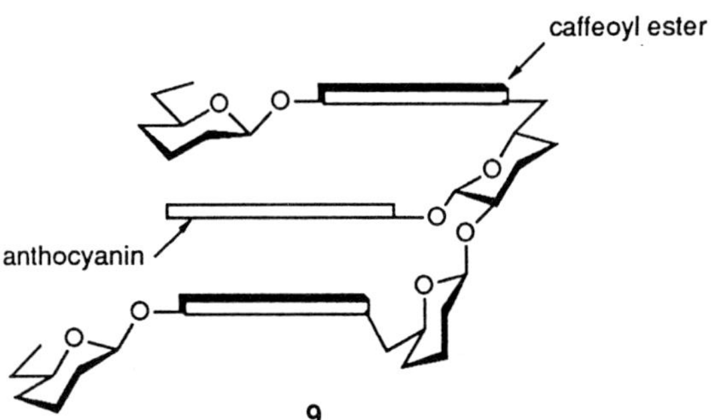

Co-pigments act, either intra- or inter-molecularly, in analogous fashion. Heavenly blue anthocyanin (HBA) (**9**) from the blue petals of the morning glory (*Ipomoea tricolor*) is composed of paeonidin and six molecules of D-glucose and three of caffeic acid (M_R = 1759) (Goto 1987; Goto and Kondo 1991). The blue colour of the flowers is caused primarily by the pH of the cell sap (~7.5). The anthocyanin is therefore probably present as the quinonoidal anhydrobase and/or its mono-anion, and is stabilized by two of the caffeoyl ester groups stacked intramolecularly with the anthocyanin nucleus.

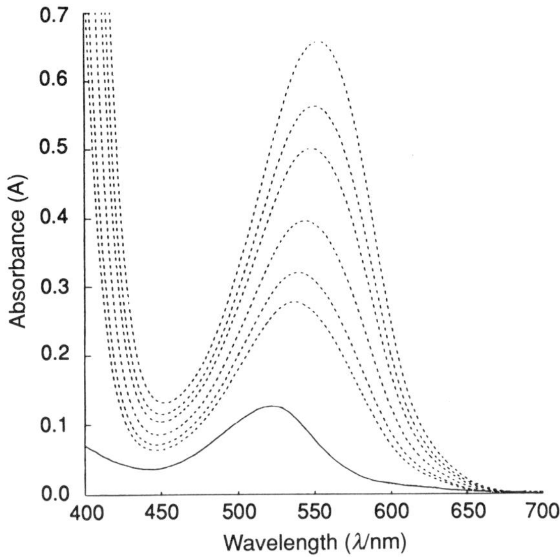

Fig. 11.9. Anthocyanin co-pigmentation—quercetin-3-β-D-galactoside. Malvin chloride (1.0×10^{-4} mol dm^{-3}) in acetate buffer (0.2 mol dm^{-3}, pH 3.65). Changes in anthocyanin visible spectrum produced by the addition of aliquots of quercetin-3-β-D-galactoside (----); final pigment to co-pigment ratio = 1:30.

Intermolecular co-pigments co-exist alongside the anthocyanin in the cell vacuole. They have little or no colour by themselves but stabilize and greatly enhance the colour of the anthocyanin by analogous *intermolecular* hydrophobically reinforced '$\pi-\pi$' stacking with the anthocyanin, either as the flavylium ion or as the quinonoidal base. Co-pigments produce an increase in visible absorbance as well as a bathochromic shift in the wavelength of maximum absorbance in the visible range. Various molecules have been identified as good natural co-pigments: hydroxycinnamoyl esters, galloyl esters ('tannins'), and flavonol glycosides. Figure 11.9 typically shows the enhancement of the red colour of a solution of malvin (pH 3.42) by the addition of aliquots of a flavonol glycoside. The effectiveness of a particular co-pigment appears to be related to its planarity and the potential surface area available for hydrophobic interactions via '$\pi-\pi$' stacking (Fig. 11.10). However this cannot be the whole explanation. Thus caffeine is an effective co-pigment but preferentially stabilizes the quinonoidal base form of the anthocyanin relative to the flavylium ion, giving violet blue solutions (Brouillard *et al.*

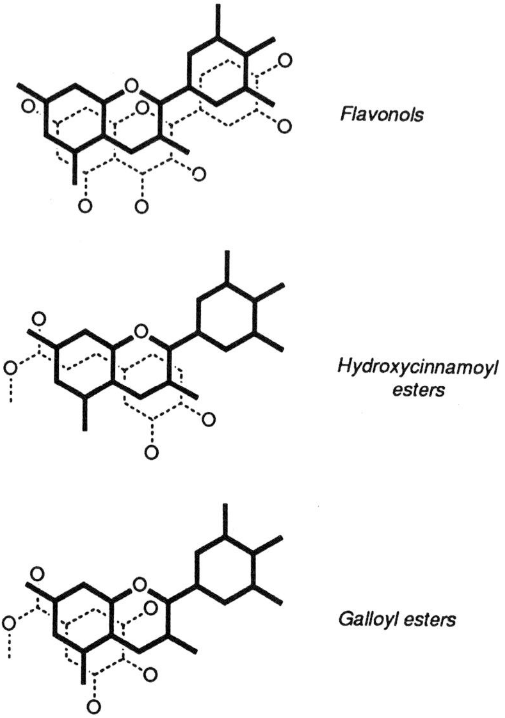

Flavonols

Hydroxycinnamoyl esters

Galloyl esters

Fig. 11.10. Anthocyanin co-pigmentation—'π-π' overlap, some comparisons.

1991; Mistry *et al.* 1991) (Fig. 11.11). Clearly electronic/dipolar effects are also probably involved in the complexation process. In the case of the metalloanthocyanins intermolecular co-pigmentation between an anthocyanin and a flavone glycoside is thought to be augmented by co-ordination of the pigment and co-pigment molecules to a central metal ion (e.g., Mg^{2+}).

It seems possible that anthocyanin co-pigmentation may well underlie other phenomena involving changes in pigmentation in plants. The brilliant red autumnal colourings, particularly of leaves of *Acer* and *Quercus* species, is generally attributed to the loss of chlorophyll and a concommitant increase in anthocyanin synthesis. However *in-vivo* appearance, change, or loss of colour, is not neccessarily related to increased anthocyanin synthesis. It may well simply be related to the release or modification of a good co-pigment. In this context it is therefore interesting to note that the leaves of many shrubs and trees which display rich autumnal

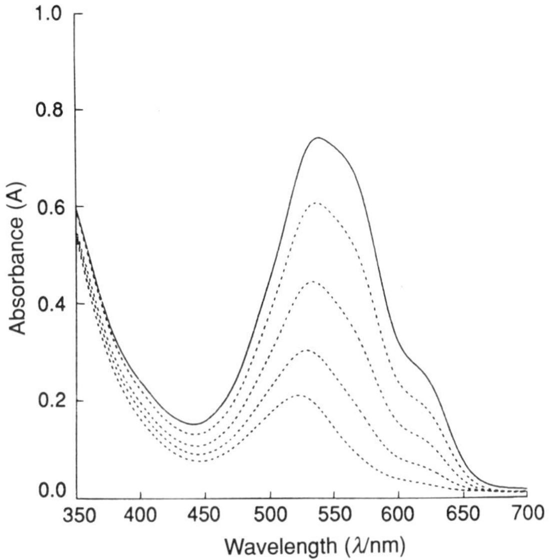

Fig. 11.11. Anthocyanin co-pigmentation—caffeine. Malvin chloride (2.6×10^{-4} mol dm^{-3}) in acetate buffer (0.2 mol dm^{-3}, pH 3.42). Changes in anthocyanin visible spectrum produced by the addition of aliquots of caffeine (----); final pigment to co-pigment ratio = 1:150.

colourings, such as *Acer* and *Quercus*, are also rich sources of galloyl esters many of which are excellent anthocyanin co-pigments.

Finally note may be made of the continuing interest of plant breeders in the development of a blue rose. Whatever the aesthetic pleasure such a rose might give and notwithstanding centuries of cross-breeding by traditional rose growers this goal remains a frustratingly distant one. However scientists in Australia claim that they are on the verge of producing the world's first blue rose. Glycosides of pelargonidin, cyanidin, and paeonidin are the most important anthocyanidins found in roses and the breakthrough lies in the possibility of transferring the gene responsible for the formation of the delphinidin glycosides in the blue petunia to the rose. Flower colour is a complex phenomenon (see above) and besides anthocyanins is dependent on pH, metal ions, and co-pigments. The observation that caffeine preferentially stabilizes the violet-blue quinonoidal base forms of anthocyanins however opens up another possibility for 'engineering' a blue rose or for that matter blue flowers in general. Transferring the genes for the synthesis of caffeine to the genus *Rosa* would truly give a hybrid tea rose!

Ageing of red wines—changes in pigmentation

New red wine is an elaborate chemical system which has the capability of sustaining a range of chemical reactions and physico-chemical equilibria. However there is little doubt that, during the ageing process, the polyphenols (flavan-3-ols, condensed proanthocyanidins) and pigments (anthocyanins) play a crucial role in the development of taste, flavour, and appearance. There is very good circumstantial evidence that many of these chemical changes occur via the initial formation and intermediacy of anthocyanin–polyphenol complexes (cf. Fig. 11.3).

The colour of red wines derives initially from the extraction of anthocyanin pigments from the skins of black grapes during vinification (Somers 1978; Somers and Evans 1979, 1986; Ribereau-Gayon *et al.* 1983; Bakker *et al.* 1986; Somers and Verette 1988). Typical concentrations of anthocyanins are around 500 mg $litre^{-1}$ (10^{-4}–10^{-3} M) in a full-bodied young red wine. In *Vitis vinifera* malvidin-3-glucoside ('oenin') and its acylated derivatives generally account for 60–80 per cent of the total anthocyanins (Somers and Verette 1988). Also leached from the grape skins, seeds, and other vascular tissue during maceration are various additional phenolic metabolites including hydroxycinnamic acid esters and flavonoids. Prominent among these flavonoids are the phenolic flavan-3-ols, (−)-epicatechin and (+)-catechin, and the structurally and biosynthetically related oligomeric procyanidins (Weinges and Piretti 1971; Haslam 1980; Da Silva *et al.* 1991) which are thought to lend astringency to young red wines (Somers and Evans 1986; Somers and Verette 1988). Estimates of the *total* phenolics in red wines range from 800–4000 mg $litre^{-1}$.

During wine conservation and maturation the pigment composition becomes progressively more complex. There is a steady decline in the concentration of the original anthocyanins with the concommitant formation of new, more stable, oligomeric pigments which maintain wine colour (Somers and Verette 1988). Most rapid changes occur during the first year or two after vinification. The structure of the products and the precise mechanisms whereby they are formed have remained speculative and no condensed pigment structure has yet been characterized (Somers and Verette 1988). However the susceptibility of anthocyanins to attack by nucleophilic reagents has long been recognized; with carbon nucleophiles substitution at C-4 is usually involved (Jurd and Waiss 1965; Jurd 1967, 1969). One of the reactions by which the colour of red wines is changed during maturation and ageing is thus considered by Somers (1971) to follow a sequence similar to that shown in Fig. 11.12. The nucleophilic species may be a phenolic flavan-3-ol or more probably, according to Somers, an oligomeric procyanidin.

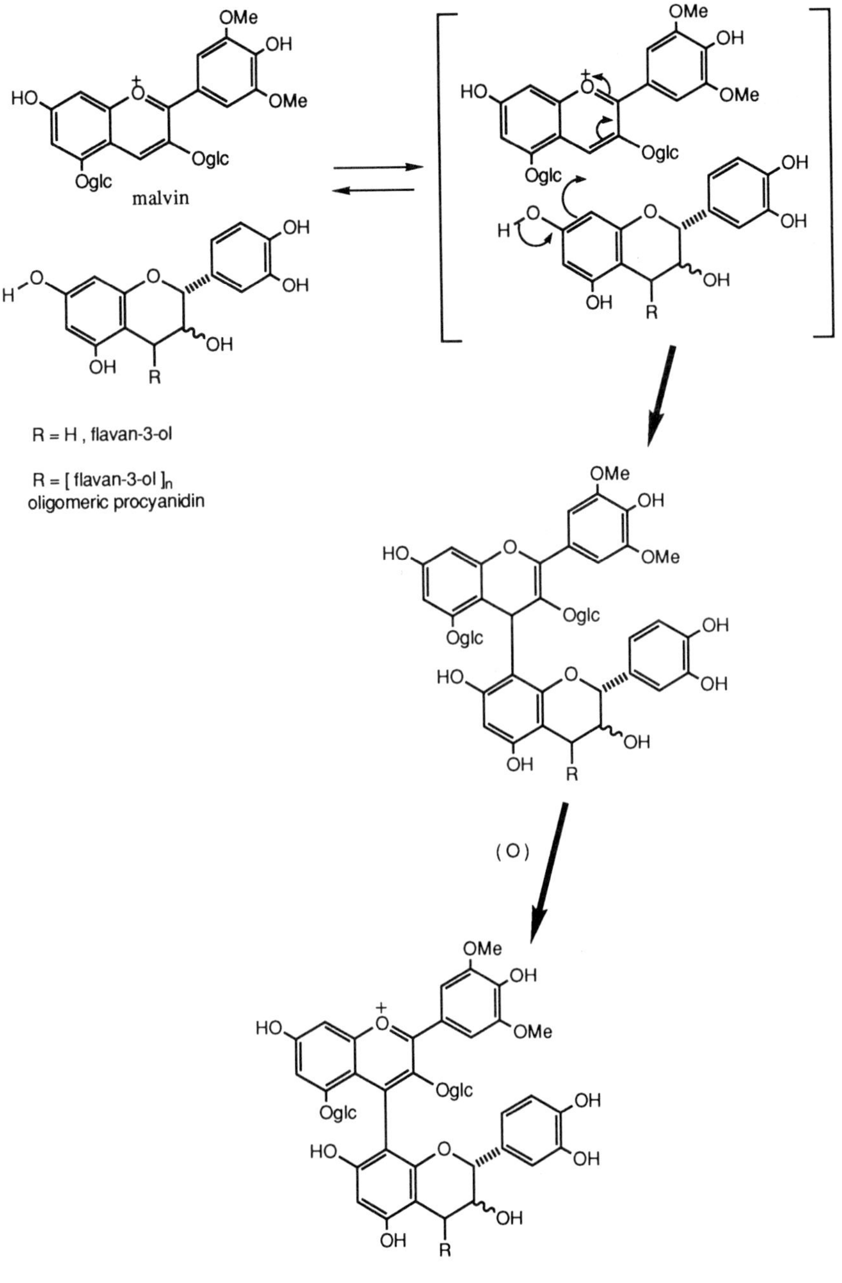

malvin

R = H , flavan-3-ol

R = [flavan-3-ol]$_n$
oligomeric procyanidin

(O)

Oligomeric / polymeric pigments

Fig. 11.12. Formation of oligomeric pigments in red wines (Somers and Verette 1988).

Young red wines have a visible absorbance maximum $\lambda \sim 520$ nm arising principally from the contributions to colour of the flavylium ion and the quinonoidal anhydro-base forms of the appropriate anthocyanins (Fig. 11.8). Between this absorbance and the one in the ultraviolet ($\lambda \sim 280$ nm) a minimum is found at $\lambda \sim 420$ nm (Ribereau-Gayon 1974). As the wine ages the maximum at $\lambda \sim 520$ nm declines in intensity, falling eventually to a shoulder in wines older than 10 years. This change corresponds to an increase in yellow colour at $\lambda \sim 420$ nm and a shift from the brightness and purple tints of young red wine to the tile-like red–orange of an old matured wine. The wine hue or tint, measured as the ratio of the absorbances A_{420}/A_{520}, typically increases from 0.4–0.5 in fresh new wines (pH 3.5–3.7) to around 0.8–0.9 in mature red wines. The ratio may well exceed 1.0 in wines older than 10 years. Since phenolic compounds are known to contribute to the oxidative browning reactions of wines (see Cheynier *et al.* 1986, 1988) Somers (1971) has suggested that the presence of additional 'quinone-like' structures, arising from the oxidation of the oligomeric pigments (Fig. 11.12), are most likely to develop as the wine ages and would therefore probably account for the browner tints of aged wines. Recent observations, albeit in model systems, underline the presumed importance of co-pigmentation in red wines, the significance of the anthocyanin–co-pigment complex as an intermediate in the development of the yellow–orange colours assumed in red wines as they mature, and the possible chemical nature of these new pigments for which alternative structures are now proposed.

Non-planar flavan-3-ols, (+)-catechin and (−)-epicatechin and related oligomeric procyanidins, are comparatively poor co-pigments for the flavylium cation (Brouillard *et al.* 1991; Mistry *et al.* 1991); in weakly acidic media (0.2 M acetate buffer, pH 4.89) (−)-epicatechin is *a priori* somewhat surprisingly a rather better co-pigment than (+)-catechin. Nevertheless the intensely bluish-red grape-berry juices derive their colour, at least in part, from co-pigmentation effects involving (+)-catechin and (−)-epicatechin. The observation that (+)-catechin is an inferior co-pigment when compared with its diastereoisomer (−)-epicatechin is presumably explicable in terms of the observations of Porter and his colleagues (1986). On the basis of high resolution NMR studies these workers described the conformational properties of the flavan-3-ols in solution principally in terms of an equilibrium between two conformational states, *E* and *A*, in which the catechol ring 'B' is respectively in a quasi-equatorial or quasi-axial position. They showed that (+)-catechin has a lower *E*:*A* ratio than (−)-epicatechin ($\sim 1.6{:}1$ compared to $\sim 6.3{:}1$) and exists in solution as an equilibrium in which the contribution from the quasi-axial conformer (*A*) is a significant one. In this conformational state (*A*) the flavan-3-ol cannot assume a shape in which

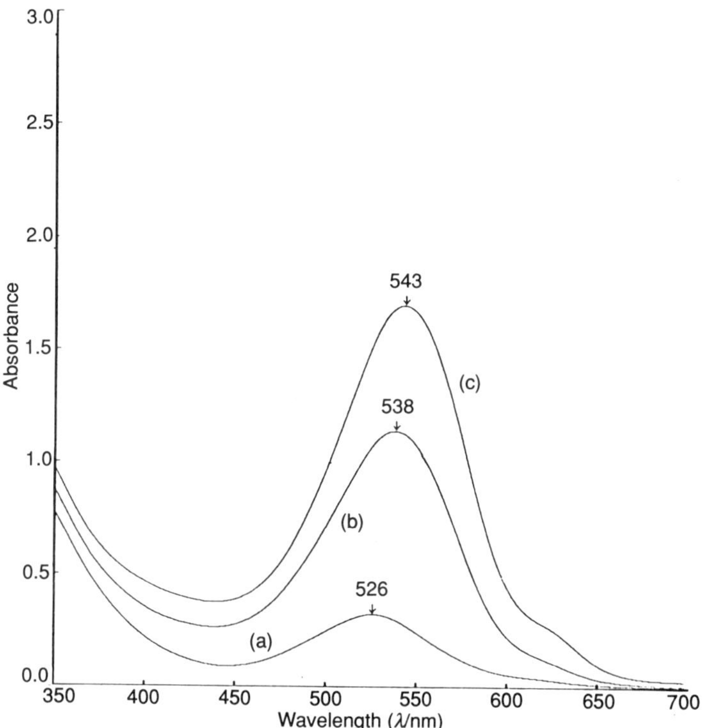

Fig. 11.13. Anthocyanin co-pigmentation—t = 24 h, 20 °C, full light. (a) Malvin chloride (1.3×10^{-3} mol dm^{-3}) in acetate buffer (0.2 mol dm^{-3}, pH 3.42) containing 10 per cent ethanol. (b) Malvin chloride (1.3×10^{-3} mol dm^{-3}) in acetate buffer (0.2 mol dm^{-3}, pH 3.42) containing 10 per cent ethanol plus (−)-epicatechin (molar ratio 1:12). (c) Malvin chloride (1.3×10^{-3} mol dm^{-3}) in acetate buffer (0.2 mol dm^{-3}, pH 3.42) containing 10 per cent ethanol plus (+)-catechin-3-*O*-gallate (molar ratio 1:9).

both aromatic nuclei are approximately coplanar as required for effective co-pigmentation, allowing both phenolic nuclei to participate simultaneously in 'π–π' stacking with the anthocyanin. Galloylation at C-3 of (+)-catechin to give (+)-catechin-3-*O*-gallate however substantially enhances the co-pigmentation of the phenolic flavan-3-ol.

Figure 11.13 depicts the result of the addition of two substrates [(−)-epicatechin and (+)-catechin-3-*O*-gallate] to solutions of malvin chloride in 0.2 M acetate buffer in 10 per cent (v/v) ethanol. Distinctive changes occur in these pigment–co-pigment solutions as they are maintained in

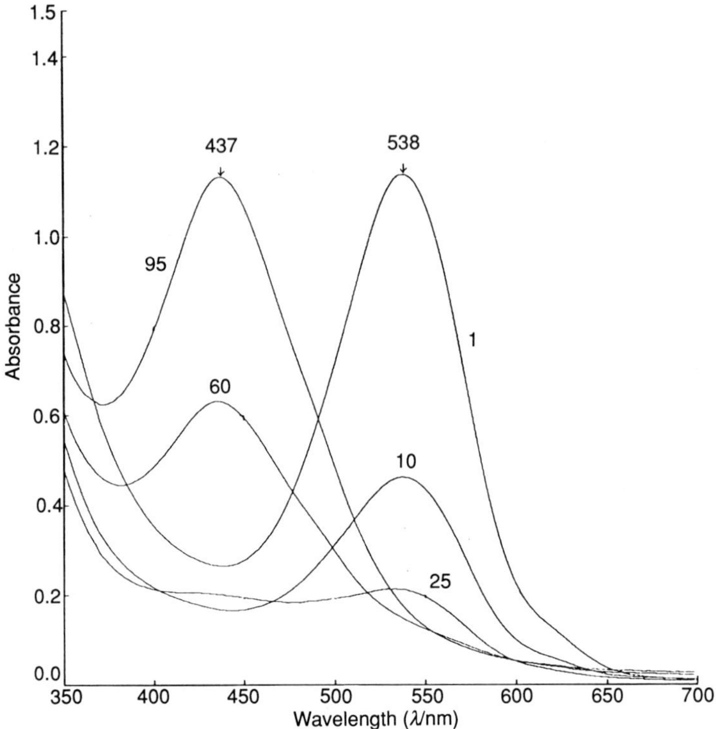

Fig. 11.14. Anthocyanin co-pigmentation—malvin chloride (1.3×10^{-3} mol dm^{-3}) in acetate buffer (0.2 mol dm^{-3}, pH 3.42) containing 10 per cent ethanol plus (−)-epicatechin (molar ratio 1:12); $t = 1$, 10, 25, 60, 95 days at 20 °C, full light.

full light at 22 °C over a period of ~ four months (Fig. 11.14 and 11.5; Hua, Cai, and Haslam, unpublished work). In solutions of malvin the principal change is simply a decrease in the magnitude of the maximum absorbance (λ ~ 525 nm) of approximately 20–25 per cent. However, when the co-pigment contains an activated phloroglucinol nucleus there are progressive changes in the colour and the absorption spectrum of the solutions (Fig. 11.14, (−)-epicatechin; Fig. 11.15, (+)-catechin-3-*O*-gallate). The absorbance maximum at λ 525–540 nm steadily declines and there is a concomitant increase in the absorption at λ 430–445 nm as the solutions change from red to yellow–orange. Moreover the rates of change of colour in the various solutions appear directly related to the effectiveness of the particular substrate as an anthocyanin co-pigment,

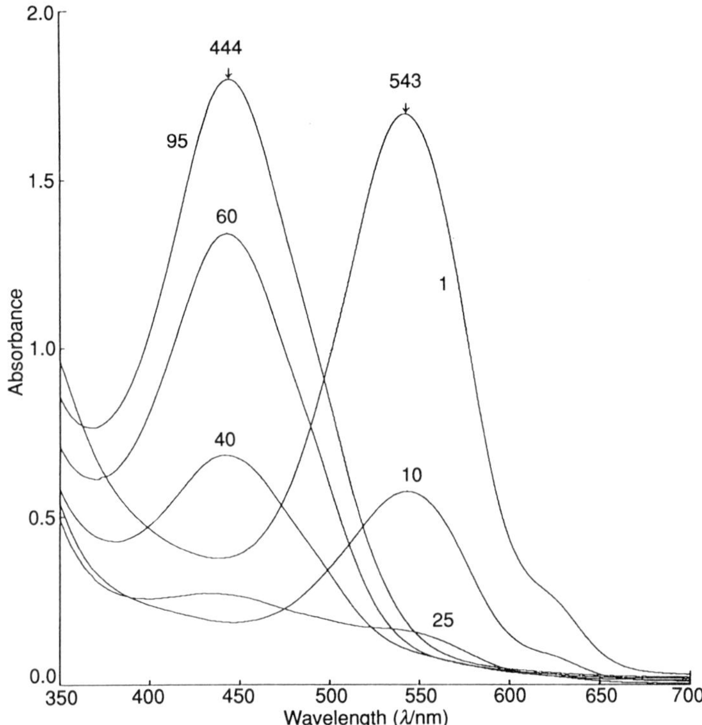

Fig. 11.15. Anthocyanin co-pigmentation—malvin chloride (1.3×10^{-3} mol dm^{-3}) in acetate buffer (0.2 mol dm^{-3}, pH 3.42) containing 10 per cent ethanol plus (+)-catechin-3-*O*-gallate (molar ratio 1:9); t = 1, 10, 25, 40, 60, 95 days at 20 °C, full light.

i.e. (+)-catechin-3-*O*-gallate > (−)-epicatechin. These observations strongly suggest that the formation of a soluble complex between the anthocyanin and the co-pigment is a kinetically significant step in the chemical transformations which follow. The wavelength of maximum absorbance of the new yellow–orange pigment forms as they appear in solution is also related to the particular co-pigment: (−)-epicatechin at $\lambda \sim 438$ nm and (+)-catechin-3-*O*-gallate at $\lambda \sim 444$ nm (Figs. 11.14 and 11.15). This observation strongly implies that the residual co-pigment remaining in solution *also* acts analogously as a co-pigment for the respective new yellow–orange pigment forms as they are produced [(−)-epicatechin < (+)-catechin-3-*O*-gallate]. It therefore indicates that, like their anthocyanin precursor, the yellow–orange pigments probably also possess an electron deficient aromatic π system (Mistry *et al.* 1991).

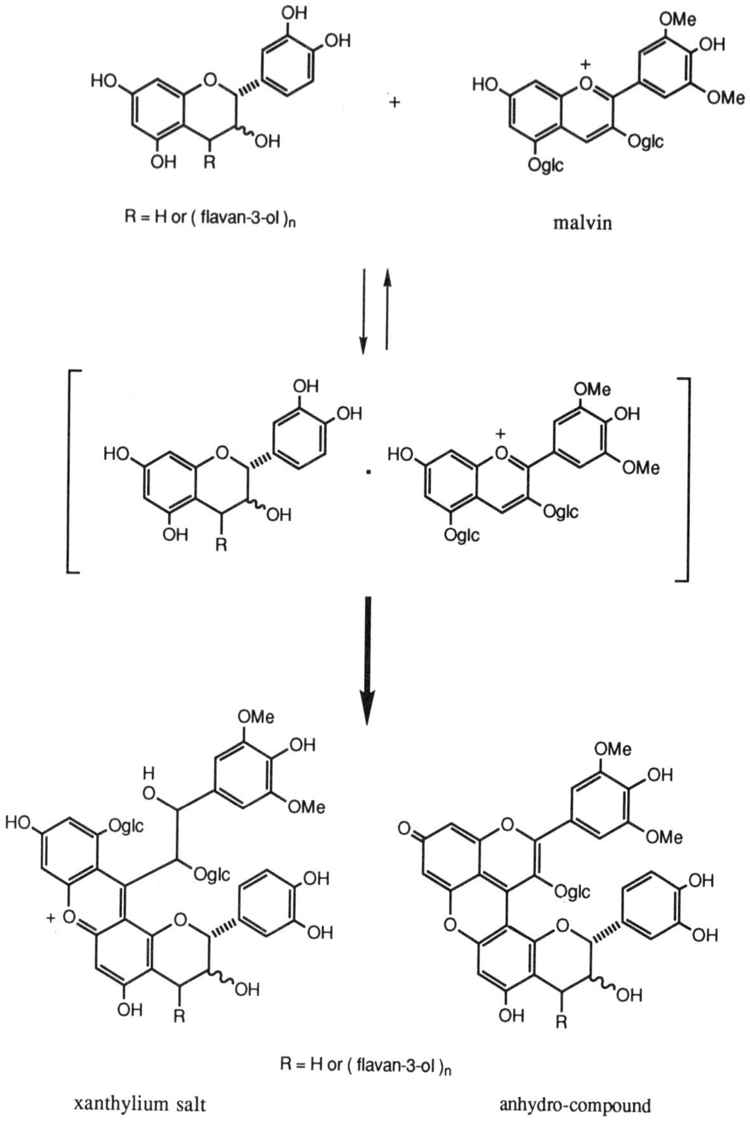

R = H or (flavan-3-ol)$_n$

malvin

R = H or (flavan-3-ol)$_n$

xanthylium salt

anhydro-compound

Fig. 11.16. Ageing of red wines—the formation of xanthylium salts.

The new pigments are highly soluble in water and are not extracted therefrom by ethyl acetate—supporting the proposition that they have a glucosidic and/or cationic structure and strongly implying that they are formed by the reaction of malvin and the co-pigment and that the final structure contains elements of both precursors. Baranowski and Nagel (1983) in an earlier paper suggested that the loss of malvin-3-glucoside from solutions containing (+)-catechin, in the presence and absence of acetaldehyde, occurred by two types of condensation reaction. They did not describe or speculate upon the nature of the products. Hrazdina and Borzell (1971) showed that a yellow compound which developed during the storage of grape juice was probably a derivative of a tetrahydroxyxanthylium salt. Timberlake and Bridle (1976) earlier discussed the formation of xanthylium compounds from malvin and (+)-catechin and presented evidence in support of this proposal and Jurd and Somers (1970) showed model proanthocyanidins when treated with 50 per cent aqueous acetic acid gave yellow xanthylium salts (λ_{max} 439 nm). These observations suggest that xanthylium salts, or anhydro derivatives, represent possible structures for the new yellow pigments described herein (Fig. 11.16). They may be envisaged as derived from the adducts (or from intermediates in the formation of these adducts) proposed earlier (Somers 1971; Fig. 11.12) as the oligomeric/polymeric pigments produced in red wines as they mature. On the basis of the evidence presented the formation of analogous yellow–orange pigments, such as have been described, in red wines themselves as they develop seems eminently reasonable. Their formation probably represents a major factor in the evolution from a young red wine to the tile-like red to orange colour of a mature wine.

References

Asen, S., Stewart, R. N., and Norris, K. H. (1971). Co-pigmentation effect of quercetin glycosides on red wing azalea. *Phytochemistry*, **10**, 171–5.

Asen, S., Stewart, R. N., and Norris, K. H. (1972). Co-pigmentation of Anthocyanins in plant tissues and its effect on colour. *Phytochemistry*, **11**, 1139–44.

Asen, S., Stewart, R. N., and Norris, K. H. (1975). Microspectrometric measurement of pH and pH effect on colour of epidermal cells. *Phytochemistry*, **4**, 937–42.

Bakker, J., Preston, N. W., and Timberlake, C. (1986). The Determination of anthocyanins in ageing red wines: Comparison of HPLC and spectral methods. *Am. J. Enol. Vitic.*, **37**, 121–6.

Baranowski, E. and Nagel, C. W. (1983) Kinetics of malvidin-3-glucoside condensation in wine model systems. *J. Food Sci.*, **48**, 419–21 and 429.

Bate-Smith, E. C. (1954*a*). Flavonoid compounds in food. *Adv. Food Res.*, **5**, 261–300.

Bate-Smith, E. C. (1954*b*). Astringency in foods. *Food*, **23**, 124.

Bate-Smith, E. C. (1962). The phenolic constituents of plants and their taxonomic significance. I. Dicotyledons. *J. Linnaean Soc. Bot. (London)*, **58**, 95–173.

Bate-Smith, E. C. and Metcalfe, C. R. (1957). Leucoanthocyanins. III. The nature and systematic distribution of tannins in plants. *J. Linnaean Soc. Bot. (London)*, **55**, 669–705.

Brouillard, R. (1988). Flavonoids and flower colour. In *The flavonoids—advances in research* (ed. J. B. Harborne), pp. 525–38, Chapman and Hall, London.

Brouillard, R. and Dubois, J.-E. (1977). Mechanisms of the structural transformations of anthocyanins in acidic media. *J. Am. Chem. Soc.*, **99**, 1359–64.

Brouillard, R., Wigand, M.-C., and Cheminat, A. (1990). Loss of colour, a prerequisite to plant pigmentation by flavonoids. *Phytochemistry*, **29**, 3457–60.

Brouillard, R., Wigand, M.-C., Dangles, O., and Cheminat, A. (1991). pH and solvent effects on the co-pigmentation reaction of malvin with polyphenols, purine and pyrimidine derivatives. *J. Chem. Soc., Perkin Trans. 2*, 1235–41.

Butler, L. G., Mehanso, H., and Carlson, D. (1987). Dietary tannins and salivary proline rich proteins—interactions, induction and defence mechanisms. *Ann. Rev. Nutrition*, **7**, 423–40

Cheynier, V., Salgues, M., Gunata, Z., and Wylde, R. (1986). Oxidation of grape juice 2-*S*-glutathionyl caffeoyl tartaric acid by *Botrytis cinerea* Laccase and characterisation of a new substance: 2,5-di-*S*-glutathionyl caffeoyl tartaric acid. *J. Food Sci.*, **51**, 1191–4.

Cheynier, V., Osse, C., and Rigaud, J. (1988). Oxidation of grape juice phenolics in model systems. *J. Food Sci.*, **53**, 1729–32 and 1760.

Collier, P. D., Bryce, T., Mallows, R., Thomas, P. E., Frost, D. J., Korver, O., and Wilkins, C. K. (1973). The theaflavins of black tea. *Tetrahedron*, **29**, 125–40.

Cramer, D. W. (1990). Caffeine and infertility. *Lancet*, **335**, 792–3.

Da Silva, J. M. R., Cheynier, V., Rigaud, J., Cheminat, A., and Moutounet, M. (1991). Procyanidin dimers and trimers from grape seed. *Phytochemistry*, **30**, 1259–64.

Goto, T. (1987). Structure, stability and colour variation of natural anthocyanins. *Prog. Chem. Org. Nat. Prod.*, **52**, 113–58.

Goto, T. and Kondo, T. (1991),Structure and molecular stacking of anthocyanins —flower colour variation. *Angew. Chem., Int. Ed. Engl.*, **30**, 17–33.

Grassman, W. and Stadler, P. (1961). Tanning effects on collagen fibrils. *Leder*, **12**, 290.

Gustavson, K. H. (1956). *The chemistry of tanning processes.* Academic Press, New York.

Hagerman, A. E. and Butler, L. G. (1981). The specificity of proanthocyanidin —protein interactions. *J. Biol. Chem.*, **256**, 4494–7.

Haslam, E. (1980). In Vino Veritas; oligomeric procyanidins and the ageing of red wines. *Phytochemistry*, **19**, 2577–82.

Haslam, E. (1989). *Plant polyphenols—vegetable tannins revisited.* Cambridge University Press.

Haslam, E. and Lilley, T. H. (1988). Natural a stringency in foodstuffs—a molecular interpretation. *C.R.C. Crit. Rev. Food Sci.*, **27**, 1–40.

Hrazdina, G. and Borzell, A. J. (1971). Xanthylium derivatives in grape extracts. *Phytochemistry*, **10**, 2211–13.

Jurd, L. (1967). Catechin–flavylium salt condensation reactions. *Tetrahedron*, **23**, 1057–64.

Jurd, L. (1969). Review of polyphenol condensation reactions and their possible occurrence in the ageing of red wines. *Am. J. Enol. Vitic.*, **30**, 289–300.

Jurd, L. and Somers, T. C. (1970). The formation of xanthylium salts from proanthocyanidins. *Phytochemistry*, **9**, 419–27.

Jurd, L. and Waiss, A. C. (1965). Anthocyanins and related compounds VI Flavylium salt—phloroglucinol condensation products. *Tetrahedron*, **21**, 1471–83.

Martin, R., Lilley, T. H., Bailey, N. A., Falshaw, C. P., Haslam, E., Magnolato, D., and Begley, M. J. (1986). Polyphenol–caffeine complexation. *J. Chem. Soc., Chem. Commun.*, 105–6.

Martin, R., Lilley, T. H., Falshaw, C. P., Haslam, E., Magnolato, D., and Begley, M. J. (1987). The caffeine–potassium chlorogenate complex. *Phytochemistry*, **26**, 273–9.

Mistry, T. V., Cai, Y., Lilley, T. H., and Haslam, E. (1991). Polyphenol interactions. Part 5. Anthocyanin co-pigmentation. *J. Chem. Soc., Perkin Trans. 2*, 1287–96.

Nonaka, G. I., Kawahara, O., and Nishioka, I. (1983). Tannins and related compounds. XV. A new class of dimeric flavan-3-ol gallates, theasinensins A and B and proanthocyanidin gallates from green tea leaf. *Chem. Pharm. Bull.*, **31**, 3906–14.

Okuda, T. (1981). Tannins of medicinal plants and drugs. *Heterocycles*, **15**, 1323–48.

Porter, L. J., Wong, R. Y., Benson, M., Chan, B. G., Vishwanadhan, V. N., Gandar, R. D., and Mattice, W. L. (1986). Conformational analysis of flavans. ^{1}H NMR and molecular mechanical (MM2) studies of the benzopyran ring of 3′,4′,5,7-tetrahydroxy-flavan-3-ols. *J. Chem. Res.*, 86–7.

Ribereau-Gayon, P. (1974). The chemistry of red wine colour. In *The chemistry of winemaking* (ed. A. D. Webb), pp. 50–87. American Chemical Society, Washington, DC.

Ribereau-Gayon, P., Pontallier, P., and Glories, Y. (1983). Some interpretations of colour changes in young red wines during their conservation. *J. Sci. Food Agric.*, **34**, 505–16.

Roberts, E. A. H. (1958). Phenolic substances of manufactured Tea. II. Their origin as enzymic oxidation products in fermentation. *J. Sci. Food Agric.*, **9**, 212–6.

Roberts, E. A. H. (1962) Economic importance of flavonoid substances: Tea fermentation. In *Chemistry of the flavonoids* (ed. T. A. Geissman), pp. 468–512. Pergamon Press, New York.

Robinson G. M. (1939). Variable colour of flower petals. *J. Chem. Soc.*, **61,** 1606–7.

Robinson, G. M. and Robinson, R. (1931). A survey of anthocyanins. I. *Biochem. J.*, **25,** 1687–705.

Sanderson, G. W. (1972). The chemistry of tea and tea manufacturing. *Rec. Adv. Phytochem.*, **5,** 247–316.

Shen, Z., Haslam, E., Falshaw, C. P., and Begley, M. J. (1986). Procyanidins and polyphenols from *Larix gmelini* bark. *Phytochemistry*, **25,** 2629–35.

Somers, T. C. (1971). The polymeric nature of wine pigments. *Phytochemistry*, **10,** 2175–86.

Somers, T. C. (1978) Interpretations of colour composition in young red wines. *Vitis*, **17,** 161–7.

Somers, T. C. and Evans, M. E. (1979). Grape pigment phenomena: Interpretation of major colour losses during vinification. *J. Sci. Food Agric.*, **30,** 623–33.

Somers, T. C. and Evans, M. E. (1986) Evolution of red wines. I. Ambient influences on colour composition during early maturation. *Vitis*, **25,** 31–9.

Somers, T. C. and Verette, E. (1988). Phenolic composition of natural wine types. In *Modern methods of plant analysis, new series*, Vol. 6. *Wine analysis* (ed. H. F. Linskens and J. F. Jackson), pp. 219–57. Springer-Verlag, Berlin, Heidelberg.

Timberlake, C. F. and Bridle, P. (1976). Interactions between anthocyanins, phenolic compounds, and acetaldehyde and their significance in red wines. *Am. J. Enol.*, **27,** 97–105.

Weinges, K. and Piretti, M. V. (1971). Isolierung des $C_{30}H_{26}O_{12}$-Procyanidins B1 aus Weintrauben. *Liebigs Ann. Chem.*, **748,** 218–20.

Wilcox, A. J. and Weinberg, C. R. (1991). Tea and fertility. *Lancet*, **337,** 1159–60.

12. *Ginkgo biloba*–cultivation, extraction, and therapeutic use of the extract

J. O'REILLY

Cara Partners, Little Island, Co. Cork, Ireland

The living fossil

The term 'living fossil' was first used by Darwin to describe *Ginkgo biloba*, as it is the sole surviving species of a group of plants, the Ginkgoales, which can be traced back more than 200 million years. Species of the order Ginkgoales were worldwide in pre-Cretaceous times and fine examples in fossil rock have been found in all parts of the world. The Ginkgoales increased to its maximum diversity during the Cretaceous period (Seward 1938). However, the great upheavals of the Quaternary period, followed by the ice age led to the destruction of all species of the Ginkgoales with the exception of *Ginkgo biloba* which survived only in the far east.

The long time span from one generation to the next, which minimizes genetic mutations, is the first reason why *Ginkgo biloba* has survived unchanged for more than 200 million years. The female tree produces its first seeds between 20 and 30 years of age and continues to produce them for over 1000 years. Secondly, the tree is resistant to all serious pests, bacteria, fungi, and viruses. Thirdly, it is well suited to its environment, being highly resilient. An example of its resilience in modern times is the ginkgo tree which was at the epicentre of the atomic bomb blast at Hiroshima (Michel 1985). All that remained was the charred tree trunk. However new shoots were produced by the roots resulting in a fine specimen today.

While claims for natural stands of *Ginkgo biloba* in China over the last 150 years have been made, this is unlikely. Fortunately, its recent history has been extended by man because of its beauty, its usefulness as a shade tree, and most importantly because the seeds are considered a delicacy in the far east. The tree was cultivated mainly around Buddhist temples throughout the ages and *Ginkgo biloba* was probably first introduced to Japan by Buddhist monks in the twelfth century. The first European to describe ginkgo was Kaempfer (1712) when he wrote of his travels in Japan at the end of the seventeeth century. The first tree was brought to Europe by Dutch merchants some 40 years later, and planted in Utrecht around 1730 (Li 1956). The plant was introduced into America in the 1780s. However, it was 1795 before the first male tree flowered in Europe. The first single female tree was found growing near Geneva in 1814. As a result fertile trees were finally secured in Europe by grafting its shoots upon male trees established elsewhere, one being in the botanic gardens in Montpellier where the first perfect seed is reported to have been produced in Europe in 1835.

On its introduction into Europe and the United States the tree was cultivated for its beauty especially during autumn when the leaves turn a beautiful yellow and also for its upright habit and resilience. It is used as a street tree in many major cities on the east coast of the United States. However, the use of the female plant as a street tree is avoided as the fruits produce large quantities of ginkgolic acids (**1**) which cause contact dermatitis similar to that of poison ivy (Sowers *et al.* 1965) and also because of their unpleasant odour. This is in contrast to the far east where the tree is planted for its seeds and the vast majority of the trees are female. While it is difficult to observe the subtle differences between young male and female plants, it is possible, for example, the buds of the male plant are larger and more round in shape than those of the female plant which are smaller and conical (Maugini and Vestri 1974). Also, when opening, the male bud on the short shoot is normally 10–11 mm, whereas the female short shoot bud is 4–5 mm.

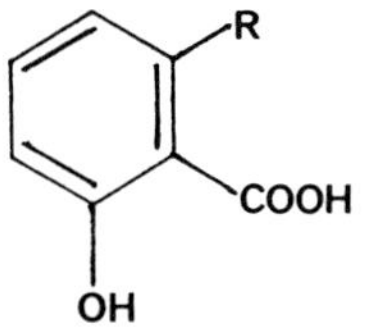

R

COOH

OH

1 R=$C_{15}H_{29}$ or $C_{17}H_{33}$

The plant is highly resistant to pests and diseases. While it is not surprising that the tree has not acquired an extensive fauna in North America (Wheeler 1975), it is unusual that in its native habitat of East Asia the plant has no predator fauna and is unusually free from insect injury. In fact, in Japan the leaves have been used as book markers to protect books from silverfish and larvae of other insects (Major 1967). Investigation into the insecticidal and fungicidal properties of *Ginkgo biloba* have shown that, on damage, the leaves produce α-hexenal by oxidation of linolenic acid (Major *et al.* 1963). Hexenal has both

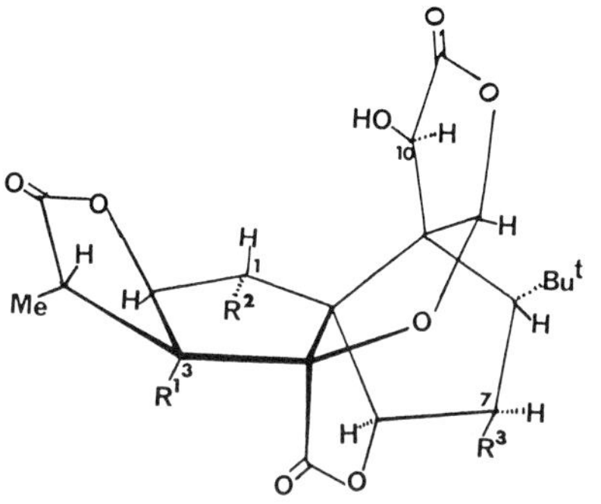

	R^1	R^2	R^3
2	OH	H	H

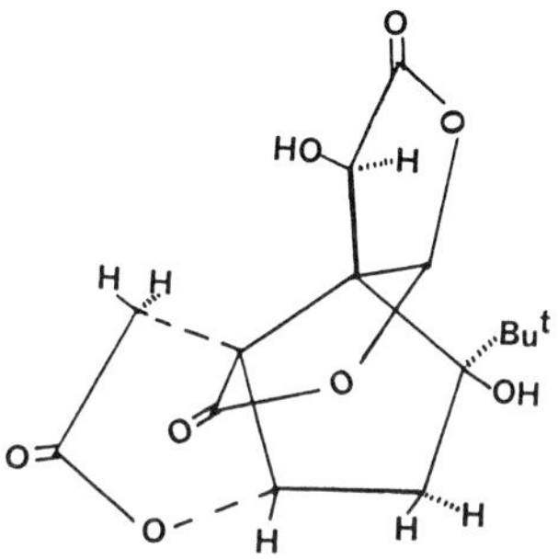

3

fungicidal (Major 1967) and insecticidal properties (Bevan *et al.* 1961). Ginkgolic acids (**1**) have known antimicrobial activity (Itokawa *et al.* 1987). Matsumoto and Sei (1987) found that ginkgolide A (**2**) was the most active of several compounds which included bilobalide (**3**) and ginkgolic acids (**1**), possessing antifeedant activity against larvae of the cabbage butterfly (*Pieris rapae crucivora*). Also the highly acidic nature of the leaves probably prevents attack by insecticidal and fauna predators.

Flavone glycosides

By far the most important group of compounds in the ginkgo extract are the flavone glycosides and acylflavone glycosides. The main flavone moieties are kaempferol (**4**) and quercetin (**5**) with small amounts of isorhamnetin (**6**) and myricetin (**7**). Flavanol-3-*O*-mono-, -di-, and -tri-glycosides have been identified (De Feudis 1991). The sugar units are generally glucose (**8**) and rhamnose (**9**). The acyl unit is coumaric acid (**10**). The acylflavone glycosides are kaempferol-3-*O*-α-(6'''-*p*-coumaroyl-glucosyl)β-1,2-rhamnoside and quercetin-3-*O*-α-(6'''-*p*-coumaroylglucosyl)-

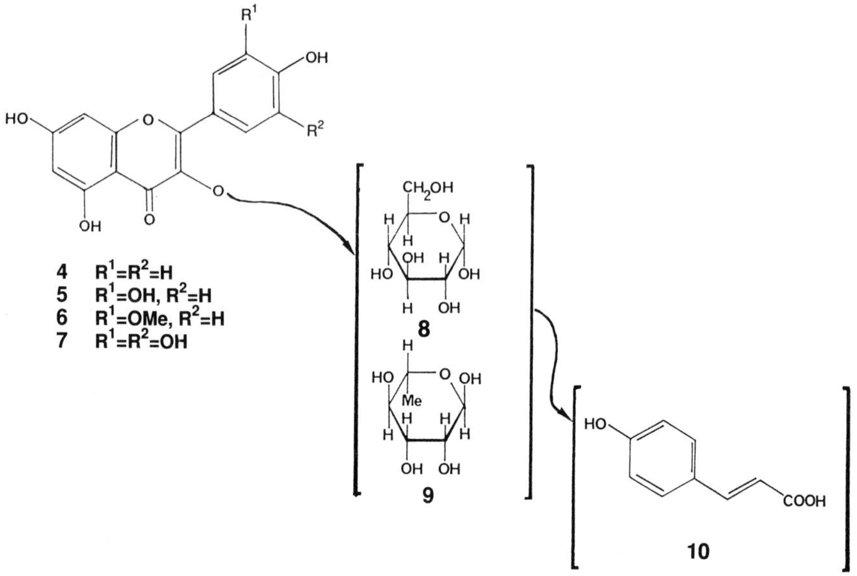

4 $R^1=R^2=H$
5 $R^1=OH, R^2=H$
6 $R^1=OMe, R^2=H$
7 $R^1=R^2=OH$

8

9

10

β-1,2-rhamnoside. The structures of the major flavone glycosides and acylflavone glycosides were identified by the Schwabe group in the late 1970s (Jaggy and Eck, personal communication). Depending on the soil type and the environmental conditions in which the tree is grown the flavone glycosides content can vary considerably. The plant grows in a wide variety of soils with the exception of wet or alkaline soils. In general, however, the flavone glycoside content in the leaves is seasonal with the highest flavone content in the leaves in spring, decreasing throughout the season to a minimum before the leaves turn yellow and then increasing after yellowing (Fig. 12.1) (Lobstein *et al.* 1991). A similar trend is observed for the acylflavone glycosides decreasing during the growing season to a minimum in October and then increasing slightly in November. However, the reverse is true for the biflavones where there is a continuous increase throughout the growing season.

Terpene lactones

The ginkgolides were first isolated as bitter substances from the leaves of *Ginkgo biloba* by Furukawa in 1932 but due to their complexity, the structures were not assigned to the compounds until 1967 by Nakanishi. He isolated four compounds from the roots and named them ginkgolide

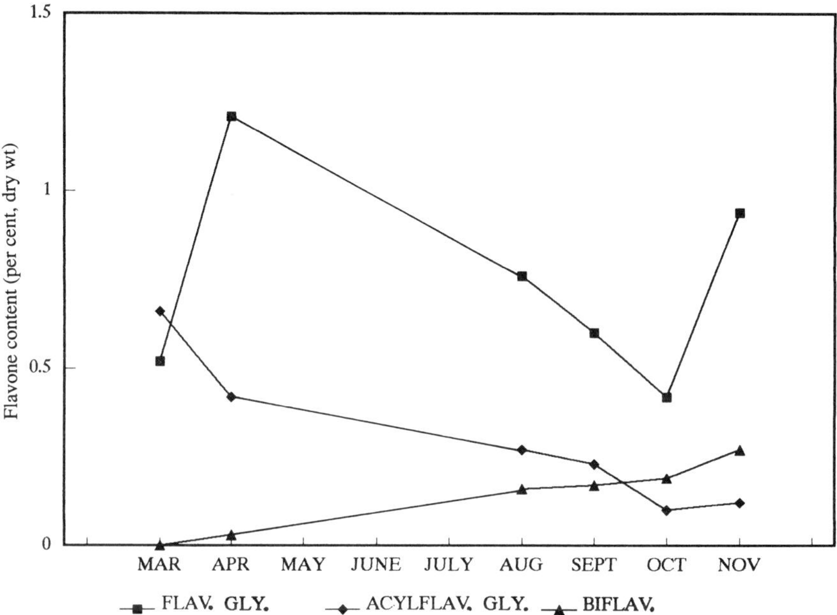

Fig. 12.1. Seasonal variations of flavone content in *Ginkgo biloba* leaves (reproduced with permission from Lobstein *et al.* 1991).

A (**2**), ginkgolide B (**11**), ginkgolide C (**12**), and ginkgolide M (**13**). The compounds are diterpene lactones with a unique cage structure and a tertiary butyl group. This tertiary butyl group is unique to the terpenes of *Ginkgo biloba*. Ginkgolide A, B, and C have 2, 3, and 4 hydroxyl groups respectively and M differs from B in the position of the hydroxy group. A third isomer of B, ginkgolide J (**14**), was isolated and identified by Weinges *et al.* in 1987. Meanwhile, a sesquiterpene bilobalide (**3**) of similar ring structure and containing the unique tertiary butyl group was isolated by Major in 1967 and the structure published in 1971 (Nakanishi *et al.*). With regard to the seasonal variation of the terpenes a maximum percentage terpene content is obtained in the leaves early in the season and decreases throughout the season. However, the trend is not as marked as that noted in the variation of the flavones.

Due to the complex framework and high oxygenation of the ginkgolides it was not obvious initially to which category of compounds they belonged. However, the diterpenoid nature was established by Nakanishi and Habaguchi (1971) (Fig. 12.2) when they successfully incorporated acetate and mevalonate into ginkgolide B. The third

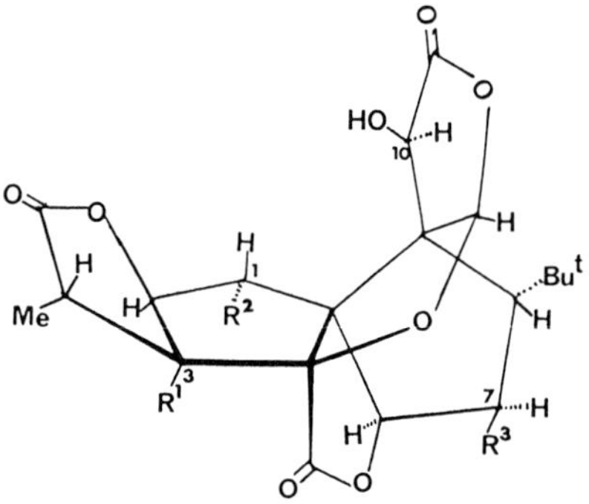

	R^1	R^2	R^3
2	OH	H	H
11	OH	OH	H
12	OH	OH	OH
13	H	OH	OH
14	OH	H	OH

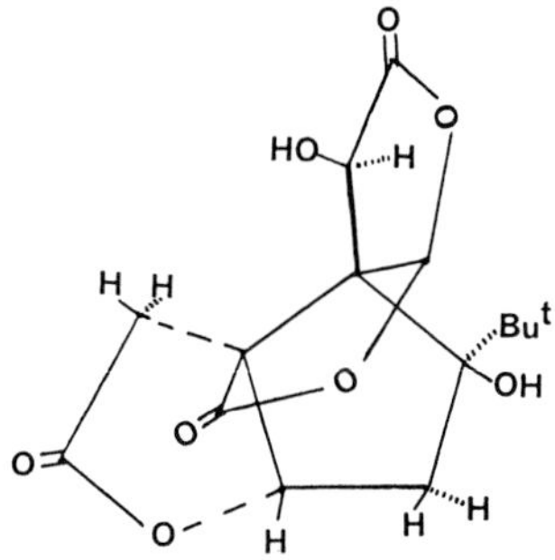

3

methyl on the tertiary butyl group was obtained by incorporation of methionine.

Interest in the ginkgolides increased significantly in the 1980s when their anti-PAF (platelet activating factor) was noted. PAF is a biologically active phospholipid whose action has been implicated in allergy and inflammation as well as in control of cardiovascular and haemodynamic systems. The most active PAF antagonist of the ginkgolides was found to be ginkgolide B. As a result of this and also the complexity of the cage structure, Professor Corey and his coworkers (1988) set about its synthesis. After a long and innovative synthetic sequence (±)-ginkgolide B was finally synthesized in 1988. This was followed by the interconversion of ginkgolide C (**12**) to ginkgolide B (**11**) by Weinges and Schick in 1991. This was achieved by formation of the silyl ether (**15**) at C-1. The thio-carbonyl (**16**) at C-7 was produced and subsequently reduced to give (**17**). Removal of the protecting silyl ether yielded ginkgolide B (**11**). The overall yield was 43 per cent.

Polyprenols

A third group of compounds in which the Japanese Company Kuraray have shown an interest, are the polyisoprenols (**18**) from *Ginkgo biloba*

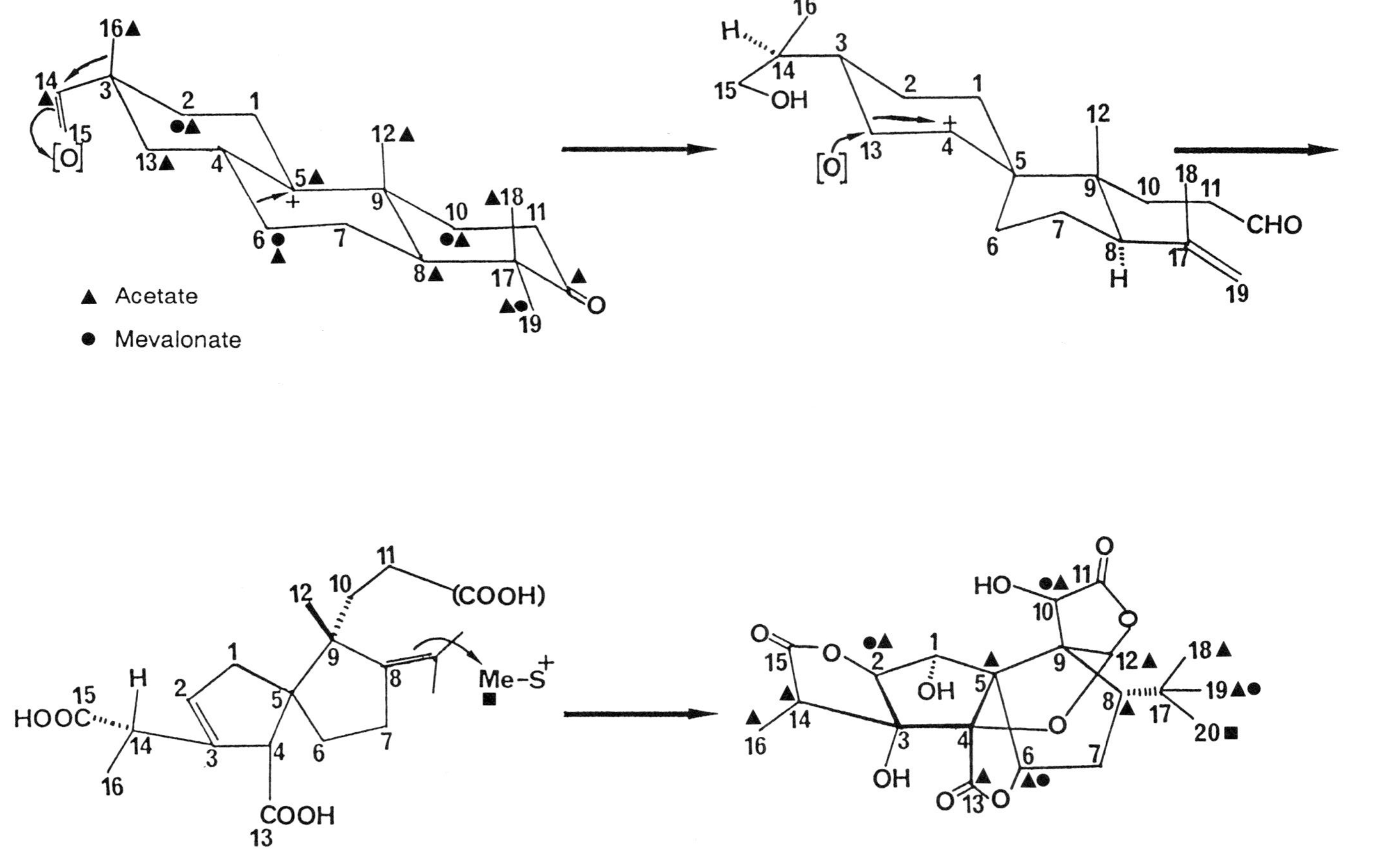

Fig. 12.2. Biosynthesis of ginkgolide B (reproduced with permission from Nakanishi and Habaguchi 1971).

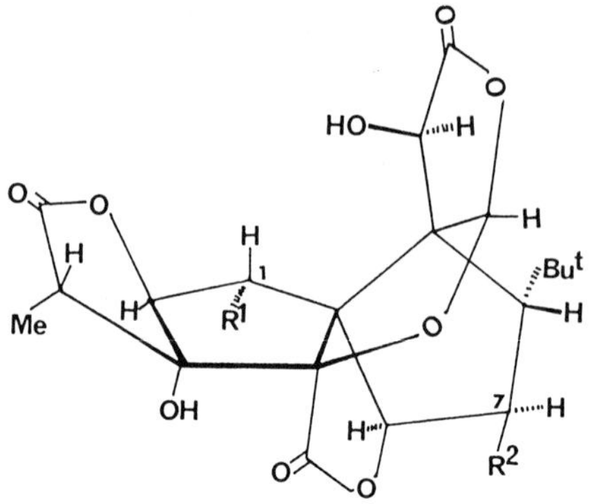

	R^1	R^2	Step recovery
12	OH	OH	
15	$OSi(Ph)_2C(Me)_3$	OH	95%
16	$OSi(Ph)_2C(Me)_3$	OCSOPh	66%
17	$OSi(Ph)_2C(Me)_3$	H	93%
11	OH	H	73%

(Ibata *et al.* 1983). These compounds have two *trans*-isoprene units and 14–16 *cis*-isoprene units. Addition of a saturated isoprene unit by Grignard reaction produces a mixture of dolichols (**19**), which are essentially identical to mammalian dolichols in alignment, number of isoprene units, and ratio of homologues in the mixture. Dolichols are

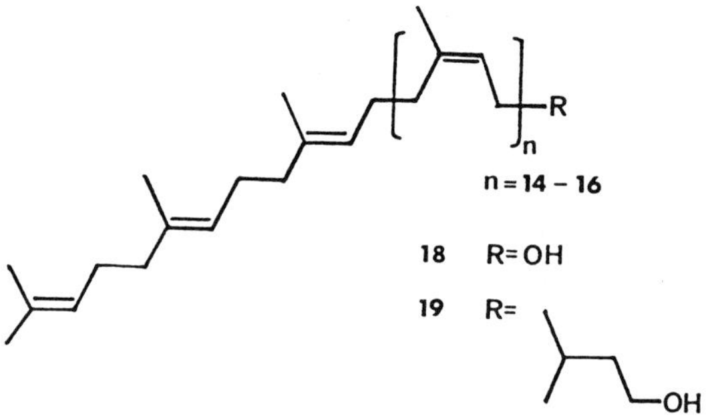

n = 14 – 16

18 R=OH

19 R=

crucial metabolites in glycoprotein biosynthesis. Kuraray (1987) have applied for patents for dolichol as a medical agent to improve a malfunctioning liver. The quantity of dolichols in animal tissue is so small that it is difficult to obtain a sufficient amount. However, *Ginkgo biloba*, which contains up to 1 per cent polyprenols on dried leaves, furnishes an ample supply of polyprenols for the production of dolichols.

Other biologically active compounds which have been isolated from ginkgo are the proanthocyanidins which are known radical scavengers, biflavonoids and organic acids which have anticarcinogenic properties (Itokawa *et al.* 1987).

Ginkgo biloba extract EGb 761

In the 1960s Dr. Willmar Schwabe, our parent company in Germany, developed an extract from the leaves of *Ginkgo biloba* for use in the treatment of cerebral insufficiency. The company formed a partnership with Laboratoires Beaufour in France who defined the extract EGb 761. This partnership in turn led to the setting up of our company, Cara Partners in Ireland, for the production of *Ginkgo biloba* extract EGb 761 and as a result, we needed a plentiful supply of quality *Ginkgo biloba* leaves. Initially our supply of leaves was obtained from intermediate 'dealers' in the different regions of the far east. However, the group very quickly realized the limitations of this method of supply. So, a strategy was formulated to ensure the plentiful supply of the high quality raw materials necessary for the production of EGb 761. This involved two phases.

Cultivation of *Ginkgo biloba*

First, we employed an agronomic engineer in the far east to organize the collection of high quality leaves in Korea, Japan, and China. Certain problems had to be overcome, most notably it was necessary to estimate the quantity and quality of leaf available and we had to convince suppliers that harvesting of the green leaves in August/September would not do irreversible damage to the trees. Then, in order to ensure a continuous supply, we had to determine an equitable return to the farmers and pay for leaves 6–8 months in advance. Finally, we had to organize a complex network for the collection of leaves from all the outlying areas and transport them to a central location within a 24 h period where we had set up factories for the drying and the baling of leaves. In the far east this strategy paid off. However, it had certain limitations as the leaves had to be manually picked from trees 5–20 m high. Also, there was a limited quantity of high-quality leaves available and with increasing sales it was necessary to diversify our source of supply.

As a result we set in motion the second stage in our strategy, i.e. the establishing of our own plantations, one in the Bordeaux region in France and the second in South Carolina. The plantations were set up in 1982. The total area for both plantations is 660 ha with 17 million ginkgo trees.

The original seeds for these plantations were obtained from China, Korea, and Japan. It was impossible to control the origin which meant the seeds obtained were genetically heterogeneous and of mediocre fertility. Also, it was necessary to sow the seeds within 10 months to ensure germination. A 50 per cent germination was considered good.

The seeds are sown in nurseries in April/May in soil which has been previously treated with methyl bromide. The seedlings are left in place for 2 years prior to planting. The first year in the nursery the plants are given a light dressing of fertilizer and sufficient irrigation and the second year the fertilization is increased. Weeding is carried out manually twice yearly and fungicide is applied in the first year to prevent rotting at the base of the young plants which is the only phytopathological problem encountered. In fact, plants which had fungal attack put forward new shoots the following year. At the end of the second year when the plants are 30 cm, the minimum height required by the planting machine, they are planted into their final position.

With regard to the plantations the trees are fertilized based on studies carried out over several years. The plantations are kept weed free with minimum use of herbicides. The only herbicides used are pre-emergent sprayed in March before the leaves come out. During the growing season there is monthly destruction of weeds with burning and also weeding by hand. We also cultivate between the rows in winter to keep down the weeds. During the growing season we have pheromone traps in place to quantify the insect population to determine if and when spraying is necessary. We have an irrigation system which is indispensable. The pruning is carried out annually in winter based on studies carried out over years of trials to optimize production. The leaves are harvested with a modified cotton picker and air dried. The moisture content of the leaves prior to drying is 70–75 per cent. When the leaves are dried to less than 10 per cent moisture, they are baled ready for extraction.

Production

The dried leaves received from the plantations and the far east are analysed for the main active ingredients, the flavone glycosides and the terpene lactones, and blended prior to extraction. *Ginkgo biloba* extract is produced to good manufacturing practice (GMP). The leaves are extracted with acetone/water (Drieu 1986). The extract is processed through multistage operations eliminating undesirable constituents and concentrating the active ingredients in the extract. The dried extract is then tested for active ingredients and the absence of solvents and other materials used in processing. The product is also analysed to ensure the elimination of unwanted substituents from the extract. The two main active ingredients, the flavone glycosides and the terpenes are analysed by HPLC. The presence of individual flavone glycosides is compared with that of a standard extract EGb 761. The extract also contains a known quantity of proanthocyanidins. These are dimers or polymers of the flavanols delphinidin (**20**) and cyanidin (**21**). Other constituents in the extract are the organic acids, shikimic acid (**22**), *p*-hydroxybenzoic acid (**23**), protocatechuic acid (**24**), vanillic acid (**25**), kynurenic acid (**26**), 6-hydroxykynurenic acid (**27**), etc. These give the extract its acidic character and play a role in the solubility of both the flavones and the terpenes. The flavones and the terpenes are not water soluble but in the chemical environment of

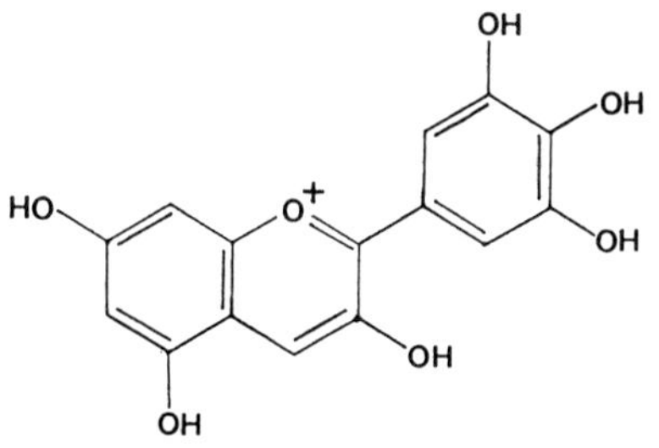

20

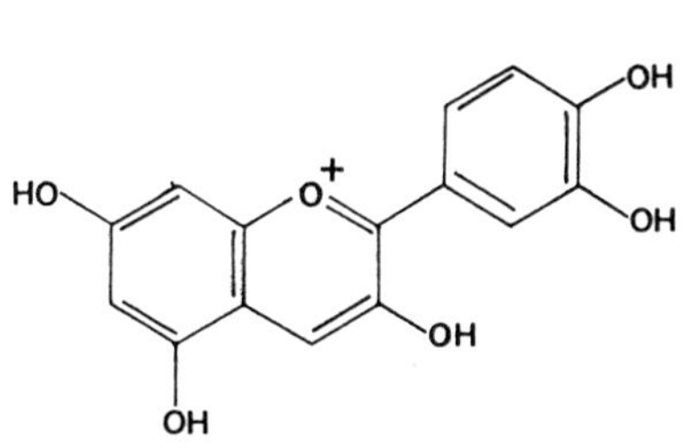

21

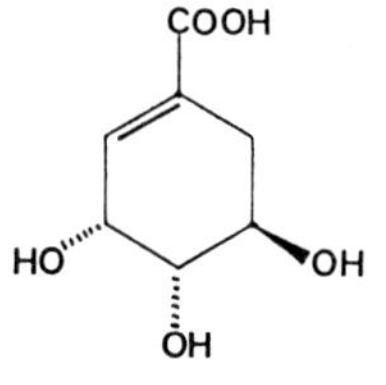

22

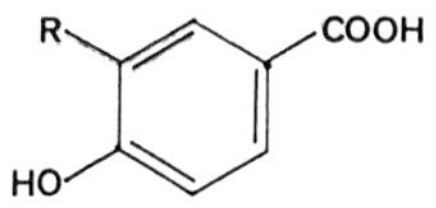

23 R= H
24 R= OH
25 R= OMe

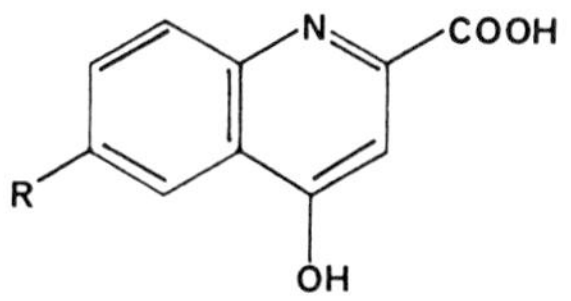

26 R= H
27 R= OH

the extract these compounds are soluble. This property increases significantly the bioavailability of these compounds in the extract. *Ginkgo biloba* extract EGb 761 is sold as a prescription drug by our group as Tanakan in France, Rökan and Tebonin Forte in Germany, and Tanakene in Spain. The extract sold by us is produced to very tight specifications and is prescribed for many different vascular ailments.

Clinical studies

At a time when we have an ever increasing ageing population, for example, in the United States the over 65s population will more than double to 50 million in the next 40 years, we are looking more at preventative medicine and improving the quality of life for the elderly. *Ginkgo biloba* extract EGb 761 is an important pharmaceutical in this category and fulfils the conditions laid down by the World Health Organisation concerning the development of drugs effective against cerebral aging. *In vitro* and *in vivo* trials indicate that EGb 761 has polyvalent activities ranging from radical scavenging, improved blood flow, vaso protection, and anti-PAF activity. Some of these activities can be attributed to specific compounds in the extract. However, it is the combined extract which gives EGb 761 its overall efficacy, and when dealing with complex diseases such as cerebral insufficiency the polyvalent activity of EGb 761 is well suited for the treatment of such disorders. This has been amply demonstrated in the many clinical trials carried out with the extract. These trials were reported in *La Presse Médicale* (La Presse Médicale 1986) and more recently reviewed by De Feudis (1991). The following are just two examples of the extensive clinical trials which indicate the efficacy of EGb 761 against cerebrovascular insufficiency.

In 1988 Halama *et al.* (Table 12.1) conducted a randomized double-blind study on the efficacy of EGb 761 in 40 out-patients ($>$ 55 years of age) with the diagnosis mild-to-medium cerebrovascular insufficiency. The Hachinski ischaemic scale was used to differentiate between vascular and primary dementia, and the Crichton scale was used to indicate the degree of severity of symptoms. Twenty patients treated with EGb 761 (120 mg/day for 12 weeks) were compared with 20 placebo controls. In the EGb 761-treated group, values for the total Sandoz clinical assessment geriatric scale (SCAG) decreased on average by 9 points, whereas they remained unchanged in the placebo group. This was a highly significant result. Evaluation of separate items (Table 12.2) showed that EGb 761 particularly affected disturbances of short-term memory, mental alertness, and dizziness. Significant effects were also shown with respect to headaches and tinnitus.

Table 12.1 Randomized, double-blind, placebo-controlled trial on the efficacy of EGb 761 in treating disturbances of cerebral performance of vascular origin: time-course of SCAG total score

	Total SCAG score	Intra-individual difference
EGB761 (120 mg/day, p.o.)		
Week 0	33.0 (30.0; 43.0)	—
Week 4	32.0 (30.0; 41.0)	0.5 (2.0; 0.0)
Week 8	23.0 (23.0; 34.0)	5.5 (9.0; 4.0)
Week 12	22.0 (22.0; 34.0)	9.0 (11.0; 6.0)*
Placebo		
Week 0	34.0 (28.0; 39.0)	—
Week 4	34.0 (30.0; 39.0)	0.0 (0.0; −1.0)
Week 8	33.0 (29.0; 38.0)	0.0 (0.0; −1.0)
Week 12	33.0 (29.0; 38.0)	0.0 (0.0; −1.0)

20 patients (8 males, 12 females) were treated with EGb 761 and 20 others (15 males, 5 females were treated with placebo. Median values for Sandoz clinical assessment geriatric scale (SCAG) scores (with 95 per cent confidence range) and their differences with respect to the initial individual values shown.

* Highly significant difference, as compared with placebo ($P < 0.00005$; one-sided χ^2 test). (Reproduced with permission from Halama *et al.* 1988.)

The second example is a clinical trial carried out by Taillandier *et al.* 1986 (Fig. 12.3) on the use of EGb 761 (Rökan) for the treatment of cerebral disorders due to ageing. The subjects were evaluated by means of a geriatric clinical evaluation scale (GCES). The study was carried out over 12 months. The greatest improvements were found in patients who were most affected. The 10 percentage point gain obtained on the behavioural scale is more than double that obtained for patients on placebo. This gain is sufficient to move the patient from a 'semi invalid' to a 'semi valid' category and therefore delaying institutionalization. This study also shows that there is a continuous improvement in the patients on EGb 761 whereas there is a levelling off of patients on the placebo.

Ginkgo biloba extract, as well as being prescribed for disturbances of the central nervous system, is indicated for other vascular complaints such as peripheral blood flow, ocular complaints, and vestibular vertigo (De Feudis 1991).

With regard to the safety of EGb 761, in clinical trials minimal side effects such as headaches, gastrointestinal disorders, and palpitations were associated with administration of EGb 761 (De Feudis 1991). In 1989 the German Health Authority were provided with a statistical

Table 12.2 EGb 761 and disturbances of cerebral performance of vascular origin. *P*-value pattern for the time-course of separate SCAG items divided according to factors: *P*-values have a purely descriptive importance

Factors/Items	Time (weeks) after beginning of EGb 761 treatment			
	0	4	8	12
Cognitive disturbances				
Confusion	2.2	2.1	1.9[c]	1.9[c]
Mental alertness	3.5	3.4	2.7[c]	2.5[b]
Impairment of recent memory	3.2	3.1	2.7[c]	2.3[a]
Disorientation	1.8	1.9	1.8	1.6
Social behaviour				
Irritability	1.1	1.2	1.2	1.2
Hostility	1.1	1.1	1.0	1.0
Bothersome	1.1	1.1	1.1	1.1
Unco-operativeness	1.5	1.5	1.4	1.4
Lack of initiative				
Motivation	2.6	2.7	2.4	2.2
Indifference to surroundings	2.6	2.6	2.2	1.9[b]
Unsociability	1.8	1.9	1.8	1.7
Self care	1.1	1.1	1.0	1.1
Affective disturbances				
Anxiety	2.6	2.6	2.3	2.0
Depressive mood	2.7	2.5	2.3[e]	2.1[c]
Emotional lability	1.3	1.3[e]	1.2[e,d]	1.2[c]
Somatic disturbances				
Fatigue	2.0	2.1	1.9	1.7[e]
Lack of appetite	1.4	1.4[d]	1.4	1.4
Dizziness	3.5	2.8[e]	2.0[a]	1.6[a]
Overall impression of patient	3.4	3.3[e]	2.5[b]	2.2[a]

[a] $P < 0.01$; [b] $P < 0.05$; [c] $P < 0.1$; [d] $P < 0.2$; [e] $P \geqslant 0.2$ (initial pathological evaluation); one-sided χ^2 test. The numbers indicated are arithmetic means of the active substance group. (Reproduced with permission from Halama *et al.* 1988.)

evaluation of EGb 761. All spontaneous reports were assembled between 1982–88 (Fig. 12.4). An important point emerges that while the number of patients increased progressively the total number of events (side effects) remained essentially the same. In fact, in a controlled double blind study (De Feudis 1991) the number of reported side effects of the EGb 761

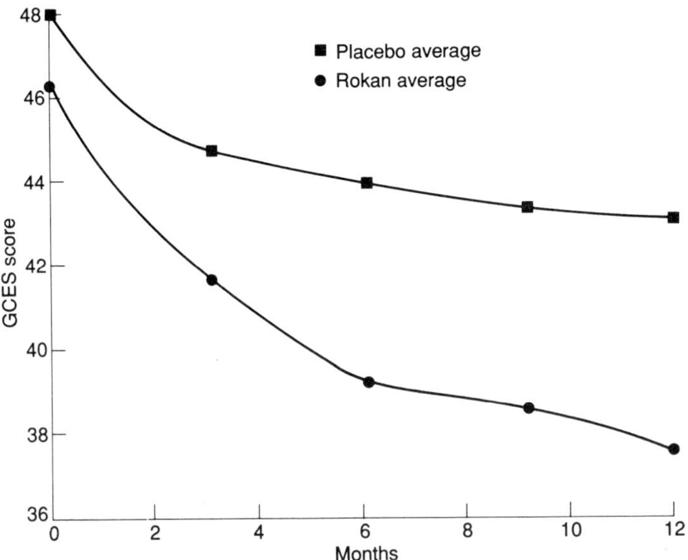

Fig. 12.3. Longitudinal study: results from geriatric clinical evaluation scale (finished cases) (reproduced with permission from Taillandier *et al.* 1986).

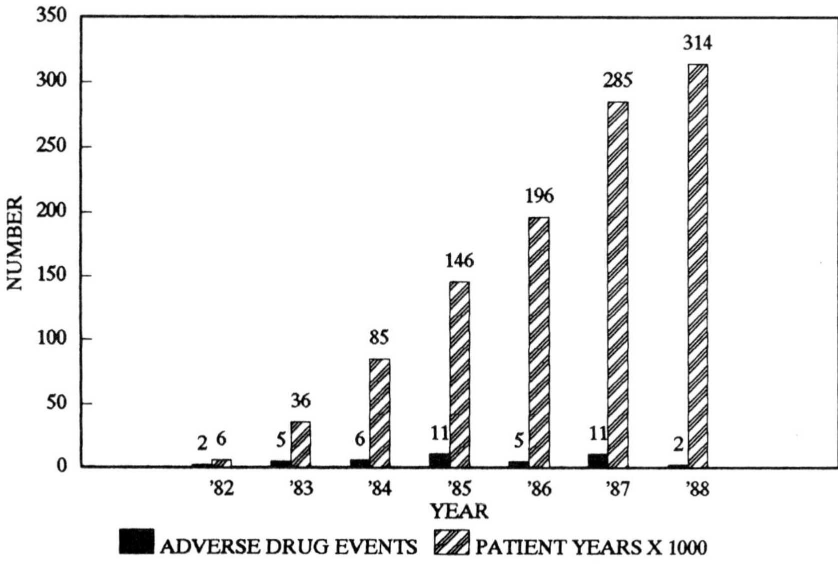

Fig. 12.4. Comparison of number of spontaneously reported events and number of patients treated 1982–1988 (reproduced with permission from Dr Willmar Schwabe Arzneimittel).

group was similar to that of the placebo group. With regard to toxicity, it was not possible to determine the LD_{50} in rats for orally administered EGb 761 and in mice it was determined to be 7.61 g/kg. Chronic toxicity studies on dogs and rats showed no evidence of organ damage at levels of 400 and 500 mg/kg/day, respectively. Oral administration of up to 1600 mg/kg/day in rats and 900 mg/kg/day in rabbits did not elicit teratogenic effects or affect reproduction and *in-vitro* and *in-vivo* assays showed that EGb 761 does not possess mutagenic activity (Drieu 1988). Therefore, concerning safety, data indicate that there is a very low risk with EGb 761-containing products.

Conclusion

In summation, the use of polyvalent therapy is necessary for the treatment of complex disease states, for example, ischemia and cerebral insufficiency and it is the combined activity of the total extract of EGb 761 that is therapeutically active and provides this polyvalent action. Finally, the ginkgo tree has given us its beauty and sustenance throughout the ages and, more recently, an exceptional pharmaceutical which improves our quality of life and in the future has possibilities for prevention of stroke. Mankind in turn has not only saved it from extinction but dispersed it throughout the world in the last 200 years and more recently our group with the planting of more than 17 million trees have increased significantly its survival into the distant future.

Acknowledgements

I wish to thank K. Drieu, Institut Henri Beaufour; H. Jaggy, Dr Willmar Schwabe, J. P. Balz, Beaufour Schwabe Group; and M. Dunne, Cara Partners for information and advice given in the preparation of this paper.

References

Bevan, C. W. L., Birch, A. J., and Caswell, H. (1961). An insect repellant from black cocktail ants. *J. Chem. Soc.*, **96**, 488.

Corey, E. J., Myung, M. C., Desai, M. C., Ghosh, A. K., and Houpis, I. N. (1988). Total synthesis of (±)-ginkgolide B. *J. Am. Chem. Soc.*, **110**, 649–51.

De Feudis, F. V. (1991). Ginkgo biloba *extract EGb 761: Pharmacological activities and clinical applications*. Elsevier, Paris.

Drieu, K. (1986). Préparation et définition de l'extrait de *Ginkgo biloba*. *La Presse Médicale*, **31**, 1455–7.

Drieu, K. (1988). Etudes de toxicité de l'extrait de *Ginkgo biloba*. Rapport de synthèse, Institut Henri-Beaufour, Paris.

Furukawa, S. (1932), Studies on the constituents of *Ginkgo biloba* leaves. *Scientific Papers of the Institute of Physical and Chemical Research*, **19**, 27–38.

Halama, P., Bartsch, G., and Meng, G. (1988). Hirnleistungsstorungen vaskularer Genese. Randomisierte Doppelblindstudie zur Wirksamkeit von *Ginkgo biloba* Extrakt. *Fortschritte der Medizin*, **106**, 408–12.

Ibata, K., Mizuno, M., Takigawa, T., and Tanaka, Y. (1983). Long chain betulaprenol-type polyprenols from the leaves of *Ginkgo biloba*. *J. Biochem.*, **213**, 305–11.

Itokawa, H., Totsuka, N., Nakahara, K., Takeya, K., Lepoittevin, J. P., and Asakawa, Y. (1987). Antitumour principles from *Ginkgo biloba* L. *Chem. Pharmaceutical Bull.*, **35**, 3016–20.

Kaempfer, E. (1712). Amoenitatum Exoticarum, Fasc. V. *Lemgo*, 811–13.

Kuraray Co. Ltd (1987). Dolichols, stimulating liver function. Jpm Kokai Tokkyo Koho JP 62–29517.

La Presse Médicale (various authors) (1986). L'extrait de *Ginkgo biloba* (EGb 761). *La Presse Médicale*, **31**, 1441–604.

Li, H. L. (1956). A horticultural and botanical history of ginkgo. *Morris Arboretum Bulletin*, **7**, 3–12.

Lobstein, A., Rietsch-Jako, L., Haag-Berrurier, M., and Anton, R. (1991). Seasonal variations of the flavonoid content from *Ginkgo biloba* leaves. *Planta Medica*, **57**, 430–3.

Major, R. T. (1967). The Ginkgo, the most ancient living tree. *Science,* **157,** 1270–3.

Major, R. T., Marchini, P., and Boulton, A. J. (1963). Observations on the production of α-hexenal by leaves of certain plants. *J. Biol. Chem.*, **238,** 1813–16.

Matsumoto, T. and Sei, T. (1987). Antifeedant activities of *Ginkgo biloba* L. components against the larva of *Pieris rapae crucivora*. *Agric. Biol. Chem.*, **51**, 249–50.

Maugini, E. and Vestri, L. (1974) Osservazioni sul ciclo di sviluppo della gemme di esemplari maschili e femmimili di *Ginkgo biloba* L. *Giornale Botanico Italiano*, **108**, 321–40.

Michel, P. F. (1985). *Ginkgo biloba—L'arbre qui a vaincu le temps*. Editions du Félin, Paris.

Nakanishi, K. (1967). The ginkgolides. *Pure Appl. Chem.*, **14**, 89–113.

Nakanishi, K. and Habaguchi, K. (1971). Biosynthesis of ginkgolide B, its diterpenoid nature and origin of the *tert*-butyl group. *J. Am. Chem. Soc.*, **93**, 3546–7.

Nakanishi, K., Habaguchi, K., Nakadaira, Y., Woods, M. C., Maruyama, M., Major, R. T., *et al.* (1971). Structure of bilobalide, a rare *tert*-butyl containing sesquiterpanoid related to C_{20}-ginkgolides. *J. Am. Chem. Soc.*, **93,** 3544–6.

Seward, A. C. (1938). The story of the maidenhair tree. *Sci. Progr.*, **32**, 420–40.

Sowers, E. F., Peyton, E. W., Collins, O. D., and Cawley, E. P. (1965). Ginkgo-tree dermatitis. *Arch. Dermatology*, **91**, 452–6.

Taillandier, J., Ammar, A., Rabourdin, J. P., Ribeyre, J. P., Pichon, J., Niddam, S., and Pierat, H. (1986). Traitement des troubles de vieillissement cérébral par l'extrait de *Ginkgo biloba*. *La Presse Médicale*, **31**, 1583–7.

Weinges, K. and Schick, H. (1991). Herstellung von Ginkgolide B aus Ginkgolide C. *Liebigs Ann. Chem.*, 81–3.

Weinges, K., Hepp, M., and Jaggy, H. (1987). Isolierung und Strukturaufklarung eines neuen Ginkgolids. *Liebigs Ann. Chem.*, 521–6.

Wheeler, A. G. (1975). Insect associates of *Ginkgo biloba*. *Entomological News*, **86**, 37–44.

13. Experiences of starting a commercial company based on plant-derived products in a developing country

ASFAQ SHEAK*

Department of Drug Administration, Ministry of Health, Thapathali, Kathmandu, Nepal

To tell a story about establishing a commercial company for plant-based products in a developing country without such experience is to proximate an object with blindfolded eyes.

Introduction

Nepal was unknown to many parts of the world until 1951, when it became a democracy and a member of the United Nations.

Nepal is mountainous except for the Terai plain that covers approximately 15 per cent of the country on the southern border with India (see Fig. 13.1). The climate varies from subtropical via subtemperate,

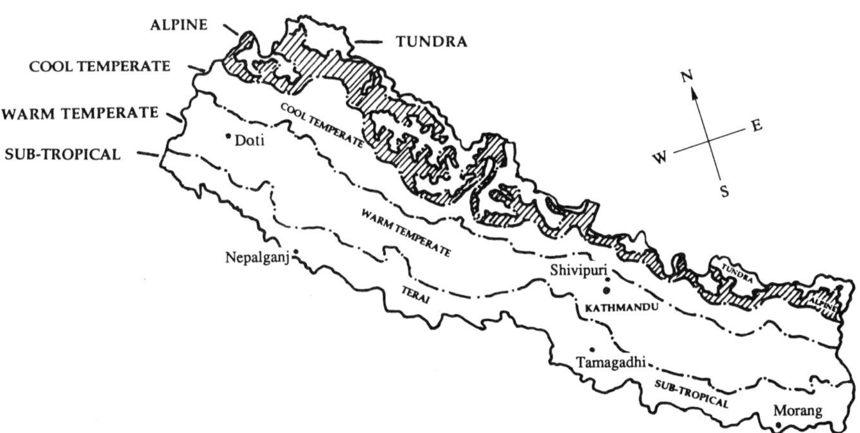

Fig. 13.1. Map of Nepal showing different climatic zones. From North to South: tundra, alpine (shaded), temperate, subtemperate, and subtropical. Further towns or regions mentioned in this chapter have been indicated.

* Formerly Director of HPPCL, Koteswore, Kathmandu, Nepal.

temperate, and alpine to tundra. Thus, the country is rich in plant resources of a diverse nature, and a very wide range of plant species can grow at almost any climatic condition desired.

Unfortunately Nepal had been selling its unprocessed natural products at a price which was too low to cover even the cost of collection. Interference by foreign traders to exploit the high quality natural resources in large quantities led towards exhaustion of the plant resources. Also an agreement had been signed to export only the crude form to India and not even the simplest processed form. Due to this pathetic condition, an attempt to reduce the bulk to suit transportation by air was made in 1981, and the venture of HPPCL (Herbs Production and Processing Company Ltd) was supported overwhelmingly. As a first step the Government of Nepal and international agencies such as UNDP, UNIDO and FAO started to make an inventory of the existing plant resources and to examine the possibility of cultivating internationally demanded sources. The agencies also assisted with developing the processing of plants. The enthusiasm gathered in this way developed confidence and determination for assured success. At the start of HPPCL some aims and objectives were set out for the company.

Aims/objectives of HPPCL

1. To sell as much processed product with added value as possible.
2. Reduction of the volume of the product at the site of collection, or at least at the HPPCL centre, before export.
3. Generation of employment and income to the people.
4. Investigation of the market for future products, mainly species exotic to Nepal.
5. Replacement of over-exploited plant resources collected in the wild by cultivated plants.

Feasibility

Before legally founding such a company, a feasibility study (Ministry of Forest and Soil Conservation 1980) on 30 naturally growing materials and 10 newly cultivated species was made for detailed costs and for selling on an economic basis. The natural herbs studied for processing and selling are listed in Table 13.1. The 10 cultivated exotic species were: four mint species (*Mentha spicata*, *M. citrata*, *M. arvensis*, *M. piperita*), citronella, lemongrass, palmarosa, pyrethrum, belladonna, and

Table 13.1 List of naturally grown herbs included in feasibility study

Medicinal plants	
1. *Adhatoda vasica*	26. *Orchis* sp. (Gamdol)
2. *Asparagus racemosus*	27. *Emblica officinalis*
3. *Berberis aristata*	28. *Curculigo orchioides*
4. *Digitalis lanata*	29. *Myria esculata* bark
5. *Rheum emodi*	
6. *Solanum xanthocarpum*	
7. *Picrorhiza kurroa*	*Aromatic plants*
8. *Parmelia nepalensis*	1. *Zanthoxyllum armatum*
9. *Mahonia nepalensis*	2. *Cinnamomum glaucescens* *
10. *Phyllanthus emblica*	3. *Nardostachys jatamansi*
11. *Terminalia chebula*	4. *Cinnamomum tamala*
12. *Terminalia belerica*	5. *Cinnamomum camphora*
13. *Valeriana wallichii*	6. *Eucalyptus* species
14. *Lycopodium clavatum*	7. *Skimea* species
15. *Bergenea ligulata*	8. *Zingiber officinale*
16. *Swertia chirata*	9. *Cardamomum subulatum*
17. *Piper longum*	10. *Pinus roxburgii*
18. *Orchis incarnata*	11. *Artemisia vulgaris*
19. *Podophyllum emodii*	12. *Acorus calamus*
20. *Rauvolfia serpentina*	13. *Gaultheria fragrantissima*
21. *Dioscorea deltoidea*	
22. *Aconitum spicatum*	
23. *Demotrichum fimbriatum*	*Insecticidal plants*
24. *Sapindus muskorossi*	1. *Chrysanthemum cinerariifolium* (pyrethrum)
25. *Rubia cordifolia*	

* Synonyms: *C. cecidodaphne* and sugandha kokila

rauwolfia. The first seven are aromatic in nature, while the last two are purely medicinal plants. Pyrethrum is a source of natural insecticides which are relatively non-toxic to human beings. The seven essential oils can be produced in an economical way, whereas the two medicinal plants though economically viable, have limited potential. The insecticide was rather difficult to obtain in commercial quantities.

Of the 43 species listed in Table 13.1, 30 were endemic and the exported products were usually the crude plants. Processing of these products seemed possible. There were some indications that similar products were sold in the international market, but this vague assumption was based only on enquiries of the herbal market and from journals of the essential oil market. A good balance between medicinal plants and essential oil-bearing plants was planned, as well as a good balance between

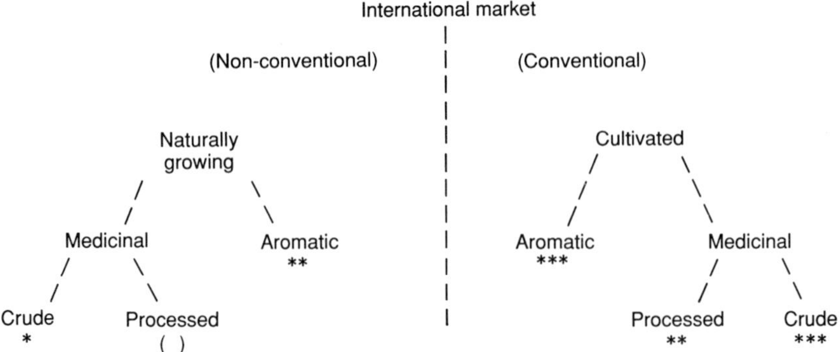

Fig. 13.2. Demarcation of the international market for plant-derived products. *–can get market; **–fairly good market; ***–very good market; ( )–no market.

natural and cultivated ones. This was deemed necessary in order to match the production with the demand. Nevertheless about 20–25 per cent of efforts on selected items remained futile, either due to lack of demand or to the uneconomical production, resulting in a too high price. Figure 13.2 shows more explicitly the type of plants that may have demand.

As long as only a feasibility study was involved the choice of which plant species were to be cultivated and put into the market was not a problem, but when a real commercial enterprise is concerned, the selection of items is the pivot for success or failure. The criteria for selection are explained further under the heading of 'Selection of items'.

Here an earlier study performed in 1983 by GATT/UNCTAD in collaboration with HPPCL to show the 150 most demanded and widely collected Nepalese plants should be mentioned. The study is based on the collection and quantity of sales of these plants by one dealer in the last five years in five different developing regions of Nepal (viz. east, central, west, midwest, and far west). The information obtained was utilized by HPPCL but the scope of the survey remained limited to crude material sales from natural resources. Thus the information was useful for private dealers to quantify the items districtwise but less useful to HPPCL who aimed to process products, at least to a primary level, before offering them for sale.

Selection of items

The first and foremost criterion for the selection of any item is the market itself, which includes the size of demand, the necessary quality of the product, and the price obtainable.

The second criterion is the technology required for cultivation and processing, together with the availability of planting material of a quality sufficient to support mass cultivation. The agricultural technique for cultivation is very important since the slightest mistake in cultivation may lead to a large drop in yield and a wide degradation of quality, both leading to a catastrophic breakdown of the turnover and the profits and thus of the company.

The third criterion is the nature of the items. Referring to the naturally growing species versus cultivated ones (Fig. 13.2), it is necessary to process the existing natural products to reduce the bulk and to achieve certain industrialization steps. However this is a problem because medicinal plants are mostly used in crude form in traditional and eastern forms of medication while Western industry needs processed medicinal plants in the form of extracts. The aromatic plants existing in nature are not used much in traditional medicine. Again it should be mentioned that one of the aims of the company was to process items before exporting. Thus a few items which were not utilized for medicinal purposes, were selected for processing as well as some of the medicinal plants, both cultivated and wild. As a compromise, to achieve viability of the new company, some unprocessed medicinal plants for the traditional Asian drug market were also cultivated and sold.

A further problem was the choice between aromatic plants from natural sources and from cultivated sources. The former class produce an oil which is non-conventional and requires greater effort to sell in the market. The use of such unconventional oils is entirely dependent on the hand of an expert perfumer. Also the use of a particular aromatic oil is dependent on its constituents. Supposing an oil is rich in a certain component, say above 70 per cent, it can be utilized for the extraction of that particular component if the natural isolate is either superior in note, preferred by the buyers for other reasons, or currently cannot be synthesized. An example to cite is wintergreen oil, which consists of 99 per cent methyl salicylate.

Greater complexity and therefore dependence on the perfumer arise when the oil contains a very large number of components in an evenly distributed ratio. Of course none of the formulated perfumes is based on a single oil, but the composition of a particular oil should be such that the addition of more components may expand the note, completely modify the perfumery characteristics, or even assist in reducing the number of components in a readily formulated product. An example is sugandha kokila (Sheak *et al.* 1989) (local name for the oil of *Cinnamomum glaucescens*). The use of such an oil in a heavily demanded perfume or flavour means that the perfumer is submitted to the mercy of the producer. Once it is included in a perfume it becomes very difficult to exclude it, thus a long term agreement is sought by both the perfumer and the producer.

The demand is also a very important criterion. Until and unless there is a reasonable demand, no commercial enterprise is viable. Yet another approach for an entirely new oil to be developed for future prospects is to produce it initially in a small quantity, then offer it for utilization, and eventually expand the market. Many commercial companies do not have such patience. In our case the oil of *Nardostachys jatamansi* can be taken as a slowly developed product, which is quite expensive, and a further investigation of its constituents would be very interesting.

The last two criteria are price and quality, which interact. According to a famous Chinese saying 'Good is not cheap and cheap is not good', i.e. if very cheaply produced the quality is likely to be low or to deteriorate in time, and when the quality is inferior, a much lower price will be offered. Take the case of eucalyptus oil where an oil containing 70 per cent of eucalyptol is more expensive than one containing a lower percentage of eucalyptol. In any case a certain minimum standard has to be maintained below which nothing will be saleable.

More specifically for the cultivated conventional oils, the quantity that can be supplied, standard of oil, and competitive price are all closely interconnected factors. The immediate generalization of price versus quality such as in eucalyptus oil plays a very important role in the international market, but sometimes the best quality is being sold at a low price in large quantities to suit bulk production and attracts the final user in his formulation. A step leading to rectification, either to eliminate lighter and heavier fractions or to concentrate the main component, improves the quality of oil, but the price is undoubtedly increased.

More will be discussed under 'Cultivation practices'. Good cultivation practices are essential for the production of large quantities, a good price, and a standard quality.

Cultivation practices

The proposed wild growing and cultivated medicinal plants have limited scope and the development and maturation of the non-conventional aromatic plants was a time-consuming operation. Full attention was therefore given to the cultivation of conventional aromatic plants. These have a large market once one has found access to the world market. Here, the ways and means adopted for the cultivation of these plants will be narrated briefly. Apart from these conventional aromatic plants the cultivation, collection, and production of non-conventional oil and medicinal plants were steadfastly continued.

The cultivation of exotic aromatic plants is neither an easy thing to do nor easy to expand. It was a big challenge to select the right material (good clones, seeds, slips, planting material), suitable land, perfect agro-techniques, and to produce a standard quality material and oil. The standard was sometimes found specified in national quality documents, but in most cases the specification of the buyer is much more important since he has the upper hand in this business and the quality is thus being dictated. Some experience in producing these products has been obtained:

(1) on company's land;

(2) by private farmers (landless and small landholder) on company's land;

(3) by private farmers on their own land.

On company's land

In the first year of this new company three different farms were visualized in order to grow crops at different altitude with varied climatic conditions such as:

(1) tropical (Tamagadhi, Terai);

(2) temperate (Shivipuri, north of Kathmandu); and

(3) subalpine (Doti, far north-west).

The last two farms were not big enough to allow commercial production. Moreover, the products could not be transported economically from the farm at far north-west Doti to a selling centre. The only remaining alternative was to go ahead with the tropically based farm at Tamagadhi, which was 350 hectare in size and which possessed a water supply and fertile soil (see Fig. 13.1 for location). The seven aromatic plant species mentioned above were a success, but also some species such as abelmoschus, basil, peppermint, spearmint, and vetiver were tried successfully. Further research on quality and market is in progress.

Out of 350 hectare, about 200 hectare is in cultivation at present. The cultivation practices were performed partly by permanent and seasonal staff of the company, while another portion was done by landless and small farmers. In addition to this a strong extension programme is being launched in the vicinity of the farm area. In order to get commercial viability shallow wells were dug and diesel power pumps were utilized to clear the land, plough, and plant. Finally, when harvested, almost all the products were distilled turn by turn in a series of distillation units owned by the company. Figure 13.3 shows a plot of land used for the production of essential oil containing grasses.

Fig. 13.3. Area of cleared jungle now used for the cultivation of essential oil bearing grasses (citronella or palmarosa). Right plot has just been harvested, left plot is to be harvested. This field is part of Tamagadhi farm in the Terai region.

It was observed that under cultivation performed in this way the yield of grass, the content of oil, and even the quality of oil was found to be high and good but the expenditures were too high to render it economical.

By private farmers (landless and small landholder) on company's land

It was realized much earlier that production carried out by company's staff is less likely to be economical as they have a guaranteed salary irrespective of crop quality and yield. They would also like to enjoy their holidays, but plants need continuous care. It is very difficult to judge whether the labour they put in equals the wages or not. It was further thought that if the income is tied more with the amount of labour both farmers and the company will benefit. The farmer will have time to work conveniently and regularly and the labour is likely to be maximized due to a feeling of ownership of the crop they produce.

Each landless farmer was provided with half a hectare of land. However, the costs of planting material, irrigation, and land preparation, if they liked to have mechanization, were provided at first and were eventually subtracted from the total income at the time of payment for the product (grass or oil). Two mechanisms for payment of crops have been developed; (a) to buy the 24 h wilt grass at a predetermined price or (b) to buy the oil, where the farmers have to pay for the cost of distillation and fuel if any is used. The first mechanism is to encourage the production, whereas the second is based on the actual oil content of the crop. All the parameters regarding production, yield, and quality of the oil are satisfactory. In this connection it seems logical to mention that a common large scale distillation unit will provide an opportunity for every farmer to distill his plants after harvesting.

By private farmers on their own land

This is a co-operative type of management,where the private farmers are ready to spare their land for cultivation of aromatic plants, but a common distillation facility is required at the site or in the vicinity of cultivation. To buy a distillation unit of small size will be difficult even for a large farmer. The company provided a small portable distillation unit to at least four farmers in the group. In future a loan is to be arranged at a low rate of interest. At least five of the farmers will decide together who will be responsible for paying the principal and interest and will collect the money at the time of payment for the oil. The crops, which attracted the cultivators attention, profiled in order of priority, are palmarosa, lemongrass, citronella, chamomile, *Mentha arvensis*, and basil. It should be mentioned here that private farmers were initially very reluctant to cultivate something other than rice or other common food crops. Only after the company *guaranteed* a certain *fixed* price (irrespective of the current world market price), which was higher than that of the food crop, for their entire production, did more and more private farmers start to cultivate the, for them novel, essential oil-bearing plants, e.g. palmarosa or citronella grasses.

Commercialization at a private level

The success of the company for a larger part depended on large scale production by the farmers on their own land, but landholding by farmers in general is very low and food production at national level is also not up to the mark. Of course food production has a high priority both for

Fig. 13.4. Example of agroforestry showing a combination of essential oil containing grasses and fast growing poplars. After approximately 8 years these trees are used as firewood. The grasses are small because the photograph was taken during the dry mid-winter period.

the individual farmer and the country as a whole. Therefore, the use of government land from waste forest as community forest land as an integrated tree/herbs production site seemed an attractive alternative (agroforestry, see Fig. 13.4). During the growing stage the trees provide some shade to the herbs. After felling the trees are either used by the company as fuel for the steam distillation during the wet summer, or sold as firewood which is in great demand in Nepal. This strategy has been tried successfully at at least three locations. Two more sites for commercial farming in east Nepal near Morang and in west Nepal at Nepalganj, have been taken into the study to see whether expansion to 110 hectare and 500 hectare, respectively, is possible (see Fig. 13.1 for location).

Processing and quality control

Establishment of commercially viable large scale production is only possible with the farmer's participation by completion of the processing operation at the site of cultivation. The simplest operation appears to be *steam distillation*. A hydro-distillation unit consisting of a single still, the lower part of which is filled with water (one-fifth) and the rest with herbal material, was observed to be working satisfactorily. A unit consisting of a separate boiler and two distillation stills, to be used sequentially, was ideal and provided a better quality oil by avoiding charring of the plant material (see Fig. 13.5). It did not need any solvent and the operation was also relatively easy for the farmers. The only knowledge required was the length of time to be set for each product and the packing conditions of the raw material such as wet, wilt, or dry and long or cut.

Fig. 13.5. Steam distillation site at Tamagadhi in Terai region south of Kathmandu, showing from left to right: water tower for cooling, spent grasses used for fuel, steam generator, two stainless steel stills, two coolers, and oil/water separator.

Fig. 13.6. Impression of central facilities of HPPCL at the Kathmandu site.

In comparison, the *extraction operation* was found to be specific for each particular material. Variables are size of apparatus, type of solvent, and type of extraction technique (continuous or batch, etc.). Furthermore, extracts need to be concentrated at low temperature and high vacuum. Even then the quality was likely to be variable. All these facilities have been developed by HPPCL at the central plant in Kathmandu (see Fig. 13.6) and are operated by experienced and qualified persons. This collection and bulking of the product is necessary in order to smooth out the variation in quality, otherwise batch dependent variations would occur. Many buyers on the international market specify their own requirements, some of them for a certain smell, others for the concentration of an active constituent only and again others for various physico-chemical characteristics. Sometimes modern techniques such as quantitative gas chromatography are required to support the olfactory evaluation by the highly trained nostril of an experienced perfumer. There are very few people who are qualified both as a perfumer and an analytical chemist, but for the selling of an essential oil on the international market sometimes gas chromatograms and physico-chemical characteristics are needed.

A wide knowledge of the chemical characteristics of essential oils that may undergo changes in association with other constituents in a formulation is of great importance. The use of lichen extract as a stabilizer for perfumes has been postulated (Sheak *et al.* 1990). The active constituent of the lichen *Parmelia nepalensis* is at a molecular level likely to bind with or to remain in close proximity of the active moieties of the components of various essential oils. These moieties consist of chemical groups such as esters, acids, aldehydes, ketones, etc. The purpose of stabilization is to elongate the stay of the perfume. The active chemical compounds in some *Parmelia* species are the cyclohexane derivatives atranorin and hematommic esters. During extraction with alcohols hematommic esters are formed from atranorin.

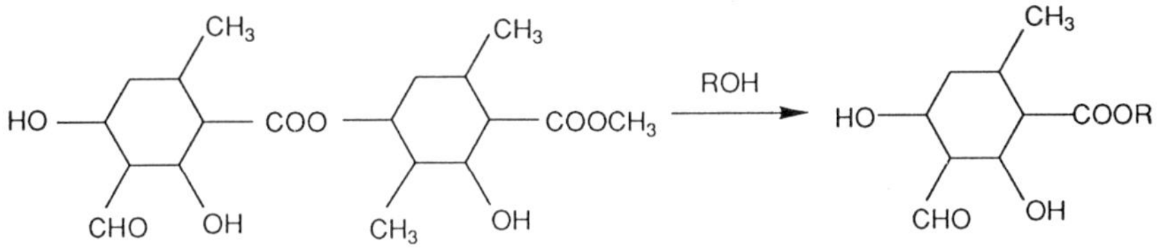

Atranorin $\xrightarrow{ROH}$ Hematommic ester

Recently the new field of natural dyes has emerged for HPPCL. There are at least four dye-containing plants which grow extensively in Nepal. These are rheum, rumex, rubia, and aloe. They contain mostly anthraquinone derivatives. The extraction may need to be carried out under low temperature and pressure. With different mordants and fillers as a clay, a powdery dye can be produced. The technique has been used locally for dyeing in the carpet industry. HPPCL undertook the extraction of *Rheum emodi* for the preparation of a dye. The dye thus obtained can be be modified chemically to obtain various shades of colour which can be further modified in fastness and shade with the use of various mordants and fillers. Apart from a dye, a laxative is also made from the *R. emodii* extract that is now extensively used.

Another important product of HPPCL which is worth mentioning is turpentine. Pine resin is collected in various medium altitude pine forests and then transported to a nearby distillation facility of HPPCL. The resulting volatile turpentine and non-volatile rosin are used by Nepalese factories for the preparation of paints and varnishes for the

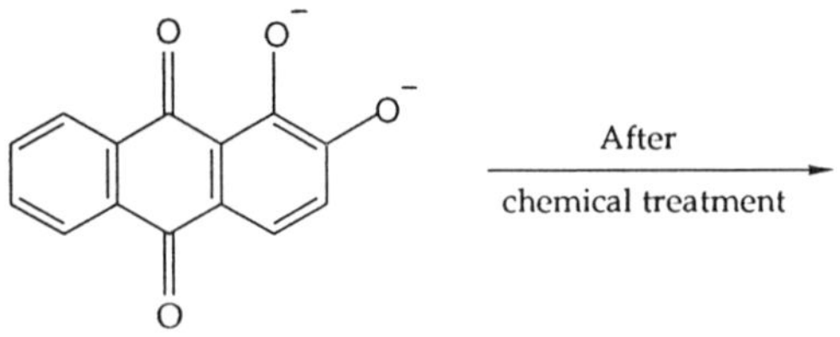

After
chemical treatment

Basic anthraquinone structure

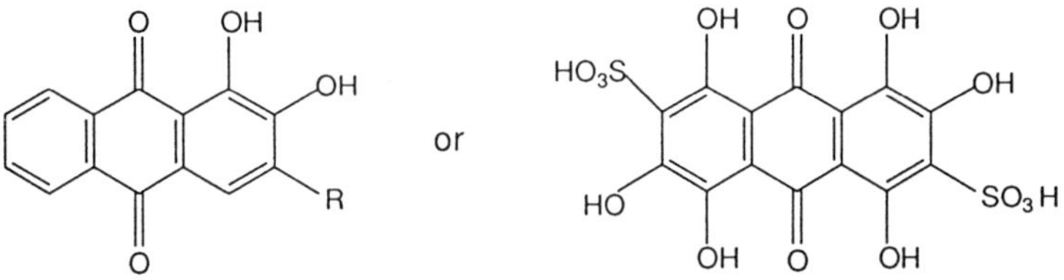

or

Alizarin

Anthracene Blue
Celliton Fast Blue B

local market thus saving the country valuable hard currency. In the same manner HPPCL also supplies the internal soap industry with essential oils and the Nepalese drug manufacturers with medicinal herbs and extracts.

Diversification of products

One of the disadvantages with this type of industry is the instability of the market. It takes quite a long time to establish a product. If it has to be removed from production at a too early stage, the management strategy, production target, and market prospect have to be abolished and a completely new setup has to be established. Thus, it is desirable that as soon as any product is likely to come on to the market, the buyer has to be fixed for a long time. This is the only way for both farmers and the company to be assured of a good future.

Funding of the company

Funding up till now is by HMG of Nepal which has a share of 90 per cent, while the other 10 per cent is shared by Royal Drug Ltd as a user, Agricultural Development Bank for the activities of cultivation, and

Forest Product Ltd for the utilization of natural resources. However, the contributions made by UNDP/FAO (1979) and UNDP/UNIDO (1980) for starting up the cultivation and the processing respectively, are highly appreciated. Although it was not a programme for capital funding, great success has been achieved through these projects, technically. Much farm equipment and machinery for extraction has come through these agencies. Many visiting experts have given technical guidance.

Transformation of HPPCL into co-operatives

In the future the development of the herbs resources in the country should be carried out by co-operatives. Eventually they should be transformed into independent units. For initial co-ordination HPPCL has been playing a significant role in providing planting material, selection of the product, developing technology, calculating cost, quality assessment and guaranteeing purchase. Once these co-operatives gear up for large scale production, they can very well exist alone. Even though progress has not been made as envisaged, the main aim remains the development of a co-operative network.

Costing and price structure

The cost calculation, including all expenditures incurred in the production to obtain the unit price, is the backbone of a commercial company. The pricing of an individual product is carried out using the following pattern:

1. *Direct costs*:
   (a) material costs;
   (b) cost of solvent after considering the loss and recovery;
   (c) direct energy used for processing;
   (d) direct manpower involvement;
   (e) packing costs and material.
2. *Indirect costs*:
   (a) administration costs;
   (b) depreciation (land, building, and equipment);
   (c) maintenance costs/quality control;
   (d) promoting costs;
   (e) interest and insurance;

(f) transportation costs;
(g) miscellaneous (unforeseen).

3. *Profit margin*:
(a) net profit (15–25 per cent) after deduction of taxes (as a public company, excessive profit of more than 25 per cent is being charged);
(b) contingency for replacement of machinery;
(c) breakage, pilferage, loss, etc.;
(d) wholesale and other commissions;
(e) dividend.

Training of manpower

For the successful running of a commercially viable company, qualified and well-trained staff are very important. In HPPCL all new staff have been employed directly from educational institutes, where they were guided in each specific subject e.g. technical, commercial, financial, and marketing. With the help of projects from FAO and UNIDO they were sent for training in various fields, which enabled them to handle the business efficiently. This was a difficult job, but now most of them are professionally better qualified, skilled, and capable. It is in the hands of the general manager how they are used.

Conclusion and summary

The biggest problem encountered during the first 10 years of HPPCL was the introduction of items in the international market. As soon as some initial contacts have been made they should be transformed to long-term contacts and preferably contracts. In this way one can stabilize one's own market. One of the most harmful mistakes made by us was to keep going from one buyer to another and thus to destabilize our market. The market itself is volatile and subject to sudden changes. We experienced this with the selling of lichen extract to India. A second mistake made by us was not to diversify our product range and not to produce in large quantities for fear of getting too large stocks, too little cash, as well as a non-expanding world market. But on the other hand it should be remembered that the process of development is always slower in developing countries due to shortage of finance and lack of promptness in decisions made by the Government. Nevertheless we say that since its founding in 1981 HPPCL has operated well and has seen tremendous growth. Sales increased from a meagre 224 000 NRps in 1981 to 11 000 000 NRps in 1990. It was considered by UNDP one of the more successful developing projects carried out in Nepal.

The present political system of Nepal neither hinders nor promotes this type of industry, although much effort is invested by the Government to elevate the living condition of the Nepalese people above the poverty level. On the above basis of 10 years experience, a programme to engage landless and small farmers in Government forest land has been planned in the Forestry Sector Master Plan, for which a soft loan of US $2.47 million from the Asian Development Bank has been allocated. The previous management had not given much attention towards this aspect, but instead gave more emphasis to developing the cultivation of private lands and to subleasing the loan to such farmers. We have observed in the past that this is a very slow process and the release of loan from private farmers will be extremely difficult. A good suggestion therefore seems to be to engage people in the cultivation of Government forest lands for the production of aromatic oils and to continue influencing private farmers.

Finally it should be stressed that selecting a good management, including Manager, Executive Board, and Chairman, is the only guarantee for continued success of HPPCL or, for that matter, any other company. Highly qualified people without any previous commercial experience can cause a rapid decrease of profits. The commercial companies in Nepal are relatively independent and the courage, ability, and judgement of the Management are the most important factors for steering a company in the right direction. Definitely the basis on which the company has run with success in the past should not be disrupted.

References

GATT/UNCTAD (1983). *Company and product profiles on medicinal plants and herbs*. Supply survey project GATT/UNCTAD.

Ministry of Forest and Soil Conservation (1980). *Feasibility of the company (HPPCL)*. Ministry of Forest & Soil Conservation, Nepal.

Sheak, A. (1990). *Sources of vegetable dye materials and their perspective in industry*. Cottage Industry Workshop, Nepal.

Sheak, A., *et al.* (1989). Essential oil of sughanda kokila. In *Proceedings of 11th International Congress of Essential Oil, New Delhi*.

Sheak, A., *et al.* (1990). Lichen resinoid of commercial importance from Nepal. In *Proceedings of International Symposium on Natural Products, 12–15 February, 1990, Nepal*. RONAST/TWAS/UNESCO.

UNDP/FAO (1979). Nep/79/007 Project.

UNDP/UNIDO (1980). Nep/80/044 Project.

14. Biosynthesis of storage fat in oil crops—today and tomorrow

STEN STYMNE

Department of Plant Physiology, Swedish University of Agricultural Sciences, PO Box 7047, S-750 07 Uppsala, Sweden

Introduction

Plant storage lipids, commonly known as vegetable oils or fats, are the second most valuable commodity in world trade and have a global annual production exceeding 50 million metric tonnes (Hatje 1989). Vegetable oil finds its main application in food manufacture, but as much as 40 per cent is used for non-food purposes, in, for example, the United States (Pryde and Rothfus 1989). The quality of the oil is governed by its fatty acid composition and this largely determines its end-use. There is a rather limited number of commercial oil crops available and each of these crops has a defined oil quality which restricts its use.

The enormous diversity in oil quality in the plant kingdom as a whole offers a gene bank for the production of vegetable oils suited to a large variety of applications. Many plant oils, such as those with substituted fatty acids containing functional groups, would be very attractive as substitutes for mineral oils in the production of products like lubricants, surfactants, varnishes, polymers, etc., provided that they could be obtained at a constant supply and at moderate costs. Today, much effort is being made to domesticate wild species containing such oils (Hirsinger 1989). The rapid advancement in recent years of molecular genetic techniques in plants, together with the increased knowledge of fatty acid and storage lipid biosynthesis in plants, offer alternative approach to the production of these oils. The possibility of making tailor-made oils for various uses from commercial oil crops by creating transgenic plants is today on the threshold of reality. An underpinning detailed knowledge of the biosynthetic pathways leading to the production of these oils is, however, an absolute prerequisite if these attempts are to be commercially viable. In this chapter an overall picture of storage lipid biosynthesis in plants will be given based on recent research results, with particular emphasis on the regulation of the fatty acid composition and the future possibility of manipulating this genetically in commercial oil seeds.

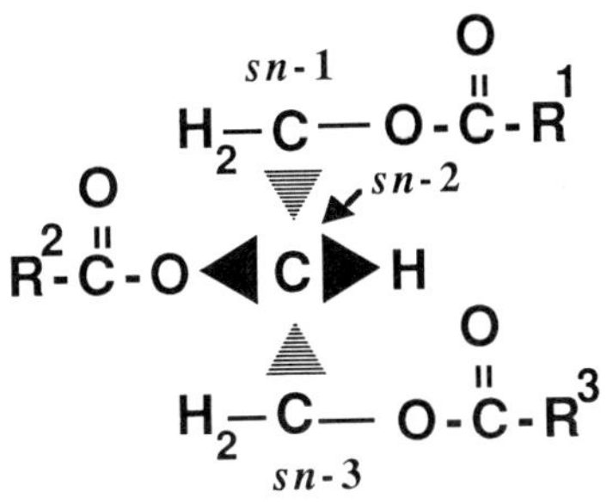

Fig. 14.1. Fischer projection of a triacylglycerol molecule. R, Fatty acid acyl chains.

Structure and functions of triacylglycerols

A common feature of many seeds is the accumulation of energy reserves in the form of oils or fats. With only one known exception, the jojoba (*Simmondsia chinensis*) seed which stores oil in the form of fatty acid alcohol esters (Yermanos 1975), these compounds are always composed of triacylglycerols where one fatty acid is esterfied to each of the three hydroxyl groups of a glycerol moiety (Fig. 14.1). Since the triacylglycerol molecule does not possess rotational symmetry, the three carbon atoms of the glycerol part are stereochemically distinguishable and are designated *sn*-1, *sn*-2, and *sn*-3 according to IUPAC-IUB standardization nomenclature. In plants, the fatty acids are not randomly distributed on the different *sn* positions; saturated fatty acids are (in most plants) almost excluded from the middle, *sn*-2, position, whereas both saturated and unsaturated acids are located at the outer positions. As shown below, the specificities and the selectivities of the glycerol acylation enzymes of the cell govern the positional distribution of the different fatty acids in the triacylglycerols.

Fatty acids in the triacylglycerols

Triacylglycerols in most plants contain a mixture of five major fatty acids, palmitic (16:0), stearic (18:0), oleic ($18{:}1^{\Delta 9}$), linoleic ($18{:}2^{\Delta 9,12}$), and linolenic acids ($18{:}3^{\Delta 9,12,15}$) (Table 14.1). These fatty acids make up over 90 per cent of the acids that are present in commercial oil crops and are also the fatty acids that are abundant in all lipid classes in all parts of the plant. Whereas the fatty acid composition of the membrane

Table 14.1 Some fatty acids occurring in plant triacylglycerols and plant sources where they are abundant

Name	Structure	Symbol	Source
'Common' fatty acids			
Palmitic acid	$CH_3(CH_2)_{14}COOH$	16:0	Oil palm mesocarp
Stearic acid	$CH_3(CH_2)_{16}COOH$	18:0	Cocoa bean
Oleic acid	$CH_3(CH_2)_7CH{=}CH(CH_2)_7COOH$	$18:1^{\Delta 9}$	Olive mesocarp
Linoleic acid	$CH_3(CH_2)_4CH{=}CHCH_2CH{=}CH(CH_2)_7COOH$	$18:2^{\Delta 9,12}$	Safflower seed
(Alpha-)Linolenic acid	$CH_3CH_2CH{=}CHCH_2CH{=}CHCH_2CH{=}CH(CH_2)_7COOH$	$18:3^{\Delta 9,12,15}$	Linseed
'Uncommon' fatty acids			
Capric acid	$CH_3(CH_2)_8COOH$	10:0	*Cuphea procumbens* seed
Lauric acid	$CH_3(CH_2)_{10}COOH$	12:0	Coconut endosperm
Petroselinic acid	$CH_3(CH_2)_{10}CH{=}CH(CH_2)_4COOH$	$18:1^{\Delta 6}$	Coriander seeds
Erucic acid	$CH_3(CH_2)_7CH{=}CH(CH_2)_{11}COOH$	$22:1^{\Delta 13}$	High erucic rape seed
Gamma-linolenic acid	$CH_3(CH_2)_4CH{=}CHCH_2CH{=}CHCH_2C{=}CH(CH_2)_4COOH$	$18:3^{\Delta 6,9,12}$	Borage seed
Crepenynic acid	$CH_3(CH_2)_4C{\equiv}CCH_2CH{=}CH(CH_2)_7COOH$	Δ^{12} acetylenic-$18:1^{\Delta 9}$	*Crepis alpina* seed
Ricinoleic acid	$CH_3(CH_2)_5CH(OH)CH_2CH{=}CH(CH_2)_7COOH$	Δ^{12} hydroxy-$18:1^{\Delta 9}$	Castor bean endosperm
Vernolic acid	$CH_3(CH_2)_4CH{-}CHCH_2CH{=}CH(CH_2)_7COOH$ (epoxide O bridging the two CH)	Δ^{12} epoxy-$18:1^{\Delta 9}$	*Euphorbia lagascae* endosperm

lipids in leaves and roots is strictly regulated and differs relatively little between plant species, there is much variation in the relative proportions of these acids in the triacylglycerols from different species. For example, linseed oil has about 60 per cent of its fatty acids as linolenic acid and safflower has 80 per cent as linoleic acid, while olive oil is essentially lacking polyunsaturated acids and has about 80 per cent of the acyl groups as oleic acid. In addition to the 'common' fatty acids, there is a large number of 'unusual' fatty acids, most of which occur in the triacylglycerols of wild species. Over 300 different types of fatty acid have been identified in seed oils so far. Their structures include a number of modifications and substitutions on the acyl chains, such as variation in chain length and double bond positions and the occurence of acetylenic bonds, functional groups, branched and cyclic fatty acids (Badami and Patil 1981). Many of these 'uncommon' acids are confined to the triacylglycerols, where they often are major fatty acids, and do not occur in the membrane lipids of the plant, not even in the seed itself. Unusual fatty acids are often restricted to one plant family or genus. A striking example is the genus *Cuphea* of the Lythraceae family. The seed fats of most members of the *Cuphea* genus contain nearly exclusively medium chain fatty acids with a chain length ranging from C_8 to C_{14}. Different subgenera have different medium chain fatty acids as the dominating acyl group (Graham *et al.* 1981).

Functions of triacylglycerols

The deposition of triacylglycerols in seeds can occur either in the embryo, including the cotyledons, as in our common annual oil crops, or in the endosperm tissue as is the case for coconut and castor bean. The only known function for the triacylglycerols in the seed is to serve as a source of energy when the seed is germinating. Thus, the triacylglycerols and their fatty acids are completely degraded and none of the fatty acids are utilized in the direct synthesis of new lipids in the germinating tissue (Kindl 1987). As discussed in more detail below, some seeds accumulate triacylglycerols with a majority of the fatty acids having a functional group, e.g. a hydroxy or an epoxy group. These oils cannot readily be digested by animals, and the laxative effect of the hydroxy fatty acid, ricinoleic acid, from castor bean is well known and has previously been exploited in pharmaceutical products. In these cases the oil might also have a function as a protective agent for the seed against predators. A large number of tropical or subtropical fruits, such as avocado and palm fruits, accumulate substantial quantities of triacylglycerols in the mesocarp tissue, the probable function of which is to make the fruits more attractive for animals which would then help to disperse their seeds.

It is possible that the great diversity in both the relative proportions and the number of different fatty acids which can occur in storage fat compared to membrane lipids in plants reflects the function of triacylglycerols as merely energy reserves. There would, in that case, be no selection pressure put on the plant to maintain a certain fatty acid composition in the triacylglycerols.

Biosynthesis of triacylglycerols

The plastid synthesizes saturated and monounsaturated fatty acids for the triacylglycerols

The *de novo* biosynthesis of fatty acids for triacylglycerol assembly occurs, as for all fatty acid synthesis in the plant, exclusively in the chloroplast (Ohlrogge *et al.* 1978) or, in the case of non-photosynthesizing tissue, in the proplastid (Browse and Slack 1985). The acetate units used in fatty acid synthesis in oil-seed tissue are mainly derived via the glycolytic breakdown of the sucrose that is channelled to the seed via the phloem. Although free acetate (Roughan *et al.* 1978) or acetylcarnetin (Masterson *et al.* 1990) seem to be the preferred substrates in fatty acid synthesis in intact chloroplasts from leaves, pyruvate could also be used as a precursor for acetyl-CoA synthesis in the plastid (Treede and Heise 1985).

The end product of the *de novo* fatty acid system is, in most cases, palmitoyl-acyl carrier protein (ACP) and stearoyl-ACP. However some seeds, like the commercially important coconut and oil palm kernel, accumulate triacylglycerols with fatty acids shorter than 16 carbons. It is likely that enzymes associated with *de novo* fatty acid synthesis are responsible for the production of these so called medium chain fatty acids. In seeds from the California bay tree (*Umbellularia californica*), an acyl-ACP thioesterase with high activities towards 12:0-ACP has been purified (Davies *et al.* 1991) and its gene cloned and expressed in *E. coli*, *Arabidopsis thaliana*, and oil-rape (*Brassica napus*) with a resulting accumulation of 12:0 fatty acids in these organisms (H.M. Davies, personal communication). Although this enzyme appears to be responsible for the production of 12:0 fatty acids in the California bay tree, no thioesterase with specificity for 10:0-ACP was found in this tissue, even though 10:0 also represents a large proportion of the acyl groups in this seed and can, together with 12:0, be synthesized *de novo* in cell-free fractions (Pollard *et al.* 1990). Attempts to demonstrate a medium chain thioesterase in coconut (Ohlrogge *et al.* 1978) have also been unsuccessful. It is therefore possible that other early fatty acid chain-termination mechanisms than medium chain specific thioesterases operate in plants.

There are at least three forms of enzymes in plants responsible for the condensation of C_2 units in the elongation of the fatty acid chain, β-ketoacyl-ACP synthetase (KAS) I, II, and III. The newly discovered KAS III catalyses the initial condensation step of a acetyl group to malonoyl-ACP and utilizes acetyl-CoA rather than acetyl-ACP (Jaworski *et al.* 1989). KAS I is able to condense acyl-chains up to C_{14} whereas KAS II has the ability to also utilize palmitoyl-ACP and thus is involved in the synthesis of stearoyl-ACP (Stumpf 1987).

The *de novo* fatty acid enzymes in plants have many similarities with corresponding bacterial enzymes and each step is catalysed by separate proteins (Stumpf 1987) and not, as in animals and yeasts, by a multifunctional polypeptide.

In most plants the majority of the stearoyl-ACP is desaturated by a soluble plastidic Δ^9 desaturase to oleoyl-ACP. The enzyme requires NADPH and an electron transport system, such as ferredoxin. It has been purified to homogeneity and its *c*-DNA gene has been isolated from castor bean, cucumber, and safflower (Shanklin and Somerville 1991; Shanklin *et al.* 1991; Thompson *et al.* 1991). There is a strong sequence homology in the genes from different plant species, but no homology with the membrane bound stearoyl-CoA desaturase from yeast or animals. This is an interesting example of a convergent evolution of enzymes catalysing the same reaction.

Once the oleoyl-ACP is formed, a thioesterase with a high specificity towards oleoyl groups catalyses the formation of free oleic acid. In oil seeds the free oleic acid, together with usually smaller proportions of palmitic and stearic acids, are transported through the plastidic envelope by yet uncharacterized mechanisms and ligated to CoA, probably in the outer part of the envelope (Roughan and Slack 1977). Further modifications of the acyl chain and the acylation to the glycerol backbone in the formation of complex lipids are catalysed by endoplasmic membrane-bound enzymes.

Triacylglycerols are assembled via the glycerol 3-phosphate pathway in the endoplasmic reticulum

The common polyunsaturated fatty acids, linoleic and linolenic acids, in the triacylglycerols are synthesized by sequential desaturation of oleate by enzymes located in the endoplasmic reticulum (ER), as described in

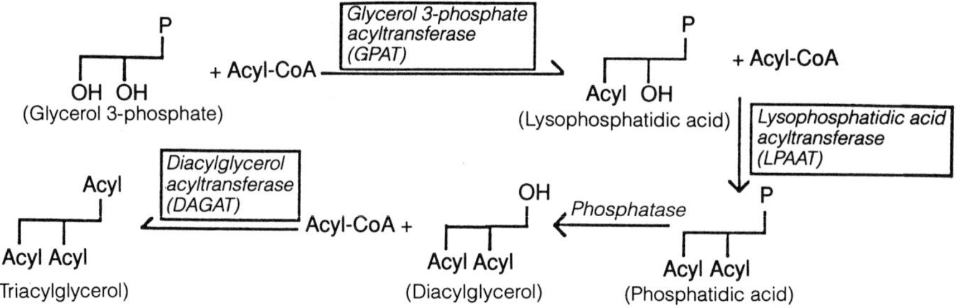

Fig. 14.2. Reactions catalysed by the three glycerol acylation enzymes involved in triacylglycerol biosynthesis via the glycerol 3-phosphate pathway.

the next section. The saturated and monounsaturated acyl groups exported from the plastid and the linoleate and linolenate synthesized in the ER are acylated from their CoA esters to the glycerol moiety in the formation of triacylglycerols by three consecutive steps (Fig 14.2). These are catalysed by enzymes located in the endoplasmic reticulum (Barron and Stumpf 1962) and follow the classical Kennedy pathway (Kennedy 1961). Glycerol 3-phosphate, generated from hydroxyacetone by glycerol 3-phosphate dehydrogenase in the cytosol (Finlayson and Dennis 1980) is acylated at the *sn*-1 position by acyl-CoA : glycerol 3-phosphate acyltransferase (GPAT) in the synthesis of lysophosphatidic acid. Oil seed GPAT has generally a broad acyl specificity and acylates both saturated and unsaturated fatty acids (Ichihara 1984; Griffiths *et al.* 1985; Bafor *et al.* 1990*a*). The acylation of the *sn*-2 position, i.e. the formation of phosphatidic acid by acylation of lysophosphatidic acid, is catalysed by a lysophosphatidic acid acyltransferase (LPAAT). Although this enzyme in, for example, turnip rape seeds can acylate both saturated and unsaturated acyl-CoAs if presented as single substrates, it shows a high selectivity for C_{18} unsaturated acyl-CoAs over saturated acyl-CoA (Griffiths *et al.* 1985; Ichihara *et al.* 1987; Bafor *et al.* 1990*a*). Thus, as a general rule, the *sn*-2 position in the plant triacylglycerols does not contain any saturated acyl groups.

Specialization in the second acylation step in seeds accumulating 'unusual' fatty acids

It has recently become apparent that the LPAAT in oil seeds accumulating medium chain acyl groups and large amounts of very long chain

($>C_{18}$) acyl groups has particular acyl specificities that enable the cell to assemble these acyl groups into triacylglycerols.

In *Cuphea* seeds, which are rich in medium chain fatty acids, the LPAAT has special properties that enable the cell to channel medium chain and long chain fatty acids in a bimodal fashion to give nearly exclusively di-medium-chain glycerols and di-long-chain glycerols (Bafor *et al.* 1990*b*; Bafor and Stymne 1992). Medium chain acyl groups are only found in the di- and tri-acylglycerols and not in the membrane lipids of the seed (Stymne *et al.* 1990) where they would probably severely impair membrane functions. Since membrane lipids and triacylglycerols share diacylglycerols as a common precursor molecule there must be enough long chain diacylglycerols available for membrane lipid synthesis. *Cuphea* seeds accumulate about 90 per cent of their acyl groups as medium chain fatty acids and the mechanism of diversion of the synthesis into di-long- and di-medium-chain diacylglycerols is probably a necessity for the proper membrane lipid assembly in the cell.

Erucic acid (22:1) is another 'triacylglycerol' specific acid which occurs in the families of Cruciferae, Limnanthaceae, and Tropaeolaceae. In Cruciferae species, the erucic acid is exclusively located in the outer, *sn*-1 and *sn*-3, positions of the triacylglycerols, whereas Limnanthaceae and Tropaeolaceae species also have this acid in position *sn*-2 (Löhden *et al.* 1990; Whitfield and Murphy 1990). It has been demonstrated that LPAAT from the oil-rape (a Cruciferae species) does not accept erucoyl-CoA whereas the corresponding enzyme from *Limnanthes douglasii* (Löhden *et al.* 1990) and *Limnanthes alba* (Cao *et al.* 1990) efficiently utilized this acyl donor. Very long chain acyl groups in triacylglycerols from Cruciferae never reach over 66 per cent, whereas Limnanthaceae and Tropaeolaceae species have 70 per cent or more, of such acyl groups in the storage lipids. It is reasonable to assume that the lack of specificity for very long chain acyl groups by the LPAAT in Cruciferae puts the upper limit for the proportion of long chain acyl groups in the triacylglycerols to approximately two-thirds of the fatty acids.

The conversion of phosphatidic acid to diacylglycerols

The conversion of phosphatidic acid to diacylglycerol is catalysed by a Mg^{2+}-dependent phosphatidic acid phosphatase, which in oil seeds is present in both an ER-bound and a soluble form (Ichihara *et al.* 1989). This might suggest that phosphatidate phosphatase, by binding and dissociation from the membranes, could play some regulatory role in triacylglycerol formation (Ichihara *et al.* 1990).

The acylation of diacylglycerol is the only unique step in triacylglycerol assembly

The ultimate acylation step in triacylglycerol formation, the acylation of diacylglycerols catalysed by diacylglycerol acyltransferase (DAGAT), is the only acylation step that triacylglycerols do not share with membrane lipids and might therefore be of regulatory importance in the diversion of lipid synthesis towards the accumulation of storage lipids. There are some conflicting reports regarding the acyl specificities of this enzyme in oil seeds. The diacylglycerol acyltransferase of developing safflower seeds showed no preference of saturated over unsaturated acyl-CoAs when measured with added diacylglycerols (Ichihara *et al.* 1988; Stymne, unpublished results). On the other hand, a strong preference for stearoyl-CoA over palmitoyl- and oleoyl-CoA by the safflower enzyme was deduced from experiments where triacylglycerols were synthesized from glycerol 3-phosphate and acyl-CoAs (Bafor *et al.* 1990*a*).

Cuphea lanceolata seeds have over 80 per cent capric (C_{10}) fatty acids in their triacylglycerols whereas the diacylglycerols of the seed only contain about 34 per cent of this type of acid, of which the majority is associated with dicaproylglycerol (Bafor *et al.* 1990*b*). This indicates a selectivity by the *Cuphea* DAGAT for medium chain diacylglycerols over long chain diacylglycerols. Further evidence for such selectivity was found in the *in-vitro* synthesis of triacylglycerols from glycerol 3-phosphate and acyl-CoAs by microsomal membranes of developing *Cuphea lanceolata* seeds (Bafor *et al.* 1990*b*). The DAGAT from both *Cuphea procumbens* and sunflower seeds acylated added dicaproylglycerols at a higher rate than dioleoylglycerols, when presented as single substrate. However, when both diacylglycerol species were presented in equimolar mixtures, the *Cuphea* enzyme acylated 13 times more dicaproylglycerols than dioleoylglycerols, whereas the sunflower enzyme only showed a twofold higher rate (Wiberg *et al.* 1992). Thus, the selectivity for medium chain diacylglycerols seems to be a specialization of the enzyme in seeds accumulating medium chain triacylglycerols.

Castor bean endosperm triacylglycerols are rich in ricinoleate (12-hydroxyoctadeca-9-enoate). Based on *in-vitro* experiments, where microsomal fractions from developing castor bean endosperm catalysed the formation of triacylglycerols from glycerol 3-phosphate and acyl-CoA, it was suggested that the castor bean endosperm possessed a DAGAT that selectively utilized ricinoleoyl-containing diacylglycerols over oleoyl-containing diacylglycerol (Bafor *et al.* 1991). When we measured the acylation of added diacylglycerols by castor bean microsomes, we found that dioleoylglycerols were utilized at a somewhat higher rate than diricinoleoylglycerols, if the acylacceptors were present as single substrates (Wiberg *et al.* 1992). When both diacylglycerol species were presented in equimolar mixtures the acylation of dioleoylglycerols was inhibited, whereas little effect was seen on the rate of triacylglycerol formation from diricinoleoylglycerols. Similar selectivity for ricinoleoyldiacylglycerols was also seen in the DAGAT from sunflower seeds and thus might be an inherent property of all oil seed DAGAT.

Polyunsaturated fatty acids are synthesized on phospholipid substrates

Oleate is converted into linoleate by a Δ^{12} desaturase and the linoleate can be further converted by a Δ^{15} desaturase to yield alpha-linolenate. The oleate desaturase in oil seeds is an endoplasmic bound enzyme requiring NAD(P)H, cytochrome b_5, and cytochrome b_5 reductase and utilizes oleate esterified to phospholipids (Stymne and Stobart 1987; Smith *et al.* 1990; Kearns *et al.* 1991) The main substrate for the oleate desaturation in oil seeds is phosphatidylcholine, and oleate attached to both *sn* positions of the lipid can be desaturated (Stobart and Stymne 1985; Griffiths *et al.* 1988). The Δ^{15} desaturase is poorly characterized due to the lability of the enzyme in cell free fractions, but available data indicate that linoleoyl-phosphatidylcholine is the substrate in oil seeds. Recent studies of the Δ^{15} desaturase activity in homogenates from developing linseed suggest that, in contrast to the Δ^{12} desaturase, position *sn*-2 is preferentially desaturated over position *sn*-1 (Stymne *et al.* 1992).

Since the synthesis of polyunsaturated fatty acids for oil-seed triacylglycerol involves desaturation of acyl groups which are esterified in phospholipids there must be mechanisms whereby the acyl substrate is channelled into the phospholipid and the polyunsaturated products are made available for triacylglycerol assembly. Two such mechanisms have been identified in oil seeds.

1. It has been shown that acyl groups from acyl-CoA can be exchanged with acyl groups of position *sn*-2 of phosphatidylcholine. The exchange, which is specific for C_{18} unsaturated acyl groups, is believed to be catalysed by an acyl-CoA:lysophosphatidylcholine acyltransferase working bi-directionally (Stymne and Stobart 1984). The oleate will, by this mechanism, enter position *sn*-2 of phosphatidylcholine, whereas the newly synthesized linoleate and linolenate from the *sn*-2 position could enter the acyl-CoA pool and be used for triacylglycerol biosynthesis via acylation to either of the three positions of the glycerol backbone.

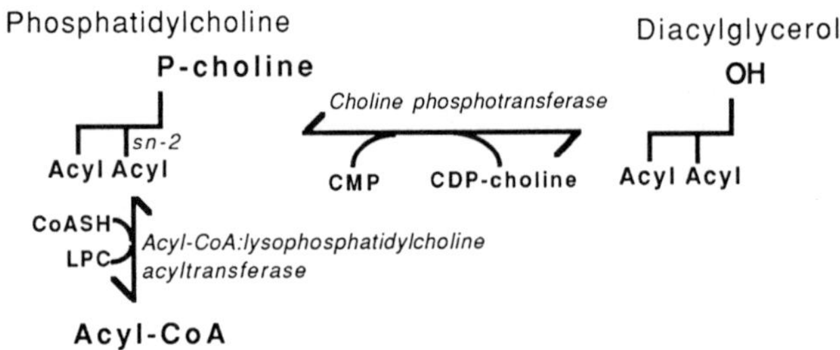

Fig. 14.3. The two equilibrium reactions for the channelling of acyl groups into phosphatidylcholine for desaturation and the transfer of desaturated acyl groups from phosphatidylcholine to the substrates in the glycerol 3-phosphate pathway for utilization in triacylglycerol biosynthesis.

2. A reversible exchange of phosphocholine groups between diacylglycerols and phosphatidylcholine mediated by cytidine monophosphate has been demonstrated in oil seed membranes (Slack *et al.* 1983, 1985). This interconversion between phosphatidylcholine and diacylglycerols offers a pathway whereby oleate at both *sn* positions can be channelled from diacylglycerols into phosphatidylcholine for desaturation and the polyunsaturated acids transferred to the diacylglycerol pool for utilization in triacylglycerol biosynthesis. The two equilibrium reactions are schematically depicted in Fig. 14.3. The relative contribution of the two reactions in providing the triacylglycerols with polyunsaturated fatty acids is not clear and might differ from species to species.

The phospholipase A_2-mediated release of ricinoleic acid from phosphatidylcholine for triacylglycerol biosynthesis is the major route for channelling this acid to triacylglycerols in the membranes of developing castor bean endosperm (Bafor *et al.* 1991) as described in more detail below. Other oil seed membranes, such as rape microsomes, have phospholipase activities that, in addition to oxygenated fatty acids, can also release some unsaturated fatty acids from phospholipids (Banas *et al.* 1992). It can therefore not be excluded that linoleate and linolenate can also be made available for triacylglycerol biosynthesis via the action of these enzymes.

Biosynthesis of 'uncommon' fatty acids occurring in triacylglycerols

Despite the large number of different fatty acids occurring in the triacylglycerols of plants and their potential economic value, few studies have been made of the biosynthesis of the more uncommon fatty acids. This section will deal with the present knowledge regarding the biosynthesis of some of the acids that are of great potential interest for the oleochemical industry. Mechanism(s) for the biosynthesis of medium chain fatty acids have been discussed in connection with fatty acid biosynthesis previously in this chapter and will not be mentioned here.

Very long chain fatty acids

A number of plant species accumulate fatty acids longer than C_{18} and these acids are confined to the storage fat of the seed. A common feature of triacylglycerols in seeds from *Cruciferae* species is the accumulation of erucic acid ($22{:}1^{\Delta 13}$). Erucic acid has been shown to be synthesized by an NAD(P)H dependent elongation of oleoyl-CoA with C_2 units from two malonyl-CoA molecules. Intermediates like keto-, hydroxy-, and transenoyl-acyl groups have been identified (Fehling and Mukherjee 1990), suggesting that the same reactions as in *de novo* fatty acid synthesis are involved. Although the elongation enzymes of *Lunaria annua* seeds were solubilized and partially purified from a 12 000 x g pellet (Fehling and Mukherjee 1990) it is likely that the enzymes are localized in the endoplasmic reticulum since an absolute correlation between the elongation enzymes and endoplasmic marker enzymes in preparations from high erucic turnip rape has been found (Bafor and Stymne, unpublished results). There appear to be separate condensing enzymes involved in the two elongation steps from 18:1, as 20:1 and 22:1 concentrations can vary independently in rape seeds (Downey and Craig 1964)

Meadowfoam (*Limnanthes alba*) seeds are extremely rich in long chain fatty acids in their triacylglycerols (about 90 per cent) and the majority of these acids are $20{:}1^{\Delta 5}$. It was shown that cell-free extracts of meadowfoam seeds could synthesize $20{:}1^{\Delta 5}$ via a NAD(P)H dependent Δ^5 desaturation of 20:0 (Moreau *et al.* 1981).

Petroselinic acid

Seeds of the *Umbelliferae* family are usually rich in petroselinate ($18{:}1^{\Delta 6}$), which is a potential raw material for the production of adipic acid (Kleiman and Spencer 1982). This isomer to oleic acid is only accumulated in the seed of the plant, but there it is present in both the triacylglycerols and the membrane lipids (Cahoon and Ohlrogge 1991). Since galactolipids of the seed also contain petroselinic acid it is likely that the synthesis of this acid can occur within the plastid. Cahoon and

Ohlrogge (1991) have reported the production of free petroselinic acid from malonyl-CoA in extracts from coriander seeds, but how the Δ^6 double bond is inserted is unknown.

Gamma-linolenic acid

Species of the Boraginaceae family have seeds with high amounts of polyunsaturated fatty acids containing a Δ^6 double bond. The most common Δ^6 acid is gamma-linolenic acid ($18:3^{\Delta 6,9,12}$), but stearidonic acid ($18:4^{\Delta 6,9,12,15}$) is also present in some seeds (Griffiths *et al.* 1989). The Δ^6 fatty acids in Boraginaceae species are present in all lipids of the seed, as well as in the leaf tissues of the plant (Stymne *et al.* 1987). The synthesis of gamma-linolenic acid in seeds occurs via a NADH dependent desaturation of linoleate esterified to position *sn*-2 in the endoplasmic reticulum (Griffiths *et al.* 1988) and thus resembles the synthesis of alpha-linolenate by the Δ^{15} desaturase in linseed (Stymne *et al.* 1992). Stearidonic acid is synthesized by Δ^6 desaturation of *sn*-2-alpha-linolenoyl-phosphatidylcholine, and can probably not be formed by Δ^{15} desaturation of gamma-linolenate (Griffiths *et al.* 1989).

Ricinoleic acid

Castor bean endosperm accumulates triacylglycerols with nearly 90 per cent of their acyl groups consisting of the hydroxylated fatty acid, ricinoleate (12-hydroxy-$18:1^{\Delta 9}$). Galliard and Stumpf (1966) showed that ricinoleate was synthesized by a NAD(P)H-dependent hydroxylation of oleate and kinetic data on ricinoleate formation *in-vitro* suggested that oleate esterfied to phosphatidylcholine was the immediate substrate for the hydroxylation (Moreau and Stumpf 1981). Recent experiments with microsomal fractions from castor bean endosperm showed unequivocally that oleate at position *sn*-2 of phosphatidylcholine was the main site for the hydroxylation (Bafor *et al.* 1991). The castor microsomes also contained a phospholipase A_2 that, with high specificity, released the newly synthesized ricinoleate from phosphatidylcholine to yield the free acid. In contrast the oleoyl-phosphatidylcholine was not hydrolysed. The ricinoleic acid could then be activated to its CoA ester by a microsomal acyl-CoA synthetase and be utilized for triacylglycerol formation by the glycerol 3-phosphate pathway (Bafor *et al.* 1991). The ricinoyl-phosphatidyl-choline specific phospholipase is not restricted to membranes from tissues accumulating ricinoleate, but is also present in microsomal fractions of safflower and rape (Bafor *et al.* 1991) which are also capable of removing fatty acid hydroperoxides at a high rate (Banas *et al.* 1992). It can be speculated that these lipases are part of a protective system which rapidly removes oxidized acyl groups from the membrane in order to maintain membrane integrity.

Vernolic acid

A few plant species accumulate fatty acids with an epoxy group. The endosperm of *Euphorbia lagascae* contains about 60 per cent vernolic acid (12-epoxy-18:1$^{\Delta 9}$) in its triacylglycerols and our laboratory has recently started to study the biosynthesis of this acid. Preliminary data show that the epoxidase is located in the microsomal fraction of the *Euphorbia lagascae* endosperm and that linoleate is the immediate substrate. The epoxidase has a requirement for NADPH, NADH being very ineffective (Stymne, unpublished results).

Production of oil crops with novel oil qualities

Domestication of wild species

The development of new vegetable oil qualities for the market could either be achieved by altering oil quality in existing oil crops, or by the introduction of novel oil-producing plant species in agriculture. Species selected for domestication in plant breeding programmes include meadowfoam (*Limnanthes alba*) for its high content of very long chain fatty acids, different *Cuphea* species for the production of medium chain fatty acids, *Euphorbia lagascae* for epoxy fatty acids, *Crepis alpina* for acetylenic acids, and *Dimorphoteca pluvialis* for fatty acids with hydroxy groups in combination with conjugated double bonds (Hirsinger 1989). There are many obstacles to overcome before these plants can be commercially viable. Although the potential yields of oil per area produced by some of these species are quite impressive, if extrapolated from small experimental trials, the harvested seed yield using common agricultural practice could be expected to be only a fraction of that. Since most of these seed oils have to compete with mineral oil-based products on the market, it is not likely that they will be able to take a market share unless the oil yield per hectare approaches that of present oil crops. The development of a new variety of a well-established crop requires usually 8–10 years of breeding efforts and it is obvious that the introduction of a wild species into agricultural use will take considerably longer (Hirsinger 1989). It is therefore probable that the introduction of new crops producing oil products in bulk quantities as replacers of mineral oil-based products lies several decades ahead. However, more highly priced seed oil qualities, such as evening primrose oil, borage oil (rich in gamma-linolenic acid), or jojoba oil (wax-esters), which are used in the pharmaceutical and cosmetic industries, will most likely continue to increase in demand and production.

Alteration of oil quality in commercial oil crops

Changing the fatty acid composition in the oil of a particular oil crop could broaden the number of end-uses. The alteration could either be a change in the relative proportions of the already existing fatty acids, or the introduction of foreign fatty acids in the oil. In the former case the new oil quality generally has to compete with vegetable oils from other plant species producing similar qualities whereas in the latter case a completely unique quality can be created. Our annual oil crops contain mainly the five common fatty acids, palmitic, stearic, oleic, linoleic, and alpha-linolenic acids. Since all these acids share parts of a common biosynthetic pathway, it is possible to manipulate their relative proportions by blocking or enhancing a few enzymatic reactions. The obvious enzymes for the manipulation of the concentration of all these acids in oil seeds are depicted in Fig. 14.4. Also manipulation of the content of other proteins which are more indirectly involved in the synthesis of a particular fatty acid, such as acyl-ACP thioesterases, lysophosphatidylcholine acyltransferase, and cytochrome b_5, might be effective.

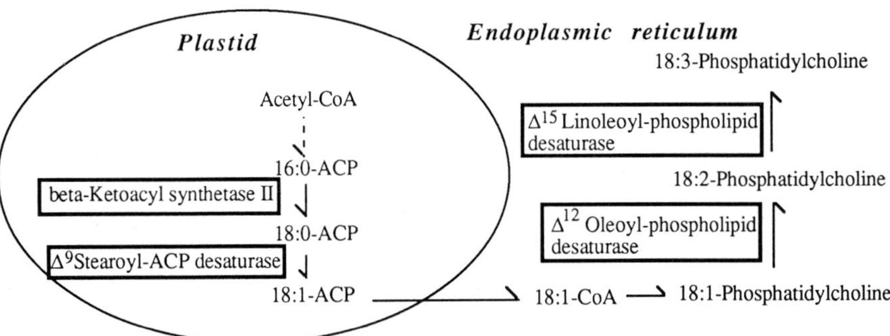

Fig. 14.4. Target enzymes for the manipulation of the relative distribution of the five 'common' fatty acids in triacylglycerols in oil seeds.

Mutation breeding

The downregulation or blocking of a particular enzyme activity in the plant can be achieved either via mutation breeding or by introducing an anti-sense construct of the gene by genetic engineering techniques (Sheehy *et al.* 1988). Mutation breeding has successfully been employed to inhibit polyunsaturated fatty acid synthesis in several oil seeds. A desired oil quality for frying purposes would, from a health point of view, be low in saturated and polyunsaturated fatty acids and high in oleic acid. Traditionally, the only commercial oil crop fulfilling these demands was olive oil. However, now sunflower (Soldatov 1976) and peanut mutants with blocked oleate desaturase (Powell *et al.* 1990) have been obtained with oil qualities very much resembling olive oil. A mutation in the linoleate desaturation in linseed has converted the oil from a high linolenic drying oil used for industrial purposes to a high linoleic edible oil (Green 1986). The effects of all these mutations are only seen in the seed lipids and the seeds show no obvious decrease in yield or viability. However this does not appear to be valid for all plant species since mutations in phospholipid desaturation in *Arabidospsis thaliana* affects the content of polyunsaturated fatty acid content in both seeds and other parts of the plant (Miquel and Browse 1990). Mutations causing changes in the amount of saturated acids in the seeds appear to be much less frequent than changes in unsaturated acids, but a soya bean mutant with elevated levels of stearic acid has been reported (Graef *et al.* 1985). The scarcity of mutants with high levels of saturated acid could be due to the fact that the mutations in lipid synthesis tend to block enzyme activities completely (Somerville and Browse 1991), and this will probably be lethal if the mutation prohibits the seed membrane lipids from containing any unsaturated fatty acids.

Alteration of oil quality by genetic engineering

The use of genetic engineering techniques to alter the fatty acid composition of the oil offers a number of advantages over mutation breeding. The enzyme activities can be both suppressed and enhanced, the latter a very rare event in mutation breeding. Normally constitutively expressed genes can be specifically suppressed in the developing seed by the introduction of anti-sense gene constructs with seed specific promotors. Finally, novel fatty acids and other traits can be introduced to the seed by insertion of genes from other species.

Mutation breeding has shown the ability of oil seeds to tolerate vast changes in the levels of polyunsaturated fatty acids. The gene coding for the Δ^{15} desaturases, which is involved in the production of linolenic acid, has now been isolated through chromosome walking techniques using an *Arabidopsis thaliana* mutant and it can be expected that the oleate desaturase gene will be obtained by the same approach (C. Somerville, personal communication). Thus, it will soon be possible to manipulate at will the levels of these fatty acids in oil seeds. Attractive phenotypes obtained through the manipulation of polyunsaturation would be, as already created by mutation breeding, high oleic varieties. An oil with a very high level of alpha-linolenic from, for example, rape seed would efficiently compete with linseed oil as a drying oil for varnishes and paints. For industrial uses of a particular fatty acid, the value of the oil increases with the percentage of this acid in the triacylglycerols. Therefore the creation of oils with only one dominating acyl group for all three of the unsaturated fatty acids would be desirable.

The first report of the alteration of acyl quality in seed oil by genetic engineering techniques has recently been published (Knutzon *et al.* 1992). Anti-sense constructs of the stearoyl-ACP desaturase gene attached to seed specific promotors have been inserted in both *Brassica rapa* and oil-rape (*Brassica napus*) with a resulting dramatic increase in stearic acids in the seeds. In oil-rape the stearic acid content was increased from 1.5 per cent to nearly 40 per cent with a near normal amount of oil and germination frequency in these seeds. If the high stearic trait could be combined with the absence of polyunsaturated fatty acid in the seed, the resulting high stearic/high oleic rape oil would serve as an excellent cocoa butter substitute, and so have a higher price than ordinary rape seed oil. Manipulation of the levels of the fatty acids that are common for both triacylglycerols and membrane lipids will, however, lead to similar changes in both these two lipid classes. It is not certain, therefore, that the seed cells could tolerate membrane lipids with very high levels of saturated fatty acids at lower temperatures, especially if it is combined with a lack of polyunsaturated acyl groups. It is pertinent to note that plants producing oils and fats with high contents of saturated acids are always of tropical origin.

The introduction of genes from foreign species into oil crops will make it possible to create totally novel oil qualities. The transfer of a lysophosphatidic acid acyltransferase gene from *Limnanthes alba* seeds to high erucic rape would allow the rape to accumulate erucic acid in all three positions of the triacylglycerols. Since the concentration of erucate in the triacylglycerols probably is directly related to the activities of the elongation enzymes, this would not cause the plant to synthesize more erucic acid *per se*. It would, however, allow further plant breeding to achieve erucic acid levels well over the 66 per cent that now seems to be the limit. There are no doubts, from the present knowledge of the biochemistry, that such transgenic seeds would be vigorous since the rape possesses efficient mechanisms to exclude erucic acid from membrane lipids and can construct trierucoyl-triacylglycerols *in vitro* (Taylor *et al.* 1991).

The transfer of a gene coding for the synthesis of a species-foreign fatty acid could severely impair cell function. Since many of the fatty acids of interest are 'triacylglycerol specific acids' in their original species, one can assume that they could severely affect membrane function if they entered the membrane lipids of the seed. Biochemical investigations regarding the biosynthesis of medium chain triacylglycerols in *Cuphea* seeds suggest that there are specializations of the glycerol acylating enzymes in these seeds, in particular the lysophosphatidic acid acyltransferase and the diacylglycerol acyltransferase, that allow the simultaneous synthesis of seed membrane lipids with exclusively long chain fatty acids and very high amounts of medium chain acids in triacylglycerols (Bafor *et al.* 1990*b*; Bafor and Stymne 1992). Medium chain fatty acids would probably compete very inefficiently with long chain acyl groups in the acylation of positions *sn*-1 and *sn*-2 of the glycerol in oil seeds which accumulate only long chain acids (Bafor *et al.* 1990*a*). Position *sn*-3 can, on the other hand, be preferentially acylated with medium chain acids (Cao and Huang 1987; Wiberg *et al.* 1992). Thus, it could be predicted that in transgenic seeds up to a third of the acyl groups could be medium chain fatty acids without interfering with membrane lipid synthesis. If the percentage of medium chain acids is higher than that, the medium chain acids have to enter the *sn*-1 and *sn*-2 positions of the glycerol, and then the lack of specialization of the glycerol acylation enzymes could either lead to an accumulation of these acids in membrane lipids, to a deficiency in long chain diacylglycerols for membrane lipid synthesis, or to both these events. Therefore, a transgenic oil crop containing high amounts of medium chain acids in their oil might have to include multiple gene transfer, including some of the glycerol acylation enzymes. The Calgene company has just announced its second successful manipulation of oil quality in transgenic plants by creating a rape with up to 25 per cent lauric acid (C_{12}) in its triacylglycerols by the insertion of a medium chain specific thioesterase gene from the California bay tree (Voelker *et al.* 1992). It will be interesting to see whether these transgenic plants have the limitations in the accumulation of laurate predicted from the biochemical studies or whether other, unknown mechanisms, might be operating under *in vivo* conditions.

The development of oil seeds with high concentrations of hydroxy or epoxy groups, by the introduction of hydroxylases or epoxidases, might cause fewer problems than medium chain acids, considering the interference of these acids to membrane lipid synthesis. It is likely that all plant membranes contain phospholipases that efficiently remove the oxygenated acyl groups in order to maintain proper membrane functions (Banas *et al.* 1992).

It is highly desirable to produce a large number of additional fatty acids in domestic oil crops. The limiting factor in the process of developing novel oil qualities in transgenic plants today is not the molecular genetic techniques, but lack of knowledge of both the enzymes involved in the synthesis of the desired oil qualities and the genes that these enzymes are coded from. Much more must also be known about the general regulation of triacylglycerol biosynthesis as well as triacylglycerol catabolism in the germinating seeds. It can therefore not be expected that transgenic plants will instantly yield commercially viable products, but these plants will aid enormously in the elucidation of regulatory mechanisms in oil synthesis in the cell. Since both private and public institutions world-wide have now realized the potential benefits of this research, the investment in both capital and manpower are considerable and can be expected to pay off in a very rapid advancement of the field in the coming years.

Economical and ecological impacts of transgenic oil seeds

The economic potential of introducing foreign fatty acids into oil crops by interspecies gene-transfer is enormous and these transgenic plants could restructure large sectors of agriculture. By molecular genetic techniques, unique oil qualities could be developed which are 'tailor-made', e.g. for polymer synthesis, lubricants, and paints. Since these products are used in bulk quantities, the area for cultivation could be a significant part of the arable land in the world. This will be of particular importance since in the present situation a surplus of food production in industrialized countries severely distorts the competitiveness of agriculture, and renewable, biodegradable resources are needed for the replacement of fossil oils. These positive effects on world economy and environment could, at least in the short term, be overshadowed by severe problems for developing countries which are dependent on their exports of coconut, palm kernel, and cocoa. If these fat qualities are produced in annual crops that can grow in temperate regions, the industrialized world would not have to depend on the tropical countries for their supply. Although it is not likely that annual crops will ever economically compete with the highly productive palm trees in the production of medium chain fatty acids, for example, the possibility cannot be excluded that the western world will be tempted to impose import restrictions on the latter commodities. Also, the possibility of modifing the oil in annual oil crops so that it contains only one particular medium chain fatty acid, as well as a more constant supply, will make the industry willing to pay a higher price than for the present palm products. There is, however, another side of the coin. Since palm trees are outstanding oil producers with a yield that is five to ten times higher than annual oil crops, genetic engineering techniques applied to these species could yield new more desired and highly priced oils and these would serve as very competitive export products for palm oil-producing countries in the third world.

The ecological consequences of altering oil quality by the creation of transgenic plants are likely to be few. Alteration of the proportions of the five 'common' fatty acids in a plant seed will not bring any advantage in survival and reproduction to the plant or, in the case of crossing or 'lateral' gene transfer, to any other plant species. The introduction of fatty acids, like hydroxy and epoxy fatty acids, that will make a previously edible seed anti-nutritional, if not toxic, could, however, have some positive effects for the seeds survival as well as negative consequences for the predators. The ecological effects of this should be evaluated before such transgenic plants are grown on a commercial scale.

Acknowledgement

The author is grateful for financial support from The Swedish Natural Science Research Council, The Swedish Council for Forestry and Agricultural Research, and Stiftelsen Svensk Oljeväxtforskning/Frö -och Oljeväxtodlarnas Service AB.

References

Badami, R. C. and Patil, K. B. (1981). Structure and occurrence of unusual fatty acids in minor seed oils. *Prog. Lipid Res.*, **19**, 119–53.

Bafor, M. and Stymne, S. (1992). Substrate specificities of glycerol acylating enzymes from developing embryos of two *Cuphea* species. *Phytochemistry*, **31**, 2873–6.

Bafor, M., Stobart, A. K., and Stymne, S. (1990*a*). Properties of the glycerol acylating enzymes in microsomal preparations from the developing seeds of safflower (*Carthamus tinctorius*) and turnip rape (*Brassica campestris*) and their ability to assemble cocoa-butter type fats. *J. Am. Oil Chem. Soc.*, **67**, 217–25.

Bafor, M., Jonsson, L., Stobart, A. K., and Stymne, S. (1990*b*). Regulation of triacylglycerol biosynthesis in embryos and microsomal preparations from the developing seeds of *Cuphea lanceolata. Biochem. J.*, **272**, 31–8.

Bafor, M., Smith, M. A., Jonsson, L., Stobart, K., and Stymne, S. (1991).

Ricinoleic acid biosynthesis and triacylglycerol assembly in microsomal preparations from developing castor-bean (*Ricinus communis*) endosperm. *Biochem. J.*, **280,** 507–14.

Banas, A., Johansson, I., and Stymne, S. (1992). Plant microsomal phospholipases exhibit preference for phosphatidylcholine with oxygenated acyl groups. *Plant Sci.*, **84,** 137–44.

Barron, E. J. and Stumpf, P. K. (1962). The biosynthesis of triglycerides by avocado-mesocarp enzymes. *Biochim. Biophys. Acta*, **60,** 329–37.

Browse, J. and Slack, C. R. (1985). Fatty acid synthesis in plastids from maturing safflower and linseed cotyledons. *Planta*, **166,** 74–80.

Cahoon, E. B. and Ohlrogge, J. B. (1991). Deposition and synthesis of petroselinic acid in seeds of Umbelliferae. *International News On Fats, Oils and Related Materials*, **2,** 342.

Cao, Y.-Z. and Huang, H. C. (1987). Acyl-CoA preference of diacylglycerol acyltransferase from maturing seeds of Cuphea, maize, rapeseed, and canola. *Plant Physiol.*, **84,** 762–5.

Cao, Y., Oo, K.-C., and Huang, A. C. (1990). Lysophosphatidate acyltransferase in microsomes from maturing seeds of meadowfoam (*Limnanthes alba*). *Plant Physiol.*, **94,** 1199–206.

Davies, H. M., Anderson, L., Fan, C., and Hawkins, D. J. (1991). Developmental induction, purification, and further characterization of 12:0-ACP thioesterase from immature cotyledons of *Umbellularia californica*. *Arch. Biochem. Biophys.*, **290,** 37–45.

Downey, R. K. and Craig, R. M. (1964). Genetic controls of fatty acid biosynthesis in rape seed (*Brassica napus* L.). *J. Am. Oil Chem. Soc.*, **41,** 475–8.

Fehling, E. and Mukherjee, K. M. (1990). Isolation and characterisation of enzymes catalyzing the synthesis of very long chain fatty acids in *Lunaria annua*. In *Plant lipid biochemistry, structure and utilization* (ed. P. J. Quinn and J. L. Harwood), pp. 151–3. Portland Press. London.

Finlayson, S. A. and Dennis, D. T. (1980). NAD^+-specific glycerol 3-phosphate dehydrogenase from developing castor bean endosperm. *Arch. Biochem. Biophys.*, **199,** 179–85.

Galliard, T. and Stumpf, P. K. (1966). Fat metabolism in higher plants. XXX. Enzymatic synthesis of ricinoleic acid by a microsomal preparation from developing *Ricinus communis* seeds. *J. Biol. Chem.*, **241,** 5806–12.

Graef, G. L., Miller, L. A., Fehr, W. R., and Hammond, E. G. (1985). Fatty acid development in a soybean mutant with high stearic acid. *J. Am. Oil Chem. Soc.*, **62,** 773–5.

Graham, S. A., Hirsinger, F., and Röbbelen, G. (1981). Fatty acids of Cuphea (*Lythraceae*) seed lipids and their taxonomic significance. *Am. J. Bot.*, **68,** 908–17.

Green, A. (1986). A mutant genotype of flax (*Linum usitatissimum* L.) containing very low levels of linolenic acid in its seed oil. *Can. J. Plant Sci.*, **66,** 499–503.

Griffiths, G.. Stobart, A. K., and Stymne, S. (1985). The acylation of *sn*-glycerol 3-phosphate and the metabolism of phosphatidate in microsomal preparations from the developing cotyledons of safflower (*Carthamus tinctorius*) seed. *Biochem. J.*, **230,** 379–88.

Griffiths, G., Stobart, A. K., and Stymne, S. (1988). Δ^6- and Δ^{12}-desaturase activities and phosphatidic acid formation in microsomal preparations from the developing cotyledons of common borage (*Borago officinalis*). *Biochem. J.*, **252,** 641–7.

Griffiths, G., Brechany, E. Y., Christie, W. W., Stymne, S., and Stobart, K. (1989). Synthesis of octadecatetraenoic acid in borage. In *Biological role of plant lipids* (ed. P. A. Biacs, K. Gruuiz, and T. Kemmer), pp. 151–3. Académiai Kiadó, Budapest and Plenum Press, New York.

Hatje, G. (1989). World importance of oil crops and their products. In *Oil crops of the world*, (ed. G. Röbbelen, R. K. Downey, and A. Ashri), pp. 1–21. McGraw-Hill Inc., New York.

Hirsinger, F. (1989). New annual oil crops. In *Oil crops of the world*, (ed. G. Röbbelen, R. K. Downey, and A. Ashri), pp. 518–32. McGraw-Hill Inc., New York.

Ichihara, K. (1984). *sn*-Glycerol 3-phosphate acyltransferase in a particulate fraction from mauring safflower seeds. *Arch. Biochem. Biophys.*, **232,** 685–98.

Ichihara, K., Asahi, T., and Fujii, S. (1987). 1-Acyl-*sn*-glycerol-3-phosphate acyltransferase in maturing safflower seeds and its contribution to the non-random fatty acid distribution of triacylglycerol. *Eur. J. Biochem.*, **167,** 339–47.

Ichihara, K., Takahashi, T., and Fujii. S. (1988). Diacylglycerol acyltransferase in maturing safflower seeds: its influences on the fatty acid composition of triacylglycerol and on the rate of triacylglycerol synthesis. *Biochim. Biophys. Acta*, **958,** 125–9.

Ichihara, K., Norikura, S., and Fujii, S. (1989). Microsomal phosphatidate phosphatase in maturing safflower seeds. *Plant Physiol.*, **90,** 413–19.

Ichihara, K., Murota, N., and Fujii, S. (1990). Intracellular translocation of phosphatidate phosphatase in maturing safflower seeds: a possible mechanism of feedforward control of triacylglycerol biosynthesis for fatty acids. *Biochim. Biophys. Acta*, **1043,** 227–34.

Jaworski, J. G., Clough, R. C., and Barnum, S. R. (1989). A cerulinin insensitive short chain 3-ketoacyl-acyl carrier protein synthetase in *Spinacia oleracea* leaves. *Plant Physiol.*, **90,** 41–4.

Kearns, E., Hugly, S., and Somerville, C. R. (1991). The role of cytochrome b_5 in Δ^{12} desaturation of oleic acid by microsomes of safflower (*Carthamus tinctorius* L.). *Arch. Biochem. Biophys.*, **284,** 431–6.

Kennedy, E. P. (1961). Biosynthesis of complex lipids. *Proc. Federation of the American Society for Experimental Biology*, **20,** 934–40.

Kindl, H. (1987). β-Oxidation of fatty acids by specific organelles. In *The biochemistry of plants. A comprehensive treatise* (ed. P. K. Stumpf), pp. 31–52. Academic Press, New York.

Kleiman, R. and Spencer, G. F. (1982). Search for new industrial oils: XVI. *Umbelliflorae*—seed oils rich in petroselinic acid. *J. Am. Oil Chem. Soc.*, **50,** 29–38.

Knutzon, D. S., Thompson, G. A., Radke, S. E., Johnson, W. B., Knauf, V. C., and Kridl, J. C. (1992). Modification of Brassica seed oil by antisense expression of a stearoyl-acyl carrier protein desaturase gene. *Proc. Natl. Acad. Sci., USA*, **89,** 2624–8.

Löhden, I., Bernerth, R., and Frentzen, M. (1990). Acyl-CoA:1-acylglycerol-3-phosphate acyltransferase from developing seeds of *Limnanthes douglasii* (R. Br.) and *Brassica napus* (L.). In *Plant lipid biochemistry, structure and utilization* (ed. P. J. Quinn and J. L. Harwood), pp. 175–7. Portland Press, London.

Masterson, C., Wood, C., and Thomas, D. R. (1990). L-Acetylcarnetin, a substrate for chloroplast fatty acid synthesis. *Plant, Cell and Environment*, **13,** 755.

Miquel, M. and Browse, J. (1990). Mutants of Arabidopsis deficient in 18:1-PC desaturation. In *Plant lipid biochemistry, structure and utilization* (ed. P. J. Quinn and J. L. Harwood), pp. 456–8. Portland Press, London.

Moreau, R. A. and Stumpf, P. K. (1981). Recent studies of the enzymatic synthesis of ricinoleic acid by developing castor beans. *Plant Physiol.*, **67,** 672–6.

Moreau, R. A., Pollard, M.-R., and Stumpf, P. K. (1981). Properties of a Δ^5-fatty acyl-CoA desaturase in the cotyledons of developing *Limnanthes alba*. *Arch. Biochem. Biophys.*, **209,** 376–84.

Ohlrogge, J. B., Shine, W. E., and Stumpf, P. K. (1978). Fat metabolism in higher plants: Characterization of plant acyl-ACP and acyl-CoA hydrolases. *Arch. Biochem. Biophys.*, **189,** 382–91.

Ohlrogge, J. B., Kuhn, D. K., and Stumpf, P. K. (1979). Subcellular localization of acyl carrier protein in leaf protoplasts of Spinacia oleracea. *Proc. Natl. Acad. Sci., USA*, **76,** 1194–8.

Pollard, M. R., Andersson, L., Fan, C., Hawkins, D. J., and Davies, H. M. (1990). A specific acyl-ACP thioesterase implicated in laurate production in immature cotyledons of *Umbellularia californica*. In *Plant lipid biochemistry, structure and utilization* (ed. P. J. Quinn and J. L. Harwood), pp. 163–5. Portland Press, London.

Powell, G., Abott, A., Knauft, D., and Barth, J. (1990). Oil desaturation in developing peanut seeds: studies of lipid desaturation in a peanut mutant that accumulates high levels of oleic acid. In *Plant lipid biochemistry, structure and utilization* (ed. P. J. Quinn and J. L. Harwood), pp. 131–3. Portland Press, London.

Pryde, E. H. and Rothfus, J. A. (1989). Industrial and nonfood uses of vegetable oils. In *Oil crops of the world* (ed. G. Röbbelen, R. K. Downey, and A. Ashri), pp. 87–117. McGraw-Hill Inc., New York.

Roughan, P. G. and Slack, C. R. (1977). Long chain acyl-coenzyme A synthetase activity of spinach chloroplast is concentrated in the envelope. *Biochem. J.*, **162,** 457–9.

Roughan, P. G., Holland, R., and Slack, C. R. (1978). Acetate is the preferred substrate for long-chain fatty acid synthesis in isolated chloroplasts. *Biochem. J.*, **184,** 565–9.

Shanklin, J. and Somerville, C. (1991). Stearoyl-acyl-carrier-protein desaturase from higher plants is structurally unrelated to the animal and fungal homologs. *Proc. Natl. Acad. Sc., USA*, **88**, 2510–14.

Shanklin, J., Mullins, C., and Somerville, C. (1991). Sequence of a complementary DNA from *Cucumis sativus* L. encoding the stearoyl-acyl-carrier protein desaturase. *Plant Physiol.*, **97**, 467–8.

Sheehy, R. E., Kramer, M., and Hiatt, W. R. (1988). Reduction of galacturonase activity in tomato fruit by antisense RNA. *Proc. Natl. Acad. Sci., USA*, **85**, 8805–9.

Slack, C. R., Cambell, L. C., Browse, J. A., and Roughan, P. G. (1983). Some evidence for the reversibility of the cholinephospho- transferase-catalysed reaction in developing linseed cotyledons *in vivo*. *Biochim. Biophys. Acta*, **754**, 10–20.

Slack, C. R., Roughan, P. G., Browse, J. A., and Gardiner, S. E. (1985). Some properties of cholinephosphotransferase from developing safflower cotyledons. *Biochim. Biophys. Acta*, **833**, 438–48.

Smith, M. A., Cross, A. R., Jones, T. G., Griffiths, T., Stymne, S., and Stobart, K. (1990). Electron-transport components of the 1-acyl-2-oleoyl-*sn*-glycerol-3-phosphocholine Δ^{12}-desaturase (Δ^{12}-desaturase) in microsomal preparations from developing safflower (*Carthamus tinctorius* L.) cotyledons. *Biochem. J.*, **272**, 23–9.

Soldatov, K. I. (1976). Chemical mutagenesis in sunflower breeding. In *Proceedings of the 7th international sunflower conference in Krasnodar, USSR*, pp. 352–57. International Sunflower Association, Vlaardingen, The Netherlands.

Somerville, C. and Browse, J. (1991). Plant lipids: metabolism, mutants, and membranes. *Science*, **252**, 80–7.

Stobart, A. K. and Stymne, S. (1985). The regulation of the fatty-acid composition of the triacylglycerols in microsomal preparations from avocado mesocarp and the developing cotyledons of safflower. *Planta*, **163**, 119–25.

Stumpf, P. K. (1987). The biosynthesis of saturated fatty acids. In *The biochemistry of plants. A comprehensive treatise* (ed. P. K. Stumpf), Vol. 9, pp. 121–36. Academic Press, New York.

Stymne, S. and Stobart, A. K. (1984). Evidence for the reversibility of the acyl-CoA:lysophosphatidylcholine acyltransferase in microsomal preparations from developing safflower (*Carthamus tinctorius* L.) cotyledons and rat liver. *Biochem. J.*, **223**, 305–14.

Stymne, S. and Stobart, A. K. (1987). Triacylglycerol biosynthesis. The biosynthesis of saturated fatty acids. In *The biochemistry of plants. A comprehensive treatise* (ed. P. K. Stumpf), Vol. 9, pp. 175–214. Academic Press, New York.

Stymne, S., Griffiths, G., and Stobart, A. K. (1987). Desaturation of fatty acid on complex lipid substrates. In *The metabolism, structure and function of plant lipids* (ed. P. K. Stumpf, J. B. Mudd, and W. D. Nes), pp. 405–412. Plenum Press, New York.

Stymne, S., Bafor, M., Jonsson, L., Wiberg, E., and Stobart, A. K. (1990). Triacylglycerol assembly. In *Plant lipid biochemistry, structure and utilization* (ed. P. J. Quinn and J. L. Harwood), pp. 191–7. Portland Press, London.

Stymne, S., Tonnet, M. L., and Green, A. G. (1992). Biosynthesis of linolenate in developing embryos and cell-free preparations of high-linolenate linseed (*Linum usitatissimum*) and low linolenate mutants. *Arch. Biochem. Biophys.*, **294**, 557–63.

Taylor, D. C., Weber, N., Barton, D. L., Underhill, E. W., Hogge, L. R., Weselake, R. J., and Pomeroy, M. K. (1991). Triacylglycerol bioassembly in microspore-derived embryos of *Brassica napus* L. cv Reston. *Plant Physiol.*, **97**, 65–79.

Thompson, G. A., Scherer, D. E., Foxall-Van Aken, S., Kenny, J. W., Young, H. L., Shintani, D. K., Kridl, J. C., and Knauf, V. C. (1991). Primary structures of the precursor and mature forms of stearoyl- acyl carrier protein desaturase from safflower embryos and requirement of ferredoxin for enzyme activity. *Proc. Natl. Acad. Sci., USA*, **88**, 2578–82.

Treede, H. J. and Heise, K.-P. (1985). The regulation of acetyl coenzyme A synthesis in chloroplasts. *Z. Naturforsch.*, **40c**, 496–502.

Voelker, A., Worrel, A. C., Hawkins, D. J., and Davies, M. (1992). Production of medium-chain fatty acids in transgenic oilseeds. In *Proceedings from the 10th international symposium on the metabolism, structure and utilization of plant lipids* (ed. A. Cherif, B. Marzouk, A. Smaoui, D. Ben Miled, and M. Zarrouk), pp. 181–4. Centre National Pédagogique, Tunis.

Whitfield, H. V. and Murphy, D. J. (1990). Storage lipid synthesis in nasturtium (*Tropaeolum majus*). In *Plant lipid biochemistry, structure and utilization* (ed. P. J. Quinn and J. L. Harwood), pp. 225–7. Portland Press, London.

Wiberg, E., Tillberg, E., and Stymne, S. (1992). Specificities of diacylglycerol acyltransferases in microsomal fractions from developing oil seeds. In *Proceedings from the 10th international symposium on the metabolism, structure and utilization of plant lipids* (ed. A. Cherif, B. Marzouk, A. Smaoui, D. Ben Miled, and M. Zarrouk), pp. 83–6. Centre National Pédagogique, Tunis.

Yermanos, D. M. (1975). Composition of jojoba seed during development. *J. Am. Oil Chem. Soc.*, **52**, 115–17.

15. Variation of flower colours by genetic manipulation of the biosynthesis of anthocyanins

JOSEPH N. M. MOL

Department of Genetics, Free University, De Boelelaan 1087, 1081 HV Amsterdam, The Netherlands

Introduction

Flavonoids constitute the main class of flower pigments. Carotenoids on the other hand are responsible for many of the bright yellows and oranges, especially in composites such as gerberas and chrysanthemums. The third class of flower pigments, the betalains are mainly found in cacti, ice plants, and beets.

The well-developed genetics and biochemistry of the flavonoid biosynthetic pathway in plants such as maize, petunia, and snapdragon make this route ideally suited for genetic engineering (for reviews see Mol *et al.* 1989*a*; Dooner *et al.* 1991; Forkmann 1991; Woodson 1991). Figure 15.1 shows a compilation of current knowledge on the flavonoid biosynthetic route in higher plants. Enzymic conversions are represented by fat arrows. In the right hand corner are listed the regulatory genes affecting multiple steps in the pathway. These genes are designated An1, An2, An10, and An11 for petunia (An stands for Anthocyanin and the numbers reflect the order in which mutations in these genes were identified) and the C1 and R family for maize. In an1an1, an2an2, an10an10, or an11an11 lines of petunia all enzyme genes downstream from F3H are silent. Therefore, these regulatory genes most likely encode proteins that are involved in the transcription of multiple enzyme genes operating in the same pathway. These proteins are called transcription factors. The entire pathway is regulated by the hormone gibberellic acid (GA) in petunia (Weiss *et al.* 1992).

The colour of a flower is mainly determined by the substitution patterns of the individual flavonoid ring structures, the pelargonidins (one B-ring substituent) being orange, the cyanidins (two B-ring substituents) being red, and the delphinidins (three B-ring substituents) being blueish (see Fig. 15.1). In addition, differences in vacuolar pH and the presence of co-pigments profoundly influence the colour of the flower.

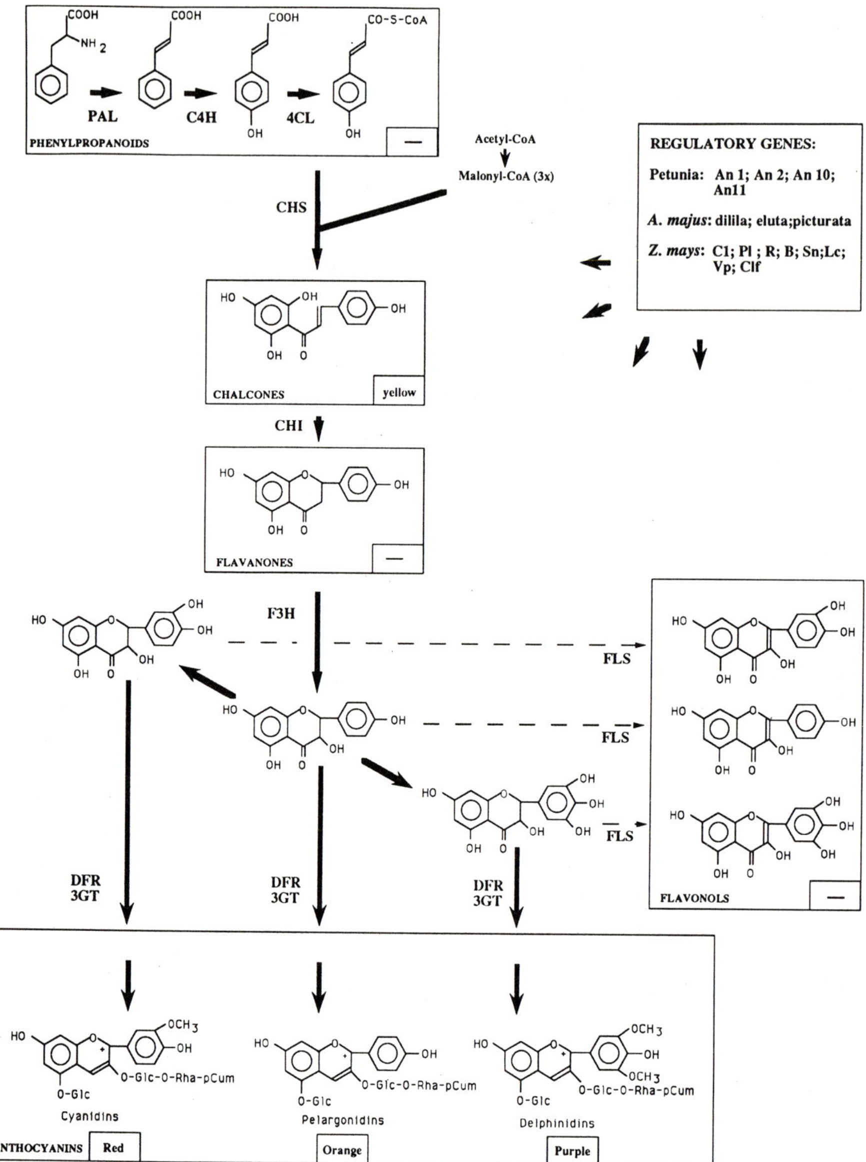

Fig. 15.1. Schematic representation of the flavonoid biosynthesis pathway in maize and petunia. PAL: Phenylalanine ammonia-lyase; C4H: cinnamate 4-hydroxylase; 4CL: 4-coumarate CoA ligase; CHS: chalcone synthase; CHI: chalcone flavanone isomerase; F3H: flavanone 3-hydroxylase; DFR: dihydroflavonol 4-reductase; 3GT: 3-glycosyl transferase. Changes in colour following modifications to the basic flavonoid ring are indicated.

The naturally available flower colour spectrum in many important ornamental plants is quite narrow. One will find neither blue roses, carnations, chrysanthemums, and tulips nor orange petunias. This is due to the absence of delphinidins and pelargonidins in these cultivars, respectively. Flowers containing colour patterns such as found in petunia red star are relatively rare in other plant species (e.g. impatiens, viola).

Recent advances in molecular biology especially in gene cloning and gene transfer have opened the possibility of engineering flower colour in a predetermined way. Two recent approaches will be discussed: (i) transfer of the pelargonidin ('orange') and delphinidin ('blue') pathways, and (ii) modulation of flower colour genes by the addition of more gene copies either in sense or anti-sense orientation.

Transfer of the pelargonidin ('orange') pathway from maize to petunia

In petunia only cyanidins and delphinidins are found, apparently because of the substrate specificity of the enzyme dihydroflavonol 4-reductase (DFR); DFR in petunia is unable to reduce dihydrokaempferol (see Fig. 15.1). The DFR enzyme from maize however is able to produce pelargonidins. This observation formed the basis for a gene transfer experiment conducted by researchers at the Max-Planck-Institute in Cologne. Meyer *et al.* (1987) transformed the petunia line RL01 that accumulates dihydrokaempferol with a gene construct composed of the maize A1 gene encoding DFR driven by a strong promoter that was derived from the cauliflower mosaic virus genome (35S promoter). This resulted in the production of salmon red pelargonidin-containing flowers. Field tests have been conducted with homogeneously coloured petunia plants that carry one copy or multiple copies of the maize A1 gene (Meyer *et al.* 1992). It turned out that the expression of the DFR transgene is highly dependent on exogenous and endogenous factors such as the age of the plant.

Transfer of the delphinidin ('blue') pathway from petunia to other cut flowers

Two genetic loci Hf1 and Hf2 control the flavonoid 3′5′-hydroxylation in petunia. These loci encode cytochrome P-450 enzymes. P-450 sequences have been isolated from a variety of organisms and these display short regions of homology. Researchers from Calgene Pacific, Victoria, have used a PCR (Polymerase Chain Reaction) approach to isolate flower-specific cytochrome P-450 sequences from petunia (Holton *et al.* 1992). Based on standard genetic mapping, putative clones for Hf1 and Hf2 could be identified. Expression of a full-size clone in yeast provided definitive proof for the identity of the P-450 clones. Transfer of Hf1 and Hf2 to commercially important cut flower species such as the rose is in progress. One important problem that may be encountered is vacuolar pH. To regulate pH in a genetic way may require the cloning of petunia Ph genes.

Down-regulation of flower pigmentation by the addition of anti-sense and sense genes

In the previous two sections strategies to alter flower colour by adding a gene for a side chain enzyme were discussed. Another strategy to modify flower colour is to reduce the expression of endogenous flower colour genes. One strategy that has proven effective in petunia is to introduce so called anti-sense genes in the plant (for a review see van der Krol *et al.* 1988*b*).

Only one of the two complementary strands of DNA is transcribed into messenger RNA (mRNA). If a gene is constructed in the test tube with the coding region inverted and transferred to an organism, the wrong strand is transcribed yielding so-called anti-sense RNA which is complementary to the mRNA. This anti-sense RNA potentially can interact with complementary mRNA to form a duplex. The mRNA in double stranded form cannot contribute to protein synthesis and the effect will be that of a mutation (Fig. 15.2). Van der Krol *et al.* (1988*a*) demonstrated for the first time that in petunia introduction of an anti-sense gene for chalcone synthase (CHS) has a dramatic effect on floral pigmentation. Since the substrates of CHS are colourless, pure white flowers are expected if the titration of CHSmRNA by its anti-sense RNA is complete and intermediate in colour if inhibition is only partial. Such phenotypes were indeed observed. Surprisingly, some transformants bore flowers with alternating coloured and colourless sectors or rings.

Expression of an anti-sense gene for DFR (van Blokland, manuscript in preparation) and that of a sugar transferase (Kroon, manuscript in preparation) operating more downstream in the pathway similarly led to substrate accumulation accompanied by a shift in flower colour. An anti-sense gene for chalcone flavanone isomerase (CHI) on the other hand failed to yield the expected yellow flowers. The transformed plants expressing the CHI gene bore flowers that were indistinguishable from

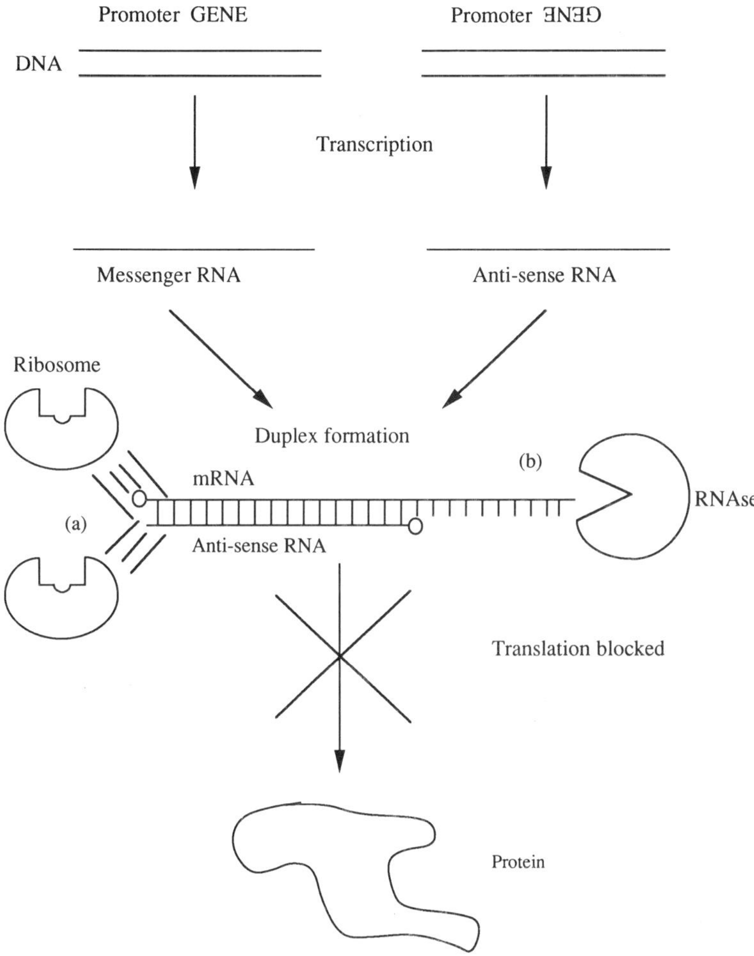

Fig. 15.2. Interactions between sense and anti-sense RNA may lead to (a) prevention of initiation of translation; (b) RNA degradation.

wild type. This clearly demonstrates that not every anti-sense gene is equally effective.

The anti-sense technology has proven very effective in down-regulating genes from other pathways, e.g. starch biosynthesis (Visser *et al.* 1991; Müller-Röber *et al.* 1992), ethylene biosynthesis (Hamilton *et al.* 1990; Oeller *et al.* 1991), oil biosynthesis (Knutzon *et al.* 1992), fruit ripening (Smith *et al.* 1988), nitrate assimilation (Vaucheret *et al.* 1992). In addition, the anti-sense technology has proven useful in the identification of so-called 'cryptic' genes for which no function is known. Anti-sense inhibition of an unknown gene may provide a mutant phenotype thereby identifying the pathway in which the gene is operating. An example of this approach was recently published by Bird *et al.* (1991). These investigators constructed a ripening-specific *c*DNA library from tomato. One particular clone, pTOM5, turned out to encode a key enzyme from the carotenoid pathway: phytoene synthase. Anti-sense expression of pTOM5 in tomato plants yielded pale-colouring fruit which had lycopene levels < 0.1 per cent.

Attempted overexpression of CHS or DFR genes by introduction of additional gene copies unexpectedly leads to reduced pigmentation (van der Krol *et al.* 1990; Napoli *et al.* 1990). Further analysis showed that reduction in expression had occurred not only of the CHS or DFR transgenes but also of their endogenous counterparts. The mechanism by which sense transgenes inhibit pigmentation is still unknown. Models which involve direct interaction of non-allelic DNA copies (Jorgensen 1991) and the synthesis of anti-sense RNA on a sense template (Grierson *et al.* 1991; Mol *et al.* 1991) have been proposed.

In collaboration with the American company DNAP, the Dutch company Florigene has developed a pure white chrysanthemum by sense CHS gene suppression. This engineered chrysanthemum will be taken through the complete procedure to achieve the status of a regulated product and to get permission to carry it to the market place.

Other strategies

Possible other strategies for suppression of gene activity have been outlined in the literature (Herskowitz 1987; Mol *et al.* 1989*b*). One approach involves competition of substrates. Plants such as peanut produce stilbenes for phytoalexins. Resveratrol synthase (RS) is the key enzyme of stilbene biosynthesis and condenses three molecules of malonyl CoA with one molecule of 4-coumaroyl CoA yielding 3′4′5′-trihydroxystilbene (resveratrol). Chalcone synthase uses exactly the same substrates. Introduction of an RS gene in an ornamental plant therefore should lead to a reduction in floral pigmentation.

Another approach involves dominant negative mutations (Herskowitz 1987). This approach can only work in multi-component systems such as multi-subunit enzymes or DNA–protein complexes. Chalcone synthase is an enzyme composed of two identical subunits. Introduction of a mutant subunit which still binds to the wild type subunit but which has lost its enzymic activity should lead to the formation of heterodimers with reduced enzyme activity. Transcription factors such as the products of An genes bind specifically to DNA and assist RNA polymerase to initiate transcription. Truncated versions of transcription factors have been described that retained the ability to bind to the DNA but lost the capability to assist in transcription initiation (Paz-Ares *et al.* 1990).

Future prospects

Genetic manipulation of flower colour genes has not only deepened our insight into the fundamental mechanisms of gene regulation but in addition has yielded some novel commercially interesting products such as new orange varieties and colour patterns. Now that Calgene Pacific has isolated the 'blue gene' the creation of blue roses, carnations, and tulips comes within reach. Stumbling blocks that still have to be removed are genetic instability (genes sometimes get modified and switch off in subsequent generations) and the legislation and public acceptance issues. The Dutch company Florigene has created a novel white-flowering variety of chrysanthemum by the sense technology and hopes to be among the first to bring a genetically modified plant into the market place. This might pave the way for edible plant products.

References

Bird, R. B., Ray, J. A., Fletcher, J. D., Boniwell, J. M., Bird, A. S., Teulieres, C., *et al.* (1991). Using antisense RNA to study gene function: Inhibition of carotenoid biosynthesis in transgenic tomatoes. *Bio/Technology*, **9**, 635–9.

Dooner, H. K., Robbins, T. P., and Jorgensen, R. (1991). Genetic and developmental control of anthocyanin biosynthesis. *Annu. Rev. Genet.*, **25**, 173–99

Forkmann, G. (1991). Flavonoids as flower pigments: The formation of the natural spectrum and its extension by genetic engineering. *Plant Breeding*, **106**, 1–26.

Grierson, D., Fray, R. G., Hamilton, A. J., Smith, C. J. S., and Watson, C. F. (1991). Does co-suppression of sense genes in transgenic plants involve antisense RNA? *Trends Biotechnol.*, **9**, 122–3.

Hamilton, A. J., Lycett, G. W., and Grierson, D. (1990). Antisense gene that inhibits synthesis of the hormone ethylene in transgenic plants. *Nature*, **346**, 284–7.

Herskowitz, I. (1987). Functional inactivation of genes by dominant negative mutations. *Nature*, **329**, 219–22.

Holton, T. A., Tanaka, Y., Kovacic, F., Lester, R. A., Hyland, C., Menting, J., *et al.* (1992). Isolation and expression of cytochrome P-450 genes controlling flower colour in Petunia hybrida. *Abstract L21, International Workshop on Flower Development and Plant Reproduction*, Amsterdam, The Netherlands.

Jorgensen, R. (1991). Beyond antisense: How do transgenes interact with homologous plant genes? *Trends Biotechnol.*, **9**, 266–7.

Knutzon, D. S., Thompson, G. A., Radke, S. E., Johnson, W. B., Knauf, V. C., and Kridl, J. C. (1992). Modification of Brassica seed oil by antisense expression of a stearoyl-acyl carrier protein desaturase gene. *Proc. Natl. Acad. Sci. USA*, **89**, 2624.

van der Krol, A. R., Lenting, P. E., Veenstra, J., van der Meer, I. M., Koes, R. E., Gerats, A. G. M., *et al.* (1988*a*). An antisense chalcone synthase gene in transgenic plants inhibits flower pigmentation. *Nature*, **333**, 866–9.

van der Krol, A. R., Mol, J. N. M., and Stuitje, A. R. (1988*b*). Modulation of eukaryotic gene expression by complementary RNA or DNA sequences. *Biotechniques*, **6**, 958–76.

van der Krol, A. R., Mur, L. A., Beld, M., Mol, J. N. M., and Stuitje, A. R. (1990). Flavonoid genes in Petunia: addition of a limited number of gene copies may lead to a suppression of gene expression. *Plant Cell*, **2**, 191–299.

Meyer, P., Heidmann, I., Forkmann, G., and Saedler, H. (1987). A new petunia flower colour generated by transformation of a mutant with a maize gene. *Nature*, **330**, 677–8.

Meyer, P., Linn, F., Heidman, I., Meyer, H., Nierdenhof, I., and Saedler, H. (1992). Endogenous and environmental factors influence 35S promoter methylation of a maize A1 gene construct in transgenic petunia and its colour phenotype. *Mol. Gen. Genet.*, **231**, 345–52.

Mol, J. N. M., Stuitje, A., Gerats. A., van der Krol, A., and Jorgensen, R. (1989*a*) Saying it with genes: molecular flower breeding. *Trends Biotechnol.*, **7**, 148–153.

Mol, J. N. M., Stuitje, A. R., and van der Krol, A. R. (1989*b*). Genetic manipulation of floral pigmentation genes. *Plant Mol. Biol.*, **13**, 287–94.

Mol, J. N. M., Van Blokland, R., and Kooter, J. (1991). More about co-suppression. *Trends Biotechnol.*, **9**, 182–3.

Müller-Röber, B., Sonneveld, U., and Willmitzer, L. (1992). Inhibition of ADP-glucose pyrophosphorylase in transgenic potatoes leads to sugar-storing tubers and influences tuber formation and expression of tuber storage protein genes. *EMBO J.*, **11**, 1229.

Napoli, C., Lemieux, C., and Jorgensen, R. (1990). Introduction of a chimeric chalcone synthase gene into petunia results in reversible co-suppression of homologous genes in trans. *Plant Cell*, **2**, 279–89.

Oeller, P. W., Wong, L. M., Taylor, L. P., Pike, D. A., and Theologis, A. (1991). Reversible inhibition of tomato fruit senescence by antisense RNA. *Science*, **254**, 437–9.

Paz-Ares, J., Ghosal, D., and Saedler, H. (1990). Molecular analysis of the C1-I allele from *Zea mays*: a dominant mutant of the regulatory C1 locus. *EMBO J.*, **9**, 315–21.

Smith, C. J. S., Watson, C. F., Ray, J., Bird, C. R., Morris, P. C., Schuch, W., and Grierson, D. (1988). Antisense RNA inhibition of polygalacturonase gene expression in transgenic tomatoes. *Nature*, **334**, 724–6.

Vaucheret, H., Kronenberger, J., Lepingle, A., Vilaine, F., Boutin, J.-P., and Caboche, M. (1992). Inhibition of tobacco-nitrite reductase activity by expression of antisense RNA. *Plant J.*, **2**, 559–69.

Visser, R. G. F., Somhorst, I., Kuipers, G. J., Ruys, N. J., Feenstra, W. J., and Jacobsen, E. (1991). Inhibition of the expression of the gene for granule-bound starch synthase in potato by antisense constructs. *Mol. Gen. Genet.*, **225**, 289–96.

Weiss, D., Van Blokland, R., Kooter, J. M., Mol, J. N. M., and Van Tunen, A. J. (1992). Gibberellic acid regulates chalcone synthase gene transcription in the corolla of Petunia hybrida. *Plant Physiol.*, **98**, 191–7.

Woodson, W. R. (1991). Biotechnology of floricultural crops. *HortScience*, **26**, 1029–33.

16. Recent advances in the production of aroma compounds in plant culture systems

BARRY V. CHARLWOOD

Plant Cell and Molecular Biology Group,
Life Sciences Division, Kings College London,
Campden Hill Road, London W8 7AH, UK

Introduction

Higher plants accumulate a wide variety of aroma compounds including the aliphatic straight-chain hydrocarbons and aldehydes (typically found in flowers and fruits), the highly odoriferous derivatives of benzoic acid and phenylpropane (characteristic especially of the Umbelliferae), and a range of nitrogen and sulphur containing volatiles. However, the ubiquitous lower isoprenoids (the mono- and the sesqui-terpenoids) form by far the largest group of natural aromas and a great deal of information has been amassed concerning the role of these secondary compounds *in planta*. This review will be restricted to a consideration of the recent advances in the biology and biochemistry of the production of lower isoprenoids (especially the monoterpenoids) in cell, organ, and tissue cultures of higher plants.

The agricultural applications of the lower isoprenoids

The most obvious areas of application in agriculture for any group of natural products are those which exploit and expand upon the natural functions of the compounds within the plant. The role of the lower isoprenoids, whether constitutively accumulated or produced as a phytoalexin response, is primarily one of defense of the plant from attack by other organisms. Details of such interactions with insects, with animals, as well as with other plants and microorganisms are readily available. For example, the monoterpenoids limonene, α- and β-pinene, and camphene, which have been found in the oleoresin of pine trees, are highly toxic to bark beetles of the genera *Dendroctonus* and *Ips*, themselves potential colonizers of loblolly and shortleaf pine (Cook and Hain 1988). Many shrubs and conifers contain lower isoprenoids which inhibit rumen microorganisms (upon which ruminants depend for the digestion of plant material) thus acting as powerful antifeedants. Other mono- and sesqui-terpenoids are potent insect repellent—for example, terpinen-4-ol has been shown to be very active (equal in activity to the commercial repellent dimethyl phthalate) against the yellow-fever mosquito *Aedes aegypti*, and is found in the oil of *Artemisia vulgaris* (mugwort) together with other repellent terpenoids such as linalool, borneol, and camphor (Hwang *et al.* 1985). The sesquiterpene hydrocarbon (*E*)-β-farnesene presents a most interesting case in that it is an aphid repellent and as such increases the movement of these insects (and hence their susceptibility to contact insecticides) when applied in very small amounts (less than 1 g/ha) to crop plants (Pickett 1991).

Lower isoprenoids also play important roles in plant–plant interactions since many (e.g. α- and β-pinene, camphene, camphor, and 1,8-cineole) have been shown to be cytotoxic to plant tissues owing to their drastic effects on respiration and photosynthesis. Some have potential application as herbicides, e.g. 1,8-cineole is the structural base of cinmethylin (Duke 1987), whilst many, particularly the sesquiterpenoid phytoalexins of the Solanaceae, exhibit potent antibiotic and antimicrobial activities (see Deans and Ritchie 1987; Fraga 1991).

The introduction of a new, or an enhancement of an existing, defence agent in an appropriate cropping plant would provide a clear possibility not only of exploiting a natural and existing biocontrol system, but also of delivering at the same time a highly marketable commodity. The improvement of product profile and yield of isoprenoids in cropping plants has, hitherto, been achieved through traditional but, for this class of compound, largely ineffectual breeding programmes. Although a number of recent attempts at manipulating secondary compound pathways *in vitro* have been successful for introducing into *Nicotiana* tissues enzymes associated with alkaloid production, for example ornithine decarboxylase (Hamill *et al.* 1990) and strictosidine synthase (McKnight *et al.* 1991), there is only one report to date (Hohn and Ohlrogge 1991) concerned with the manipulation of the pathway to the lower isoprenoids. One of the bottle-necks to this research area is the lack of availability of appropriate *in-vitro* systems which accumulate mono- and sesqui-terpenoids and in which manipulation of the biosynthetic pathway at the genomic level may be carried out and the results studied in detail at the biochemical level.

The accumulation of lower isoprenoids by undifferentiated cultures

Early observations with cultures of anise (Becker 1970) indicated that undifferentiated tissues produce only very low amounts of essential oils, and similar findings have since been reported for a considerable number of essential oil producing plants (for reviews see Koch-Heitzmann and Schultze 1989; Charlwood and Charlwood 1991*a*). The gradual de-differentiation of plant tissue cultures from the original explant to, typically, the fifth sub-culture is generally accompanied by a partial or, in many cases, a complete loss of accumulation of the lower isoprenoids. Where a measurable level of product accumulation is retained during subsequent passages, it typically falls to within the range 0.0001 to 0.15 per cent of the fresh weight of the culture; the corresponding oil accumulation in a cropping plant would be *c.* 0.1–5 per cent of the fresh weight of the aerial parts.

There is no obvious correlation between oil accumulation in undifferentiated tissue and the genus, species, variant, or even cell line of a particular plant. Thus essential oil accumulation in different callus lines of *Pelargonium tomentosum* ranged from 0.001–0.67 per cent of that of the parent plant, whilst callus from other *Pelargonium* variants certainly accumulated measurable quantities of mono- and sesqui-terpenoids, and in one instance this reached a level of 5 per cent of that found in the parent plant (Charlwood and Charlwood, 1991*b*). For certain other genera, however, there appear to be no recorded instances of essential oil accumulation in undifferentiated cultures. Thus, Segura and Calvo (1991) describe the formation of callus and suspension cultures of *Lavandula angustifolia* by a number of research groups employing a wide range of experimental conditions, but in none of these cultures have monoterpenoids been detected. On the other hand plantlets regenerated from a number of these cell lines did accumulate lavender oil (albeit at a level of only 12 per cent of that of the mature plant) indicating that the suppression of monoterpenoid accumulation in unorganized tissue is reversible (Banthorpe *et al.* 1986). In their review Segura and Calvo (1991) state that there 'appears to be no explanation for the absence of terpenoids in undifferentiated tissues of *Lavandula*, despite the presence of the enzymes responsible for monoterpenoid anabolism'. There are, however, several possible hypotheses that can now be put forward to rationalize these findings.

Isoprenoid accumulation as a function of development

Much evidence has been amassed to suggest that an element of differentiation is an absolute requirement in many culture systems before the full biosynthetic capacity of the cell can be elaborated (for a review see Charlwood *et al.* 1986). More recent examples confirm this hypothesis. In *P. fragrans*, shoot organogenesis leads to a 200-fold increase in essential oil accumulation (with significant quantities of α- and β-pinene and farnesene) compared with the unorganized callus (Charlwood and Charlwood 1991*a*), whilst the accumulation of nootkatone (a sesquiterpenoid which exhibits the odour of grapefruit) is associated with specific morphological differentiation in the callus cells of *Citrus paradisi* (del Rio *et al.* 1991). A particularly interesting case is reported by Kennedy *et al.* (1991) in which a progressive increase in the complexity of the essential oil profile of tissue during embryo development in *Daucus carota* cultures was observed. The initial suspension culture accumulated no lower terpenoids, whereas developing embryos at the globular stage accumulated β-pinene, those at the heart-shaped stage contained also terpinolene, β-curcumene, and β-caryophyllene, and at the torpedo stage α-pinene, myrcene, and β-bisabolene could be detected as well. At the final stages of development, the plantlets contained the same range of lower isoprenoids as the whole plant. The essential oil components of carrot may play an important role in the interaction of the plant with *Psila rosae* (the carrot root fly).

In view of findings such as those indicated above, several studies have focused on the expression of essential oil accumulation in fully differentiated shoot organ cultures. Shoot cultures of *P. tomentosum*, *P. fragrans*, and *P. graveolens* accumulate about 50 per cent of the amount of oil found in the intact plant (Katagi *et al.* 1986; Charlwood and Moustou 1988), whilst this value rises to almost 75 per cent for proliferating shoot cultures of *Rosmarinus officinalis* (Jain *et al.* 1991). On the other hand, shoot cultures of *Lippia dulcis* accumulate some 2.9 per cent (based on tissue dry weight) of the sweet sesquiterpenoid hernandulcin after 28 days culture on hormone-free medium (Sauerwein *et al.* 1991*b*). Since hernandulcin is reported to be present only in very small amounts (0.004 per cent dry weight) in the oil of the aerial parts of the whole plant, the *in vitro* production would appear to be remarkably high.

An explanation of the non-accumulation of essential oil in undifferentiated systems

The likely reasons why mono- and sesqui-terpenoids may not be accumulated in undifferentiated cultures have been listed in detail in a previous review (Charlwood *et al.* 1990). One obvious possibility is that undifferentiated cultures do not express genes for the enzymes responsible for the essential steps in the pathway. Banthorpe *et al.* (1986), however, demonstrated that callus cultures of *Rosa damascena* and *Lavandular angustifolia* that did not accumulate monoterpenoids, possessed high activities of enzymes able to convert mevalonic acid (MVA) and isopentenyl pyrophosphate (IPP) into geraniol and nerol. In particular, cell-free extracts from callus of the former achieved a 33.6 per cent incorporation of radioactivity from labelled IPP into mono- and sesqui-terpenoids, and this compares with an incorporation of only 0.1 per cent obtainable using a cell-free extract from the intact plant

More recently Croteau and his co-workers (Falk *et al.* 1990) have determined the catalytic activities of the enzymes associated with the synthesis of (+)-camphor in a cell suspension culture of *Salvia officinalis*. Although the leaves of this species accumulate more than 1.5 per cent of (on a dry weight basis) of essential oil, no camphor could be detected in the suspension culture, and biosynthetic activity (as determined by the trapping of tracer from [U-^{14}C]-sucrose in the monoterpenoid pool) appeared to be extremely low. Camphor is synthesized from geranyl pyrophosphate (GPP) via bornyl pyrophosphate (BPP) according to the route shown in Fig. 16.1, the key enzymes involved being BPP cyclase, BPP phosphohydrolase, and borneol dehydrogenase. The activities of enzymes have been measured throughout the growth cycle of the culture as shown in Fig. 16.2. BPP phosphohydrolase and borneol dehydrogenase activities were present throughout the culture cycle and reached levels which were between 100 and 50 per cent (respectively) of those extractable from the intact leaves. Generally, cyclase activity (BPP cyclase, 1,8-cineole cyclase, and monoterpene olefin synthases) was present only at the end of the log phase of growth (day 13) for a period of just 48 h, and the maximum level of BPP cyclase was just 5 per cent of that present in the intact leaves. It would appear, therefore, that it is the monoterpene cyclase which is limiting camphor production in the *in vitro* system. However, it may be calculated that each culture still possessed sufficient cyclase to produce 3.3 μg of camphor, a level well above the detection limit (3 ng per culture), even though monoterpenoid accumulation could not be demonstrated in this culture.

bornyl pyrophosphate cyclase
bornyl pyrophosphate phosphohydrolase
borneol dehydrogenase
CH_2OPP
OPP
OH
O
bornyl pyrophosphate
borneol
camphor
geranyl pyrophosphate

Fig. 16.1. Biosynthesis of camphor in sage (after Falk *et al.* 1990).

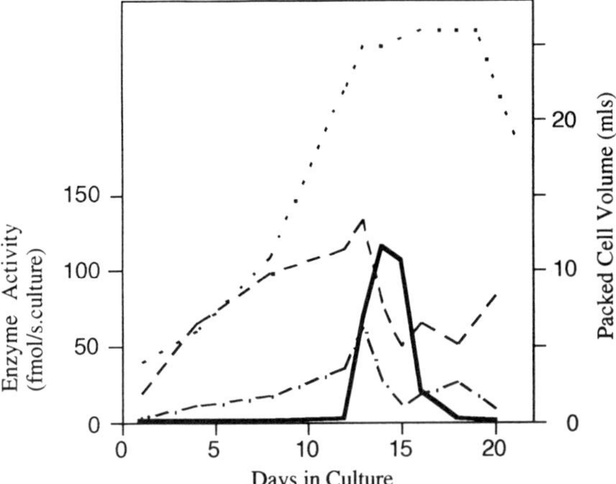

Fig. 16.2. Characteristics of cell suspension culture of sage. ..., Growth of culture (packed cell volume); —, bornyl pyrophosphate cyclase activity; ––, bornyl pyrophosphate phosphohydrolase activity × 10^{-3}; — · —, borneol dehydrogenase activity × 10^{-2}. (From Falk *et al.* 1990; redrawn with permission.)

The actual sites of synthesis of the lower isoprenoids in the whole plant still remain somewhat unclear. In *Mentha spicata*, all of the enzymes responsible for carvone formation (namely limonene cyclase, limonene hydroxylase, and carveol dehydrogenase) were found to be localized in the glandular trichomes themselves (Gershenzon *et al.* 1989), and synthesis appeared to be located in the secretory cells of the trichomes rather than the stalk or basal cells. Whether the anabolic processes are associated with the smooth endoplasmic reticulum or with the leucoplasts (or both) in the secretory cells has still to be determined. It is evident, however, that morphological differentiation of trichome structures in culture is not an absolute requirement for the expression of the enzymes associated with lower isoprenoid biosynthesis.

If the observed non-accumulation of lower isoprenoids in undifferentiated culture is not occasioned by the lack of appropriate enzymes then it may be associated with the inability of the cultured cells to sequester the product. This process requires both a mechanism by which the secondary compound can be transported, and a segregated storage location in which the product may be compartmentalized. It is widely believed that monoterpenoids are transported for both anabolic and catabolic purposes via their water soluble glycoside derivatives, and much evidence exists to support this view (for a review see Stahl-Biskup 1987).

There are numerous reports to indicate that undifferentiated cultures generally retain their ability to glycosylate exogenously-supplied substrates, for example, Berger and Drawert (1988) have shown that cultures of *Mentha*, *Rosmarinus*, *Pyrus*, and *Vitis* are able to convert linalool, geraniol, menthol, farnesol, and even benzyl alcohol into their respective glycosides with conversion rates of up to 70 per cent. Monoterpenoid glycosides have been identified in a number of undifferentiated cultures (see for example Lang and Hörster 1977), but it is widely reported that they are not generally accumulated in this form. However, the potential for transportation of product via the glycoside certainly remains operative *in vitro*.

In intact plants, lower isoprenoids are stored in specialized secretory structures such as glandular hairs (as in *Pelargonium* and *Mentha*), resin ducts (as in *Thuja* and *Pinus*), or lipid vesicles (as in roots of *Fedia* or *Valeriana*). Such structures are clearly not available in undifferentiated cultures, and it is this factor which appears to militate against accumulation of product *in vitro*. In order for a cell culture to accumulate a product, it is necessary (i) that such an accumulation is not detrimental to the continued survival of the producing cell and (ii) that the rate of synthesis of the product is greater than the rate of degradation (Brown *et al.* 1987). The lower isoprenoids are certainly cytotoxic to plant cells causing drastic reductions in the number of mitochondria and Golgi bodies, inhibiting respiration and photosynthesis, and generally decreasing cell membrane permeability. Experiments with *Pelargonium* suspension cultures have shown that α-terpinene, limonene, geraniol, and carvone are especially toxic to cells causing typically a 90 per cent loss of viability when applied at the level of 100 mg/litre of medium during the lag phase of the culture cycle (Brown *et al.* 1987). During the stationary phase of the cycle, plant cells can withstand somewhat higher concentrations of monoterpenoids, but it may be calculated that a typical 50 ml culture containing 10 g wet weight of cells could not accumulate more than 5 mg of monoterpenoid before autotoxicity would occur. This corresponds to approximately 3–4 per cent of the oil content of the parent plant and is in the order of magnitude of accumulation that is maximally observed in unorganized cultures of *Pelargonium* and, indeed, in other species as well.

Such autotoxicity is not restricted to lower isoprenoids, however. Quinine is reported to be lethal to suspension cells of *Cinchona ledgeriana* at a concentration of 2 g/litre of medium whereas the unmethoxylated alkaloids (cinchonine and cinchonidine) are much less toxic (Walton *et al.* 1987). It is possible to select for high alkaloid-yielding cell lines of *C. ledgeriana* but these typically accumulate cinchonine and cinchonidine, with quinine being present in amounts equivalent to only 10 per cent of the unmethoxylated precursor. In the case of berberine, Sato *et al.* (1990) have demonstrated that those cell cultures which accumulate this cytotoxic alkaloid have the ability either to transport it safely to storage vacuoles or to excrete it into the medium from which it cannot be re-absorbed by the cells.

For cultures of sage, however, it would appear that even sub-lethal levels of camphor are not accumulated, and this would suggest the operation of a rapid and unregulated breakdown of the newly synthesized product. It has long been recognized that lower isoprenoids have a fairly rapid turnover *in planta*. Burbott and Loomis (1969) showed that the half life of monoterpenoids in *Mentha piperita* was in the range 5–30 h, whilst in *Rosa dilecta* radiolabel, which had been incorporated into monoterpenoids and their glycosides after 1 h of feeding, was almost completely lost within 4 h (Francis and O'Connell 1969). A series of epoxidase and epoxide hydratase activities have been characterized in cell-free extracts of *Tanacetum vulgare*, *Artemisia annua*, and *Santolina chamaecyparissus* which convert GPP and neryl pyrophosphate (NPP) into terpene epoxides and ring opened products (Fig. 16.3), and presumably to acetyl CoA (Banthorpe *et al.* 1977). These seasonally dependent

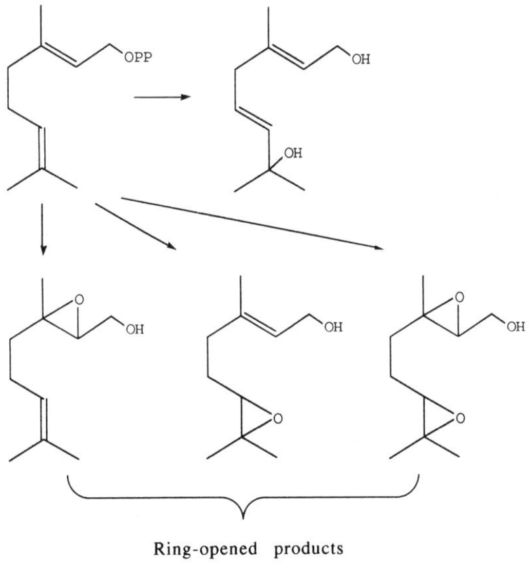

Fig. 16.3. The catabolism of geranyl pyrophosphate by cell-free extracts derived from plants of *Tanacetum*, *Artemisia*, and *Santolina* (after Bantrorpe *et al.* 1977).

'salvage' enzymes are able to catabolize up to 95 per cent of exogenous substrate into water-soluble products that cannot be readily extracted into organic solvents. In the plant such salvage enzymes are under strict spatial and temporal control but it is probable that, in undifferentiated culture systems, this control is lost. Indeed, there have been numerous reports concerning the biotransformation of exogenous monoterpenoids into (mainly) unidentified water-soluble products by undifferentiated plant cells of, for example *Vitis vinifera* (Cormier and Ambid 1987) and *Lavandula angustifolia* (Lappin *et al.* 1987). More recent and detailed studies of the biotransformation of geraniol by cell suspension cultures of *Citrus limon* (Berger *et al.* 1990) showed the rapid formation (within 30 min) of nerol, citronellol, and geranic acid and, within 3 to 4 h, of citronellic acid. During this time cleavage of the geraniol skeleton also occurred and both branched and regular chain fatty acids were formed (Fig. 16.4) resulting ultimately in an enlarged acetyl CoA pool.

Suspension cultures of sage were able efficiently to catabolize exogenous camphor at a rate that was estimated to be at least 1000-fold greater than the rate of synthesis of the monoterpenoid calculated on the

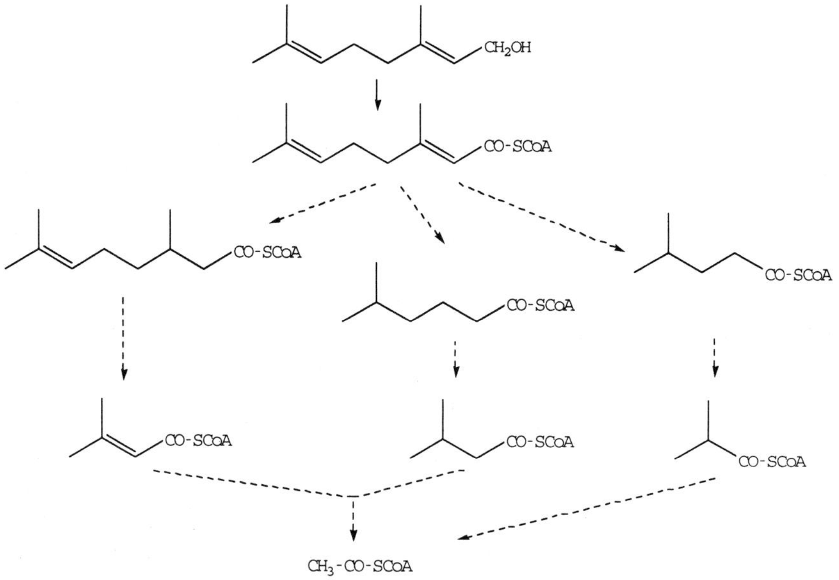

CH_2OH

CO-SCoA

CO-SCoA

CO-SCoA

CO-SCoA

CO-SCoA

CO-SCoA

CO-SCoA

CH_3-CO-SCoA

Fig. 16.4. Proposed scheme of formation of products identified following the catabolism of geraniol by suspension culture of *Citrus limon* (after Berger *et al.* 1990).

basis of BPP cyclase activity (Falk *et al.* 1990). Interestingly, camphor catabolism in culture did not follow the route (via the glucoside–glucose ester of 1,2-campholide) that has been established to operate in the intact plant (Croteau *et al.* 1987), but rather it was suggested that the cells used camphor as an energy source.

It is concluded that there are two general limitations which may prevent the accumulation of lower isoprenoids in undifferentiated cultures:

1. Where products are toxic to the producing cells, accumulation will be limited to a sub-lethal level: such a level would be dependent upon the environmental conditions under which the culture is maintained and the physiological state of the cells during isoprenoid synthesis.
2. Where the undifferentiated culture produces catabolic enzymes the activities of which are greater than those of the synthetic enzymes, then no accumulation will be observed irrespective of the cytotoxicity of the product. It may very well be that such catabolic enzymes are indeed induced as part of the strategy of the cell to deal with potentially toxic materials.

The accumulation of lower isoprenoids in transformed organ cultures

Since most monoterpenoids appear to be more or less cytotoxic, whilst a number of commercially important essential oil-producing plants also seem to yield cultures with high catabolic activities, some method must be employed *in vitro* to remove the product from the site of synthesis (to prevent autotoxicity) or even out of the culture environment entirely (to prevent degradation). A number of techniques have been employed by which product removal could be achieved, for example application of two-phase systems, use of polymeric adsorbants, operation of a continuous flow column, etc., but these are largely empirical solutions and often fail to produce the expected advantage. An ideal solution would exploit the natural mechanisms employed by the intact plant for product storage, whilst at the same time retaining the fermentation characteristics and the synthetic potential of an *in-vitro* system. To some extent the use of *Agrobacterium*-transformed organ cultures offers an approach towards this ideal, whilst also permitting the possibility of manipulating the biosynthetic pathways in the culture by gene transfer techniques.

In terms of secondary compound production, 'hairy root' cultures obtained via transformation using *A. rhizogenes* have been the most successful so far. Such cultures are genetically and biochemically very stable, often possess the same biosynthetic capacity as the root of the intact plant, and typically exhibit very high growth rates (as high as 10 times the growth rate of their untransformed counterparts). Furthermore, 'hairy roots' can be cultured in the absence of plant growth regulators, the presence of which is often detrimental to secondary compound accumulation *in vitro*. A detailed review of the characteristics of transformed root cultures as applied to secondary compound production is available (Rhodes *et al.* 1990). Unfortunately, of course, few lower isoprenoids are actually accumulated in root tissues, and this places an important limitation on the technique. However, neryl isovalerate, neryl butyrate, and other, as yet not fully identified, monoterpenoids accumulated in 'hairy root' cultures of *Artemisia absinthium* (Kennedy *et al.* 1993), whilst a range of lower isoprenoids (including camphene, (−)-limonene, terpinolene, β-caryophyllene, and hernandulcin) were accumulated by 'hairy root' cultures of *Lippia dulcis* which had been cultured in a 16 h light: 8 h dark photoperiod (Sauerwein *et al.* 1991*a*). Furthermore, transformed root cultures retain the capacity of their untransformed counterparts to synthesize sesquiterpenoid phytoalexins when the appropriate elicitor is provided (Furze *et al.* 1991 and references therein).

A useful model system in which to investigate the possibilities of genetic manipulation of the lower isoprenoid pathway has been established with valerian plants, in particular species of the genera *Fedia*, *Valeriana*, *Valerianella*, and *Centranthus* (Charlwood and Charlwood 1991*a*). Plants of the Valerianaceae readily produce 'hairy root' outgrowths upon infection of the sterile plant with *A. rhizogenes* (LBA 9402 harbouring the plasmid pRi 1855), and excised transformed roots may be grown in axenic culture on half-strength Gamborg's B5 (GB5) or Murashige and Skoog (MS) medium in the absence of plant growth regulators. A clonal line from *Centranthus macrosiphon* achieved a 12.5-fold increase (based on dry weight of root tissue) over a 15 day culture period with a minimum doubling time of 3.6 days (Fig. 16.5; Caetano 1993). This is equivalent to an increase of 1.0 to 1.5 g dry weight/litre of culture/day; the untransformed roots of this species grow at a rate of *c.* 0.4–0.5 g dry weight/litre/day. The roots of many valerian plants accumulate a class of non-glucoside, monoene- and diene-cyclopentanoid monoterpenoids (the valepotriates). These compounds are used as mild, non-addictive sedatives in parts of Europe and South America and, recently, have also been shown to have high cytotoxic activity which might be of value in the treatment of autoimmune diseases and some types of cancer. Transformed root cultures of valerian plants retain the capacity to synthesize valepotriates and accumulate a wide range of monoene and diene compounds (Fig. 16.6) similar to those of the parent plants. However, the concentration of valepotriates found in transformed root

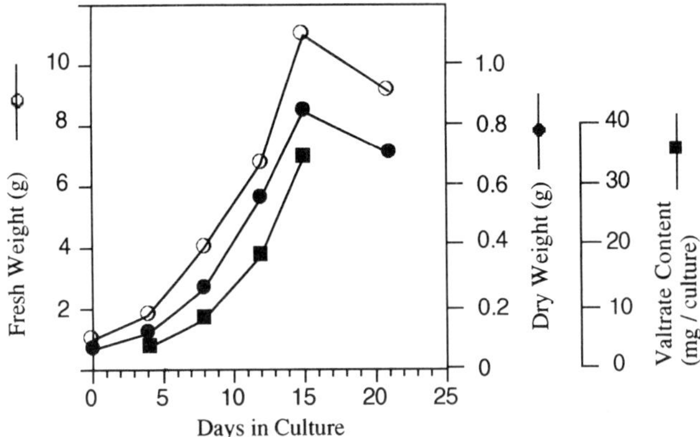

Fig. 16.5. Growth and valtrate accumulation in a 'hairy root' culture of *Centranthus macrosiphon* (Caetano 1993).

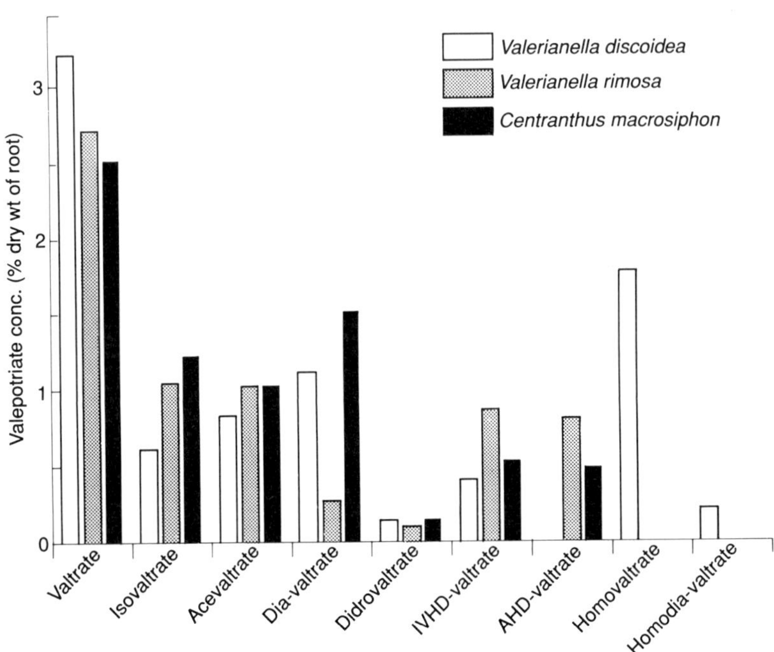

Fig. 16.6. Accumulation of valepotriates by transformed root cultures of some valerian plants (Caetano 1993).

cultures are typically very much higher than those associated with the roots of intact plants; for example a 'hairy root' clone of *Valerianella rimosa* accumulated the major diene valepotriate, valtrate, at a level of 4.6 per cent (based on dry weight) which is some seven times higher than the amount found in the roots of the whole plant (Caetano 1993).

The light regime under which a 'hairy root' clone is maintained can have a marked effect on the capacity of the culture to accumulate product: dark grown cultures of *Valerianella discoidea* accumulated 1.63 per cent (dry weight) of valtrate but this increased to 2.22 per cent upon continuous illumination, and to 2.74 per cent when the culture was subjected to a 16 h light: 8 h dark photoperiod (Caetano 1993). It is of interest to note that the intact roots of *V. discoidea* accumulate only 0.27 per cent of valtrate. In transformed and untransformed roots, the cyclopentanoid monoterpenoids are stored in lipid vesicles which are located throughout the cortex. Valepotriate accumulation *in vitro* appears to be correlated with the formation of the oil vesicles and these droplets could only be observed upon morphological differentiation of callus cultures (Violon *et al.* 1984). The numbers of lipid vesicles within transformed root cultures certainly proliferate as the accumulation of valepotriates increases. This phenomenon is readily observable when, for example, valepotriate accumulation is stimulated in transformed roots of *Fedia cornucopiae* by the addition of gibberellic acid (GA3) to the medium (Caetano 1993). In these experiment the density of lipid vesicles increased in proportion to the increase in valtrate (diene) and isovaleroxy-hydroxydidrovaltrate (monoene) produced, the maximum increases in accumulation (110 and 40 per cent, respectively, compared with control cultures) being achieved at a GA3 concentration of 10^{-3} μg/litre of medium. This finding suggests that the accumulation of valepotriates in transformed root cultures is not ultimately controlled by the availability of specialized sites in which to store product, as appears to be the case for many lower isoprenoids which are sequestered in glandular trichomes (Charlwood *et al.* 1989). The transformed root system described above clearly permits the possibility of manipulating the pathway to the iridoid monoterpenoids at the DNA level. In this context we have already succeeded in bringing about the stable integration of foreign DNA sequences (for example GUS, NPT II, and 3-hydroxy-3-methylglutaryl CoA reductase) into 'hairy roots' of *F. cornucopiae* using binary vector systems based on *A. rhizogenes*.

Unfortunately, there are no *Agrobacterium* species which can normally be expected to give rise to shoot cultures following plant infection, although such cultures would be of considerable value for the study of the accumulation of lower isoprenoids. Some nopaline strains of the tumour-inducing *A. tumefaciens* (namely T37, C58, and N273) produce, infrequently, shooty teratomas on *Nicotiana tabacum*. Similarly 'shooty mutant' strains of *A. tumefaciens*, which bear mutations at the 'shooty locus' (*tms1* and *tms2* genes), often produce shooty teratomas upon infection of tobacco plants. Alternatively, it is possible to produce shooty teratomas of tobacco following infection with a disarmed strain of *A. tumefaciens* LBA 4404 harbouring a pBIN19 plasmid containing an *ipt* gene under the control of its own promoter (cf. Rhodes *et al.* 1992). Although these techniques appear to produce shooty phenotypes with *N. tabacum*, Rhodes and co-workers have recently shown that similar results are not necessarily obtained following infection of fennel, sage, parsley, angelica, mint, and even *N. rustica* with the above mentioned bacterial strains (Rhodes *et al.* 1992). Indeed for *Mentha piperita citrata* (Bergamot mint) and *M. piperita vulgaris* (Scotch Black Mint), shooty teratomas formed only upon infection with the nopaline strains of *A. tumefaciens*, or with LBA4404 harbouring a plasmid containing the *ipt* gene behind a powerful promoter (i.e. CaMV35S or CaMV35S with duplicated up-stream enhancer sequences). Even in these cases the shooty teratomas did not form immediately but only upon excision of the initially formed gall from the original explant, decontamination from *Agrobacterium*, and subculture on to a medium containing no plant growth regulators.

The T37-derived shooty culture of *M. piperita citrata* showed a minimum doubling time of 3.2 days when grown on GB5 medium and this is very similar to the growth rate of untransformed shoot proliferation cultures (Payne *et al.* 1991). The monoterpenoid content of the transformed shooty culture was qualitatively and quantitatively similar to that of the parent plant, the major components (>85 per cent of the oil) being linalool and linalyl acetate (Spencer *et al.* 1990). The T37-derived shooty culture of *M. piperita vulgaris* grew less well (minimum doubling time 5.7 days) and the oil differed from that of the intact plant particularly with respect to the amounts of menthone and menthofuran, although menthol was the major component in the oil from both sources. The transformed shooty cultures of *M. piperita vulgaris* obtained following infection with *A. tumefaciens* harbouring a plasmid containing the *ipt* gene grew significantly more slowly than T37-derived cultures (minimum doubling time 8.4 days) although the accumulation of monoterpenoids was higher (Rhodes *et al.* 1992).

It may be concluded that both 'hairy root' cultures and *Agrobacterium*-transformed shoot cultures provide opportunities to study the complexities of lower isoprenoid synthesis, storage, and turnover as well as to manipulate the biosynthetic pathway at the DNA level. Such potential may not be presently available by the alternative approach of attempting to introduce foreign genes directly into an intact plant.

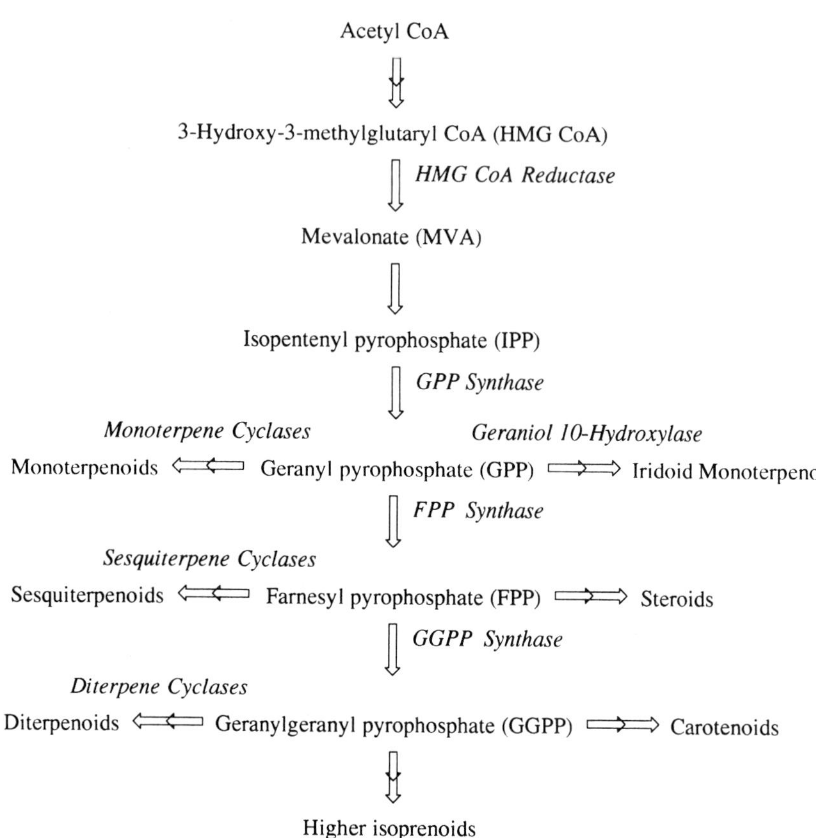

Fig. 16.7. The isoprenoid pathway.

Regulation of the biosynthetic pathway to the lower isoprenoids

The main biosynthetic route and the key branch points of the isoprenoid pathway are shown in Fig. 16.7. 3-Hydroxy-3-methylglutaryl CoA reductase (HMGR) catalyses the formation of mevalonate (MVA), the first committed step in the isoprenoid pathway, and controls the flux of carbon to steroids in mammals and yeast. In higher plants, factors which alter the accumulation of lower isoprenoids (for example light regime, plant growth regulators, fungicides, etc.) also have an effect on HMGR activity. However, the specific regulatory role of this enzyme is difficult to determine as HMGR may be located in as many as three compartments of the plant cell namely mitochondria, plastids, and cytosol. Furthermore, it is probable that the HMGR in each compartment is responsible only for the regulation of synthesis of the specific classes of isoprenoids associated with that compartment. Much of the data presently available derives from studies on wound- or elicitor-induced phytoalexin accumulation in *Solanum tuberosum*. A large, but temporary, increase in HMGR activity in both microsomal and organelle fractions is observed when steroidal glycoalkaloid accumulation is induced in potato tubers by the action of wounding, whereas elicitation of sesquiterpenoid phytoalexin synthesis by treatment with arachidonic acid lead to a longer term increase in HMGR activity but only in the microsomal fraction (Stermer and Bostock 1987). Yang *et al.* (1991) have shown that increases in HMGR activity in potato tubers that have been wounded or pathogen-activated are correlated with increases in HMGR mRNA levels, indicating that the response is due to increased gene transcription. However, kinetic studies and the use of a gene-specific probe suggested that one isogene of HMGR is activated following sesquiterpenoid elicitation by a pathogen, and that this is different from the isogene (or isogenes) activated by wounding. A small HMGR family has been identified in potato, whilst two HMGR genes have been found in tomato, in rubber, and in *Arabidopsis* (Stermer *et al.* 1991; Yang *et al.* 1991 and references therein). In the latter case it has been suggested (Learned and Fink 1989) that differential localization of isoforms of HMGR may not necessarily imply multiple genes but could be brought about by post-translational modification.

The precise regulatory nature of HMGR, compared with those of the sesquiterpene cyclases and even squalene synthase, in determining the production of lower isoprenoids has been somewhat indistinct since several early studies indicated that the latter enzymes may be more strictly controlled in tobacco tissue than HMGR itself. However, most compelling evidence has recently emerged in support of a fundamental role for HMGR. Chappell *et al.* (1991) have shown that cellulase or cell wall extracts of *Phytophthora parasitica* elicited the accumulation of the sesquiterpenoid phytoalexin capsidiol in tobacco suspension cells and that this was correlated with an increase in HMGR activity together with activities of enzymes further along the pathway. However, treatment of tobacco cells with pectolyase produced no induction of HMGR and no increase in capsidiol accumulation, but did give rise to increased activities of those enzymes able to catalyse MVA into sesquiterpenoids. The authors conclude that HMGR is limiting for sesquiterpenoid phytoalexin synthesis and suggest that the constitutive form of the enzyme may not be able to substitute for a lack of activity of the induced form.

Little information is available about the regulatory roles, if any, of the other enzymes of the main pathway to the lower isoprenoids. Geranyl pyrophosphate synthase has been partially purified from *Lithospermum erythrorhizon* (Heide 1988) although the generality of the independent existence of this enzyme in essential oil-accumulating plants still awaits confirmation. Farnesyl pyrophosphate synthase has been purified to homogeneity from *Capsicum annuum* (Hugueney and Camara 1990) and appears to be located entirely in the extraplastidic compartment; this contrasts with the situation for geranylgeranyl pyrophosphate synthase which is localized in plastids.

The branching of the isoprenoid pathway towards the lower isoprenoids is mediated by mono- and sesqui-terpenoid cyclases or, for the iridoid monoterpenoids, geraniol 10-hydroxylase. In sage cultures, the monoterpene cyclase appears to be the rate limiting enzyme (Falk *et al.* 1990). This view is supported by the finding that monoterpenoid accumulation correlated well with monoterpene cyclase activity (and, incidentally, the anatomical complexity of the resin-secreting structures) in a range of conifers (Lewinsohn *et al.* 1991). Wound-induction of monoterpene cyclase activity is not a general phenomenon: it does not occur, for example, in *Pinus* species which possess an advanced system of resin ducts, but can be demonstrated in *Abies grandis* in which only a few isolated resin cells appear (Gijzen *et al.* 1991). Here, not only was there an increase in consitutive cyclase activity but new monoterpene cyclases formed giving rise to an accumulation of oleoresin that was enriched in α- and β-pinene and limonene compared to that formed by the non-wounded plant.

Exploration of the regulatory role of monoterpene cyclases has been hampered by the lack of appropriate probes. However, new techniques have recently been developed for isolating reasonable quantities of secretory cells in a highly purified form (Gershenzon *et al.* 1992), and these preparations are reported to provide excellent sources of both monoterpene cyclases and hydroxylases. Furthermore, γ-terpinene synthase from *Thymus vulgaris* has recently been purified to homogeneity (Alonso and Croteau 1991) and major advances in this area of research are to be expected in the near future.

Geraniol 10-hydroxylase (G 10-H) may play a regulatory role in the biosynthesis of iridoid monoterpenoids and the enzyme has recently been purified from *Catharanthus roseus* cultures (Pennings *et al.* 1991). It is quite possible that G 10-H is also involved in the regulation of iridoid alkaloid biosynthesis *in vitro* (Schiel *et al.* 1987) although further information is awaited concerning this possibility.

That sesquiterpene cyclases have a role in the regulation of carbon flux along the C15 pathway is demonstrated by the finding that the activity of these enzymes is increased markedly upon phytoalexin-elicitation of tobacco suspension cells (Vögeli and Chappell 1988). 5-Epi-aristolochene cyclase activity was induced in the cultured cells 2 h after treatment with cellulase, reached a maximum within 15 h, and remained at that level for 40 h (Vögeli and Chappel 1990). It was clearly demonstrated that the induced cyclase was synthesized *de novo* and that such synthesis was correlated with the synthesis of mRNA coding for the enzyme. The authors concluded that sesquiterpenoid-phytoalexin production was regulated in tobacco cultures by transcriptional control of the cyclase gene. It is of interest to note that the increase in sesquiterpene cyclase activity following elicitation of tobacco cells is accompanied by a decrease in squalene synthase thus providing a further directional factor to ensure that carbon flux is diverted towards sequiterpenoids at the expense of sterols.

Future prospects for genetic manipulation of the isoprenoid pathway

Preliminary attempts aimed at altering the profile of lower isoprenoids produced by a plant cell line might reasonably involve inserting further copies of genes for those enzymes that are flux-limiting, and down-regulating the message for those enzymes responsible for non-target pathways. Currently, studies concerned with the insertion of genes associated with secondary compound biosynthesis have concentrated on tobacco. In other species one may expect that it will be possible to bring about the over-expression of constitutive genes or the *de novo* expression of foreign genes, although the product of the latter may be subjected to novel forms of protein processing by the host-plant (McKnight *et al.* 1991). Hohn and Ohlrogge (1991) have inserted the gene for a sesquiterpene cyclase (trichodiene synthase), isolated from the fungus *Fusarium sporotrichiodes*, into tobacco plants and obtained expression of the cyclase at levels (2 to 3 nmol/mg protein/h) which are very similar to those of the sesquiterpene cyclases induced in elicited tobacco cultures. Furthermore, the transgenic plants accumulated trichodiene although only at a rather low concentration (5 to 10 ng/g fresh weight of tissue). A very plausible explanation for this result is, of course, that the HMGR level in the unelicited tissue was insufficient to supply the enhanced requirement for MVA, a point which might be addressed by conducting elicitation experiments on the transgenic plants (Hohn and Ohlrogge 1991).

It is clear that simply up-regulating one step in a pathway cannot be expected to increase carbon flux through to the desired product, as such a simplistic approach ignores the existence of complex regulatory mechanisms designed specifically to protect the plant cell from over-production of one metabolite at the expense of others. Furthermore, there will always be the danger that up-regulating one key enzyme will just move the bottle-neck to the next regulatory stage in the pathway. However, there is no *a priori* reason why appropriate, multiple up-regulation together with the application of anti-sense techniques to avoid unwanted side products, should not provide the desirable result in favourable circumstances. If it is only required that low levels of a bioactive phytoalexin be continuously produced in a transgenic plant (perhaps in order to maintain an effect on the behaviour of an insect pest or pollinator for example) then sequestration of the foreign product may not pose a problem. If, however, it is desired to introduce into a plant the ability to accumulate significant amounts of a new metabolite, then attention will have to be given to directing the newly introduced genes to the correct cell and compartment, and to ensuring that the product can be safely transported and stored. This implies that, for the *de novo* synthesis of monoterpenoids, trichome specific promoters will be required and these should be available in the near future (Wagner 1991). Currently, however, one of the most widely used promoters for expression of foreign genes in transgenic plants is the TR promoter for mannopine synthase (Saito *et al.* 1991), and in transgenic tobacco plants this promoter appears specific for expression in the phloem tissue of stem and roots. Furthermore, such expression may be significantly enhanced by wounding or treatment of the tissue with the plant growth regulators indoleacetic acid and 2,4-dichlorophenoxyacetic acid. Saito *et al.* (1991) suggest that this promoter could be used to introduce enzymes which might biotransform naturally formed secondary compounds on-the-fly (i.e. whilst they are being transported from their synthetic site, via the vascular system, to their storage site). The further suggestion that plant growth regulators might be used as potential elicitors of genes controlled by the TR promoter may be fraught with problems since such regulators show dramatic effects on secondary compound synthesis in their own right. It is perhaps worthwhile noting that in the phenylpropanoid pathway, 5′-upstream (N-terminus) sequences to a heterologous gene have proved sufficient to confer elicitor inducibility on the gene product, and this may provide an appropriate strategy for introducing novel phytoalexins into a plant.

As an alternative to the direct introduction of foreign genes, the rather more empirical technique of selecting for over-producing mutant cell lines has recently been applied to the isoprenoid pathway. Björk and Nasiri (1991) have obtained cell lines of tobacco which are resistant to high levels of HMGR inhibitors (citrinin, compactin, lovastatin, and synvinolin) by virtue of maintaining enhanced pools of MVA. One citrinin-resistant cell line contained 0.614 mg MVA/g dry weight, whilst the level of MVA in the control line was below the limit of detection. Whether an enhanced MVA pool is conserved in plants regenerated from the resistant lines, and what effect this might have on isoprenoid accumulation and phytoalexin production (particularly following elicitation with pectolyase) remains to be reported.

Clearly many of the tools required for the insertion of new enzymes into essential oil-producing plants are already available or may be expected to be so in a relatively short time. Our lack of detailed knowledge of the regulation of isoprenoid synthesis and accumulation still presents a real problem, but innovative use of gene transfer techniques together with follow-up biochemical studies in suitable *in-vitro* systems should hasten the elucidation of these mechanisms. The era of the 'designer plant' is fast approaching.

References

Alonso, W. R. and Croteau, R. (1991). Purification and characterization of the monoterpene cyclase γ-terpinene synthase from *Thymus vulgaris*. *Arch. Biochem. Biophys.*, **286**, 511–17.

Banthorpe, D. V., Bucknall, G. A., Gutowski, J. A., and Rowan, M. G. (1977). Epoxidation and hydration of prenyl pyrophosphates by plant extracts. *Phytochemistry*, **16**, 355–8.

Banthorpe, D. V., Branch, S. A., Njar, V. C. O., Osborne, M. G., and Watson, D. G. (1986). Ability of plant callus cultures to synthesize and accumulate lower terpenoids. *Phytochemistry*, **25**, 629–36.

Becker, H. (1970). Volatile oils in *Pimpinella anisum*. *Biochem. Biophys. Pflanz.*, **161**, 425–41.

Berger, R. G. and Drawert, F. (1988). Glycosylation of terpenols and aromatic alcohols by cell suspension cultures of peppermint (*Mentha piperita* L.). *Z. Naturforsch.*, **43c**, 485–490.

Berger, R. G., Akkan, Z., and Drawert, F. (1990). Catabolism of geraniol by cell suspension cultures of *Citrus limon*. *Biochim. Biophys. Acta*, **1055**, 234–9.

Björk, L. and Nasiri A. (1991). Selection of strains of *Nicotiana tabacum* overproducing 3-hydroxy-3-methyl-glutaryl-coenzyme A reductase (E.C. 1.1.1.34 mevalonate-$NADP^+$-oxidoreductase). I. The selection method. *Biochem. Biophys. Pflanz.*, **187**, 173–5.

Brown, J. T., Hegarty, P. K., and Charlwood, B. V. (1987). The toxicity of monoterpenes to plant cell cultures. *Plant Sci.*, **48**, 195–201.

Burbott, A. J. and Loomis, W. D. (1969). Evidence for metabolic turnover of monoterpenes in peppermint. *Plant Physiol.*, **41**, 173–9.

Caetano, L. C. (1993). The accumulation of secondary metabolites in *Agrobacterium rhizogenes* transformed root cultures of valerian plants. PhD Thesis, University of London.

Chappell, J., Von Lanken, C., and Vögeli, U. (1991). Elicitor-inducible 3-hydroxy-3-methylglutaryl Coenzyme A reductase activity is required for sesquiterpene accumulation in tobacco cell suspension cultures. *Plant Physiol.*, **97,** 693–8.

Charlwood, B. V. and Charlwood, K. A. (1991*a*). Terpenoid production in plant cell cultures. In *Ecological chemistry and biochemistry of plant terpenoids* (ed. J. B. Harborne and F. A. Tomas-Barberan), pp. 95–132. Clarendon Press, Oxford.

Charlwood, B. V. and Charlwood, K. A. (1991*b*). *Pelargonium* spp. (Geranium): *In vitro* culture and the production of aromatic compound. In *Biotechnology in agriculture and forestry* (ed. Y. P. S. Bajaj), Vol. 15, pp. 339–52. Springer-Verlag, Berlin.

Charlwood, B. V. and Moustou, C. (1988). Essential oil accumulation in shoot-proliferation cultures of *Pelargonium* species. In *Manipulating secondary metabolism in culture* (ed. R. J. Robins and M. J. C. Rhodes), pp. 187–94. Cambridge University Press.

Charlwood, B. V., Hegarty, P. K., and Charlwood, K. A. (1986). The synthesis and biotransformation of monoterpenes in plant cells in culture. In *Secondary metabolism in plant cell cultures* (ed. P. Morris, A. H. Scragg, A. Stafford, and M. W. Fowler), pp. 15–34. Cambridge University Press.

Charlwood, B. V., Moustou, C., Brown, J. T., Hegarty, P. K., and Charlwood, K. A. (1989). The regulation of accumulation of lower isoprenoids in plant cell cultures. In *Primary and secondary metabolism of plant cell cultures II* (ed. W. G. W. Kurz), pp. 73–84. Springer-Verlag, Berlin.

Charlwood, B. V., Charlwood, K. A., and Molina-Torres, J. (1990). Accumulation of secondary compounds by organized plant cultures. In *Secondary products from plant tissue culture* (ed. B. V. Charlwood and M. J. C. Rhodes), pp. 167–200. Clarendon Press, Oxford.

Cook S. P. and Hain, F. P. (1988). Toxicity of host monoterpenes to *Dendroctonus frontalis* and *Ips calligraphus. J. Entomol. Sci.*, **23,** 287–92.

Cormier, F. and Ambid, C. (1987). Extractive bioconversion of geraniol by a *Vitis vinifera* cell suspension employing a two-phase system. *Plant Cell Rep.*, **6,** 427–30.

Croteau, R., El-Bialy, H., and Dehal, S. S. (1987). Metabolism of monoterpenes: Metabolic fate of (+)-camphor in sage (*Salvia officinalis*). *Plant Physiol.*, **84,** 649–53.

Deans, S. G. and Ritchie, G. (1987). Antibacterial properties of plant essential oils. *Int. J. Food Microbiol.*, **5,** 165–80.

Duke, S. O. (1987). Terpenoids from the genus *Artemisia* as potential pesticides. *Abs. Papers Am. Chem. Soc.*, **194,** AGRO 104.

Falk, K. L., Gershenzon, J., and Croteau, R. (1990). Metabolism of monoterpenes in cell cultures of common sage (*Salvia officinalis*). *Plant Physiol.*, **93,** 1559–67.

Fraga, B. M. (1991). Sesquiterpenoids. In *Methods in plant biochemistry* (ed. B. V. Charlwood and D. V. Banthorpe), Vol. 7, pp. 145–85. Academic Press, London.

Francis, M. J. O. and O'Connell, M. (1969). The incorporation of mevalonic acid into rose petal monoterpenes. *Phytochemistry*, **8**, 1705–8.

Furze, J. M., Rhodes, M. J. C., Parr, A. J., Robins, R. J., Whitehead, I. M., and Threlfall, D. R. (1991). Abiotic factors elicit sesquiterpenoid phytoalexin production but not alkaloid production in transformed root cultures of *Datura stramonium*. *Plant Cell Rep.*, **10**, 111–14.

Gershenzon, J., Maffei, M., and Croteau, R. (1989). Biochemical and histochemical localization of monoterpene biosynthesis in the glandular trichomes of spearmint (*Mentha spicata*). *Plant Physiol.*, **89**, 1351–7.

Gershenzon, J., McCaskill, D., Rajaonarivony, J. I. M., Mihaliak, C., Karp. F., and Croteau, R. (1992). Isolation of secretory cells from plant glandular trichomes and their use in biosynthetic studies of monoterpenes and other gland products. *Anal. Biochem.*, **200**, 130–8.

Gijzen, M., Lewinsohn, E., and Croteau, R. (1991). Characterization of the constitutive and wound-inducible monoterpene cyclases of grand fir (*Abies grandis*). *Arch. Biochem. Biophys.*, **289**, 267–73.

Hamill, J. D., Robins, R. J., Parr, A. J., Evans, D. M., Furze, J. M., and Rhodes, M. J. C. (1990). Over-expressing a yeast ornithine decarboxylase gene in transgenic roots of *Nicotiana rustica* can lead to enhanced nicotine accumulation. *Plant Mol. Biol.*, **15**, 27–38.

Heide, L. (1988). Geranylpyrophosphate synthase from cell cultures of *Lithospermum erythrorhizon*. *FEBS Lett.*, **237**, 159–62.

Hohn, T. and Ohlrogge, J. B. (1991). Expression of a fungal sesquiterpene cyclase gene in transgenic tobacco. *Plant Physiol.*, **97**, 460–2.

Hugueney, P. and Camara, B. (1990). Purification and characterization of farnesyl pyrophosphate synthase from *Capsicum annuum*. *FEBS Lett.*, **273**, 235–8.

Hwang, Y.-S., Wu, K.-H., Kumamoto, J., Axelrod, H., and Mulla, M. S. (1985). Isolation and identification of mosquito repellents in *Artemisia vulgaris*. *J. Chem. Ecol.*, **11**, 1297–306.

Jain, M., Banerji, R., Nigam, S. K., Scheffer, J. J. C., and Chaturvedi, H. C. (1991). *In vitro* production of essential oil from proliferating shoots of *Rosmarinus officinalis*. *Planta Med.*, **57**, 122–4.

Katagi, H., Takahashi. E., Nakao, K., and Inui, M. (1986). Shoot-forming cultures of *Pelargonium graveolens* by jar fermentation. *Nippon Nogeikagaku Kaishi*, **60**, 15–17.

Kennedy, A. H., Chamberlain, D., Wilson, G., and Ryan, M. F. (1991). Volatiles identified from five stages of embryo development separated from a heterogeneous suspension culture of *Daucus carota*. *Plant Cell. Rep.*, **10**, 435–8.

Kennedy, A. I., Deans, S. G., Svoboda, K. P., Gray, A. I., and Waterman, P. G. (1993). Volatile oils from normal and transformed roots of *Artemisia absinthium*. *Phytochemistry.*, **32**, 1449–51.

Koch-Heitzmann, I. and Schultze, W. (1989). Flüchtige Inhaltsstoffe in pflanzlichen Zellkulturen—eine Übersicht. *Biochem. Physiol. Pflanz.*, **184**, 3–30.

Lang, E. and Hörster, H. (1977). An Zucker Gebundene Reguläre Monoterpene. *Planta Med.*, **31**, 112–18.

Lappin, G. L., Stride, J. D., and Tampion, J. (1987). Biotransformation of monoterpenoids by suspension cultures of *Lavandula angustifolia*. *Phytochemistry*, **26**, 995–7.

Learned, R. M. and Fink G. R. (1989). 3-Hydroxy-3-methylglutaryl-coenzyme A reductase from *Arabidopsis thaliana* is structurally distinct from the yeast and animal enzymes. *Proc. Natl. Acad. Sci. USA*, **86**, 2779–83.

Lewinsohn, E., Gijzen, M., Savage, T. J., and Croteau, R. (1991). Defense mechanisms of conifers. *Plant Physiol.*, **96**, 38–43.

McKnight, T. D., Bergey, D. R., Burnett, R. J., and Nessler, C. I. (1991). Expression of enzymatically active and correctly targeted strictosidine synthase in transgenic tobacco plants. *Planta,* **185**, 148–52.

Payne, G. F., Bringi, V., Prince, C., and Shuler, M. L. (1991). *Plant cell and tissue culture in liquid systems*. Hanser, Munich.

Pennings, E. J. M., Meijer, A. H., Stevens, L. H., and Verpoorte, R. (1991). The applications of hydrophobic interaction chromatography to the purification of plant proteins. *Phytochem. Anal.*, **2**, 191–8.

Pickett, J. A. (1991). Lower terpenoids as natural insect control agents. In *Ecological chemistry and biochemistry of plant terpenoids* (ed. J. B. Harborne and F. A. Tomas-Barberan), pp. 297–313. Clarendon Press, Oxford.

Rhodes, M. J. C., Robins, R. J., Hamill, J. D., Parr, A. J., Hilton, M. G., and Walton, N. J. (1990). Properties of transformed root cultures. In *Secondary products from plant tissue culture* (ed. B. V. Charlwood and M. J. C. Rhodes), pp. 201–25. Clarendon Press, Oxford.

Rhodes, M. J. C., Spencer, A., Hamill, J. D., and Robins. R. J. (1992). Flavour improvement through plant cell culture. In *Bioformation of flavours* (ed. R. L. S. Patterson, B. V. Charlwood, C. MacLeod, and A. A. Williams), pp. 42–64. Royal Society of Chemistry, Cambridge.

del Rio, J. A., Ortuño, A, García Puig, D., Iborra, J. L., and Sabater, F. (1991). Accumulation of the sesquiterpenes nootkatone and valencene by callus cultures of *Citrus paradisi*, *Citrus limonia* and *Citrus aurantium*. *Plant Cell Rep.*, **10**, 410–13.

Saito, K., Yamazaki, M., Kaneko, H., Murakoshi, I., Fukuda, Y., and Van Montagu, M. (1991). Tissue-specific and stress-enhancing expression of the TR promoter for mannopine synthase in transgenic medicinal plants. *Planta*, **184**, 40–6.

Sato, H., Kobayashi, Y., Fukui, H., and Tabata, M. (1990). Specific differences in tolerance to exogenous berberine among plant cell cultures. *Plant Cell Rep.*, **9**, 133–6.

Sauerwein, M., Yamazaki, T., and Shimomura, K. (1991*a*). Hernandulcin in hairy root cultures of *Lippia dulcis*. *Plant Cell Rep.*, **9**, 579–81.

Sauerwein, M., Flores, H. E., Yamazaki, T., and Shimomura, K. (1991*b*). *Lippia dulcis* shoot cultures as a source of the sweet sesquiterpene hernandulcin. *Plant Cell Rep.*, **9**, 663–6.

Schiel, O., Witte, L., and Berlin, J. (1987). Geraniol-10-hydroxylase activity and its relation to monoterpene indole alkaloid accumulation in cell suspension cultures of *Catharanthus roseus*. *Z. Naturforsch*, **42c**, 1075–81.

Segura, J. and Calvo, M. C. (1991). *Lavandula* spp. (Lavender): *In vitro* culture, regeneration of plants, and the formation of essential oils and pigments. In *Biotechnology in agriculture and forestry* (ed. Y. P. S. Bajaj), Vol. 15, pp. 283–310. Springer-Verlag, Berlin.

Spencer, A., Hamill, J. D., and Rhodes, M. J. C. (1990). Production of terpenes by differentiated shoot cultures of *Mentha citrata* transformed with *Agrobacterium tumefaciens* T37. *Plant Cell Rep.*, **8**, 601–4.

Stahl-Biskup, E. (1987). Monoterpene glycosides: State-of-the-art *Flavour Fragrance J.*, **2**, 75–82.

Stermer, B. A. and Bostock, R. M. (1987). Involvement of 3-hydroxy-3-methylglutaryl Coenzyme A reductase in the regulation of sesquiterpenoid phytoalexin synthesis in potato. *Plant Physiol.*, **84**, 404–8.

Stermer, B. A., Edwards, L. A., Edington, B. V., and Dixon, R. A. (1991). Analysis of elicitor-inducible transcripts encoding 3-hydroxy-3-methylglutaryl coenzyme A reductase in potato. *Physiol. Mol. Plant Pathol.*, **39**, 135–45.

Violon, C., Dekegel, D., and Vercruysse, A. (1984). Relation between valepotriate content and differentiation level in various tissues from Valerianaceae. *J. Nat. Prod.*, **47**, 934–40.

Vögeli, U. and Chappell, J. (1988). Induction of sesquiterpene cyclase and suppression of squalene synthetase activities in plant cell cultures treated with fungal elicitor. *Plant Physiol.*, **88**, 1291–6.

Vögeli, U. and Chappell, J. (1990). Regulation of a sesquiterpene cyclase in cellulase-treated tobacco cell suspension cultures. *Plant Physiol.*, **94**, 1860–6.

Wagner, G. J. (1991). Secreting glandular trichomes: More than just hairs. *Plant Physiol.*, **96**, 675–9.

Walton, N. J., Parr, A. J., Robins, R. J., and Rhodes, M. J. C. (1987). Toxicity of quinoline alkaloids to cultured *Cinchona ledgeriana* cells. *Plant Cell Rep.*, **6**, 118–21.

Yang, Z., Park, H., Lacy, G. H., and Cramer, C. L. (1991). Differential activation of potato 3-hydroxy-3-methylglutaryl Coenzyme A reductase genes by wounding and pathogen challenge. *Plant Cell*, **3**, 397–405.

17. Production of foreign proteins and peptides in transgenic plants

ENNO KREBBERS,* DIRK BOSCH,*† and JOEL VANDEKERCKHOVE‡

Plant Genetic Systems, J. Plateaustraat 22, B-9000 Gent, Belgium; ‡University of Gent, Department of Physiological Chemistry, Ledeganckstraat 35, B-9000 Gent, Belgium

Introduction

Other chapters in this volume have focused on natural plant products and the manipulation of their synthesis either to alter plant characteristics or to produce these substances for other purposes. In this chapter we will focus on the possible use of plants as production systems for completely foreign (to the plant) products. Plants offer several potential advantages as biological production systems. Chief among these is the low cost of growing plants and the availability of technology for harvesting and initial processing at medium and large scales. Disadvantages include the speed of transformation systems; the period from the initiation of an experiment with a suitable explant to the harvest of seed from the primary transformant is usually measured in months.

The production of lipids, carbohydrates, or other non-protein products would require the transfer to the plant of sets of genes for complete biochemical pathways, and given the current state of plant genetic engineering techniques this is still a highly ambitious goal. Instead, efforts in this area have focused on the production of foreign proteins or peptides in plants, since this can generally be accomplished by the transfer of only one or two genes to the plant genome. Peptides can be expressed as parts of fusion products with natural plant proteins (see below). This review will summarize some of the principles involved and recent progress; a more extensive review including previous work has been published elsewhere (Krebbers *et al.* 1992*a*).

Principles

Successful expression of foreign proteins in plants requires that both biological and production criteria be met. Some of the most important are summarized as follows.

†Present address: CPRO-DLO, Postbus 16, NL-6700 AA Wageningen, The Netherlands.

Biological criteria

1. The gene of interest must be transcribed, meaning that it will usually be necessary to place the protein coding sequence behind a gene promoter which has been shown to work in plants.
2. The transcript must be suitably processed. As there may be some differences in plant and mammalian splicing mechanisms (Barta *et al.* 1986), this may entail removal of introns from genomic clones.
3. The transcript must be translatable; it is known that some (but by no means all) foreign genes are inefficiently translated in plants, due to codon usage or other factors. This may require changes in the coding sequence (Perlak *et al.* 1991).
4. The translation product must be directed to the appropriate intracellular location, or secreted. Appropriate targeting signals must thus be included. It may not always be known where a foreign protein can be stably accumulated without interfering with cellular metabolism, and a protein may not fold correctly, form disulphide bridges, or be stably stored if expressed in the wrong intracellular compartment.
5. As proteins ordinarily made as propeptides may not be correctly processed in a heterologous system, where necessary constructs expressing only the mature portions of the product must be used.
6. As alluded to above, expression of the protein must not reduce the fitness of the plant to the point where it grows so poorly as to no longer function as an efficient production tool.

Production criteria

1. The expression level of the introduced gene must be high enough to ensure an economically viable production system.
2. The product must not carry any undesirable modification. For example, there are some differences in the details of glycosylation in plants and animals (Johnson and Chrispeels 1987).
3. The part of the plant in which the protein accumulates must be easy to harvest and process at large scale.
4. While not absolutely necessary, the production economics of the system would be enhanced if the material could be stored for some time before processing. For example, seeds and some kinds of tubers can be stored, while leaves generally cannot be for any length of time.
5. The final product must be free of contaminating substances. These may be different depending on what tissues of the plant, or the plant species itself, the product is isolated from.

The biological criteria will in some cases overlap, but in some cases may conflict with production criteria. For example, a suitable promoter may be chosen which is expressed in plants and confers high expression levels on the introduced gene. However, cases may arise where the tissue in which the protein is accumulated to the highest level may not be one which can be stored long before processing. The choice of promoters (which confer tissue specificity, or the lack of it) and intracellular targeting signals (which direct proteins to different locations within the cell) will thus always be a compromise made to match as many as possible of the above criteria.

Practice

Thus far, these issues have not always been considered at length in practice. For a variety of reasons most workers have chosen to use 'constitutive' promoters. None of these experiments are yet at the point where actual commercial production is underway; studies thus far have concentrated on feasibility studies aimed at demonstrating the possibility of accumulating a particular protein in plant cells (usually leaves), and none of the resulting plants have been tested under field conditions. Even from these preliminary studies, however, it is clear that intracellular location is important, and that expression levels vary greatly depending on the details of the constructs used and other circumstances. Some of the lessons from older work are reviewed briefly again here, but the reader is referred to the original references and to Krebbers *et al.* (1992*a*) for more detail.

Hiatt *et al.* (1989) expressed the heavy and light chains of immunoglobulin separately in two transgenic plant lines using a constitutive promoter. In leaves they observed low levels (0.14 per cent) of the proteins present. When they crossed the two lines, progeny expressing both chains contained antibodies at up to 1.3 per cent of total protein. These results suggest that proteins are more likely to be stably accumulated if they can fold correctly, as would be expected when both the heavy and light chains are present. Intracellular targeting was also shown to be important in this study; expression without a signal peptide, which would result in the protein being retained in the cytoplasm, gave almost no accumulation of either chain. Düring *et al.* (1990) carried out similar experiments expressing both chains from the same construct using a different constitutive promoter and obtained significantly lower levels of expression for reasons not completely understood.

Human serum albumin (HSA) has also been expressed in transgenic plants (potato) using a constitutive promoter, and the protein was detected in leaf cells (Sijmons *et al.* 1990). Two major points were raised by these experiments. First, it was again shown that expression of a protein normally entering the secretory pathway without a signal peptide resulted in little or no accumulation of the product. Secondly, it is in principle possible to express a protein usually made as a propeptide without the extra fragment normally processed out in the homologous system. HSA could be expressed without its natural amino-terminal pro-sequence (but with a signal peptide). When the pro-sequence was left in, the protein was made but not processed. In both cases, expression levels reached only 0.02 per cent, however.

The same group (Pen *et al.* 1992) has also expressed alpha-amylase (from *Bacillus licheniformis*) in plants, in this case in tobacco using a constitutive promoter supplemented with an enhancer. Alpha-amylase is an industrial enzyme, which as the authors point out is an example of an industrial product used in bulk for which purity standards are not stringent. Expression levels of up to 0.3 per cent in leaves and 0.2 per cent in seeds were achieved. Two observations of particular interest to this chapter were made. The product is glycosylated, which it is not in the homologous system. Despite this, the enzyme is biologically active, and its activity appears to be completely normal and can liquify starch, which is the industrial application. This demonstrates that different modifications need not always be a problem. Furthermore, there appeared to be no effect on endogenous starch. The starch is localized within the cells, while the enzyme was secreted and found in the extracellular space, thus segregated from the potential substrate. This shows that production can be done in such a way as to avoid interference with other metabolic processes in the plant, and the authors raise the possibility of using this system to liquify starch upon processing, when crushing of the cells would bring enzyme and substrate, both produced in the same cell, together.

The example of alpha-amylase demonstrates that some products can be accumulated in different tissues. Work in the authors' laboratories (Bosch *et al.* submitted) has indicated that this need not always be the case. Trout growth hormone (tGH), of interest to the fish farming industry, was expressed using two different tissue specific promoters, one directing expression in the seeds, the other in the leaves. While RNA was detected in both tissues, protein could be detected only in the leaves of transformed tobacco plants, not in the seeds of the transformed *Arabidopsis* plants. In this case, when the protein is expressed in leaves without a signal peptide, a band can still be observed on western blots. However, the band is a smear instead of the distinct band observed when a signal peptide is included. This suggests that while tGH may be more resistant than immunoglobulin chains to degradation in the cytoplasm, it is still partially degraded. tGH was found to be partially but not completely glycosylated in tobacco leaves. This was of interest as the glycosylation status of the protein in trout itself is not completely understood. Both the glycosylated and unglycosylated forms of the protein appeared to form correct disulphide bridges, while the cytoplasmic tGH obtained when the signal peptide was omitted was insensitive to DTT treatment, suggesting that no disulphide bridges were formed. Unfortunately, the amounts accumulated when the signal peptide was included were not sufficient (about 0.1 per cent of total protein) to do *in vivo* bioactivity tests.

These experiments lead to the conclusion that a given protein may not be stably accumulated in all tissues. Seeds contain large amounts of storage proteins which are stably accumulated, but also proteases and protease inhibitors. Perhaps the tGH was expressed at a time, or targeted to a place, where it was not protected from proteases as the storage proteins are. Changing either or both of these parameters might solve the problem. Another cautionary note was provided by the *m*RNA levels. While the two promoters used are not equally active in the relevant tissues (Bosch *et al.* submitted), *m*RNA levels were similar, suggesting that properties of the coding region or termination signals may be important.

'Transient' expression of foreign proteins using viral systems falls outside the scope of this review; see De Zoeten *et al.* (1989) and Grill *et al.* (1989) for details.

In summary it may be said that expression of foreign proteins for production purposes is still in its infancy. The results summarized above make it clear that a large number of parameters, involving transcriptional, translational, and post-translational factors, are still not well enough understood to allow creation of a chimeric gene guaranteed to provide stable high level expression of any protein. Until this is so, it will be a matter of case by case testing of a variety of different chimeric constructs.

Production of peptides as fusion proteins

Given the current inability to design chimeric genes with the necessary characteristics at will, one solution is to leave part of the work to nature by making fusion proteins, something which seems particularly promising for peptides. As the list of peptides with interesting biological functions has increased (see references in Krebbers and Vandekerckhove 1990), interest has grown in alternative production systems. By making fusion proteins in which peptides are attached to or embedded in larger natural plant proteins, issues of where (in which tissue) and when to express the gene are essentially taken care of. Work in the authors' laboratories has focused on the seed storage proteins for this purpose. Many laboratories (reviewed in Krebbers *et al.* 1992*b*) have considered modifications of seed storage proteins not only for production purposes, but in order to alter seed amino acid composition. Discussion of the latter application falls outside the scope of this review, but the technical issues raised are essentially the same.

The essential problem is how to determine which parts of a protein can be modified without interfering with the biosynthesis or transport of that protein. These issues are discussed at length elsewhere (Krebbers *et al.* 1992*b*) but can be summarized as follows. Seed storage proteins are thought to serve as a source of nutrition for the seedling, and appear to have no enzymatic function (Higgins 1984). Despite this their biosynthesis is highly complex (Chrispeels 1991). They are synthesized on the rough endoplasmic reticulum (ER), where they fold, form disulphide bridges where relevant, and in some cases form higher order aggregates (Dickinson *et al.* 1990). With the exception of the prolamins, which assemble into protein bodies directly in the ER (Lending and Larkins 1989), they pass through the Golgi, where glycosylation side chains, if any, are modified. Then, on the basis of targeting information whose nature is understood only in a minority of the cases (Chrispeels 1991), seed proteins are segregated from proteins destined for secretion and instead are transported to the protein storage vacuoles (PSV, also referred to as protein bodies). Most of these steps are incompletely understood, but at least one example (Hoffman *et al.* 1988) has clearly shown that if inappropriate modifications are made, the protein is degraded.

In the absence of protein structural information for most storage proteins (phaseolin being the exception; Lawrence *et al.* 1990), comparison of protein sequences of analogous proteins from different species in order to define highly conserved regions, presumably important to the structure of the protein, has been one approach to identify regions of a protein which might tolerate modifications. Vandekerckhove *et al.* (1989) applied this technique to identify such a loop in the 2S albumins. In subsequent work (De Clercq *et al.* 1990*b*; Krebbers *et al.* 1991; and De Clercq, Van Der Klei, Krebbers, and Vandekerckhove, unpublished) a series of modified 2S albumins have been expressed in the seeds of transgenic plants. These contained either added essential amino acids or biological peptides, including leuenkephalin, a neuropeptide, and magainin, a 25 amino acid peptide with antibacterial activity isolated from the skin of *Xenopus leavis*.

2S albumins are, as the name implies, small water soluble proteins. In different species they represent from 20 to 60 per cent of the total seed protein (Youle and Huang 1981). *In vivo* they are found as two subunits of 9 and 3 kDa linked by disulphide bridges, but they are synthesized as complex precursors which undergo significant post-translational processing, including at least three proteolytic cleavage steps (Crouch *et al.*

1983). Early work (De Clercq *et al.* 1990*a*) had shown that these steps were correctly carried out in transgenic plants, as well as that the proteins were correctly targeted to the PSV. However, on the basis of the highly conserved positions of the eight cysteine residues in 2S albumins from several species, it was possible to identify a region between the sixth and seventh cysteine which was less highly conserved in length and which was not involved in a post-translational processing step. The successful modifications referred to above were all made in this region, using a gene from *Arabidopsis thaliana* as the basis for the modifications. In both *Arabidopsis*, which was used as a model plant for reasons of convenience (faster transformation and regeneration), and *Brassica napus* (oilseed rape), yields of isolated peptide reached 100–200 nmol/g of seed, which would correspond to 1 kg of a 25 amino acid peptide per hectare of *Brassica* plants. Preliminary data (A. Conceição and E. Krebbers) suggest that the use of the at2S2 promoter instead of the at2S1 promoter (Guerche *et al.* 1990) used in the initial experiments should give higher levels of expression of chimeric storage protein genes.

The peptides were cleaved from the storage protein using protease cleavage sites built into the chimeric protein. The 2S albumins were first purified on the basis of their size and solubility, then treated with the appropriate protease. Clearly the technique will become more broadly applicable as more specific proteases become available. The identification of restriction proteases such as factor Xa and certain viral proteases (e.g. Dougherty *et al.* 1989) serve as examples. Current work in the authors' laboratories is directed towards developing conditions for 2S albumin purification and cleavage which can be applied at larger scales on a routine basis. Some peptides require modifications such as carboxyl terminal amidation in order to be active, but here again current developments should allow progress (Beaudry *et al.* 1990). A major limitation, as with other storage proteins, remains the lack of a crystal structure, so that each new modification must be made with little idea as to its chances of success. The development of reliable *in vitro* or *in vivo* transient expression systems would be a major step forward. Unfortunately no such system has proved to be uniformly applicable for 2S albumins in particular or storage proteins in general (Krebbers *et al.* 1992*b*). Experiments are underway to explore the limits of the 2S albumin system, in terms of the size and nature of the peptides which can be inserted or substituted into the molecule.

Future prospects

A series of technical, economic, and regulatory questions remain for the use of plants as a production system for either proteins or peptides. The costs of purification will be a major factor, and this will in part depend on expression levels. As has been seen, the results obtained so far have varied widely. When the time comes to register products made using the new technology, standards will have to be set for the different contaminants found in plants relative to more traditional production systems. Of course, there are advantages here too; clearly animal viruses will not be a problem in plant systems. It is unclear whether a pharmaceutical product produced in plants but already produced by other means will have to go through a new set of clinical trials, or whether it will suffice to demonstrate that the product is chemically identical to that already approved. The regulatory issues related to growing such plants in the field instead of the greenhouse, which will be necessary to gain some of the economic benefits of scale the system promises, need to be resolved. Appropriate precautions to prevent the unwanted spread of introduced genes, such as those now used in field trials of other transgenic plants (National Research Council 1989), will be necessary. These concluding comments are not intended to sound pessimistic, but to point out that a great deal of work remains at various levels to build on the technical feasibility studies done thus far to establish a practical production system. There is no fundamental reason to believe that this will not be possible.

References

Barta, A., Sommergruber, K., Thompson, D., Hartmuth, K., Matzke, M. A., and Matzke, A. J. M. (1986). The expression of a nopaline synthase—human growth hormone chimeric gene in transformed tobacco and sunflower callus tissue. *Plant Mol. Biol.*, **6**, 347–57.

Beaudry, G. A., Mehta, N. M., Ray, M. L., and Bertelsen, A. H. (1990). Purification and characterization of functional recombinant alpha-amidating enzyme secreted from mammalian cells. *J. Biol. Chem.*, **265**, 17 694–9.

Bosch, D., Smal, J., and Krebbers, E. A growth hormone is expressed, correctly folded, and partially glycosylated in the leaves but not the seeds of transgenic plants. Submitted.

Chrispeels, M. J. (1991). Sorting of proteins in the secretory system. *Annu. Rev. Plant Physiol. Plant Mol. Biol.*, **42**, 21–53.

Crouch, M. L., Tenbarge, K. M., Simon, A. E., and Ferl, R. (1983). *c*DNA clones for *Brassica napus* seed proteins: evidence from nucleotide sequence analysis that both subunits of napin are cleaved from a precursor polypeptide. *J. Mol. Appl. Gen.*, **2**, 273–83.

De Clercq, A., Vandewiele, M., De Rycke, R., Van Damme, J., Van Montagu, M., Krebbers, E., and Vandekerckhove, J. (1990*a*). Expression, processing and targeting of an *Arabidopsis* 2S albumin in transgenic tobacco. *Plant Physiol.*, **92**, 899–907.

De Clercq, A., Vandewiele, M., Van Damme, J., Guerche, P., Van Montagu, M., Vandekerckhove, J., and Krebbers, E. (1990*b*). Stable accumulation of modified 2S albumin seed storage proteins with higher methionine contents in transgenic plants. *Plant Physiol.*, **94**, 970–9.

De Zoeten, G. A., Penswick, J. R., Horisberger, M. A., Ahl, P., Schultze, M., and Hohn, T. (1989). The expression, localization, and effect of a human interferon in plants. *Virology*, **172**, 213–22.

Dickinson, C. D., Scott, M. P., Hussein, E. H. A., Argos, P., and Nielsen, N. C. (1990). Effect of structural modifications on the assembly of a glycinin subunit. *Plant Cell*, **2**, 403–13.

Dougherty, W. G., Cary, S. M., and Parks, T. D. (1989). Molecular genetic analysis of a plant virus polyprotein cleavage site: a model. *Virology*, **171**, 356–64.

Düring, K., Hippe, S., Kreuzaler, F., and Schell, J. (1990). Synthesis and self-assembly of a functional monoclonal antibody in transgenic *Nicotiana tabacum*. *Plant Mol. Biol.*, **15**, 281–93.

Grill, L., Erwin, R. L., Berliner, D. L., and Hubbard, E. T. (1989). Non-nuclear chromosomal transformation. PCT patent application WO 8908145, Biosource Genetics Corporation.

Guerche, P., Tire, C., Grossi De Sa, F., De Clercq, A., Van Montagu, M., and Krebbers, E. (1990). Differential expression of the *Arabidopsis* 2S albumin genes and the effect of increasing gene family size. *Plant Cell*, **2**, 469–78.

Hiatt, A., Cafferkey, R., and Bowdish, K. (1989). Production of antibodies in transgenic plants. *Nature*, **342**, 76–8.

Higgins, T. J. V. (1984). Synthesis and regulation of major proteins in seeds. *Annu. Rev. Plant Physiol.*, **35**, 191–221.

Hoffman, L. M., Donaldson, D. D., and Herman, E. M. (1988). A modified protein is synthesized, processed, and degraded in the seeds of transgenic plants. *Plant Mol. Biol.*, **11**, 717–29.

Johnson, K. D. and Chrispeels, M. J. (1987). Substrate specificities of *N*-acetylglucosaminyl-, fucosyl-, and xylosyltransferases that modify glycoproteins in the Golgi apparatus of bean cotyledons. *Plant Physiol.*, **84**, 1301–8.

Krebbers, E. and Vandekerckhove, J. (1990). Production of peptides in plant seeds. *Trends Biotechnol.*, **8**, 1–3.

Krebbers, E., Rudelsheim, P., De Greef, W., and Vandekerckhove, J. (1991). Laboratory and field performance of transgenic *Brassica* plants expressing chimeric 2S albumin genes. In *Proceedings of GCIRC 1991*, Saskatoon, Sask., Canada, pp. 716–21. GCIRC, Paris.

Krebbers, E., Bosch, D., and Vandekerckhove, J. (1992*a*). Prospects and progress in the production of foreign proteins and peptides in plants. In *Plant protein engineering* (ed. P. R. Shewry and S. Gutteridge), chapter 17, pp. 315–26. Cambridge University Press.

Krebbers, E., Van Rompaey, J., and Vandekerckhove, J. (1992*b*). Expression of modified seed storage proteins in transgenic plants. In *Transgenic plants* (ed. A. Hiatt), chapter 3, pp. 37–60. Marcel Dekker, New York.

Lawrence, M. C., Suzuki, E., Varghese, J. N., Davis, P. C., Van Donkelaar,

A., Tulloch, P. A., and Colman, P. M. (1990). The three-dimensional structure of the seed storage protein phaseolin at 3 Å resolution. *EMBO J.*, **9**, 9–15.

Lending, C. R. and Larkins, B. A. (1989). Changes in the zein composition of protein bodies during maize endosperm development. *Plant Cell*, **1**, 1011–23.

National Research Council, (1989). *Field testing genetically modified organisms.* National Academy Press, Washington, DC.

Pen, J., Molendijk, L., Quax, W. J., Sijmons, P. C., van Ooyen, A. J. J., van den Elzen, *et al.* (1992). Production of active *Bacillus licheniformis* alpha-amylase in tobacco and its application in starch liquefaction. *Bio/technology*, **10**, 292–6.

Perlak, F. J., Fuchs, R. L., Dean, D. A., McPherson, S. L., and Fischhoff, D. A. (1991). Modification of the coding sequence enhances plant expression of insect control protein genes. *Proc. Natl. Acad. Sci. USA*, **88**, 3324–8.

Sijmons, P. C., Dekker, B. M. M., Schrammeijer, B., Verwoerd, T. C., Van den Elzen, P. J. M., and Hoekema, A. (1990). Production of correctly processed human serum albumin in transgenic plants. *Bio/technology*, **8**, 217–21.

Vandekerckhove, J., Van Damme, J., Van Lijsebettens, M., Botterman, J., De Block, M., Vandewiele, M., De Clercq, A., Leemans, J., Van Montagu, M., and Krebbers, E. (1989). Enkephalins produced in transgenic plants using modified 2S seed storage proteins. *Bio/technology*, **7**, 929–32.

Youle, R. and Huang, A. H. C. (1981). Occurrence of low molecular weight and high cysteine containing albumin storage protein in oilseeds of diverse species. *Am. J. Bot.*, **68**, 44–8.

18. Industrial production of fine chemicals —plants versus chemical synthesis

ALLE BRUGGINK

DSM Andeno, PO Box 81, 5900 AB Venlo, The Netherlands

Introduction

The emergence of new technologies does not immediately imply that existing industrial processes become old-fashioned and must be replaced. When research in biotechnology grew exponentially in the early 1980s predictions were made that several established petrochemical-based bulk chemical products would be substituted by bio-based equivalents. With the introduction of enzymes in the (industrial) synthesis of organic products many symposia were organized to discuss the sad future of organic synthesis going to be replaced by cheaper, milder, and more selective biosynthesis.

Even the debates on the production of chemicals from plants did not entirely escape these euphoric predictions. New biotechnological techniques would both improve the production and isolation of various products from plants. Special government sponsored programmes were developed to stimulate 'green chemistry' and to find new uses for new fine chemicals from plants.

By now we all know reality is much more harsh. Development of new products and processes always takes a lot more time than anticipated. Particularly in industry changes are very time-consuming before commercial success is reached. It is a well-known economic fact that both in petrochemicals and bulk chemicals products and processes have a lifetime of 25–50 years.

In the fine chemical and specialties industries the turnover time of products and processes is much shorter. Lifetimes of 5–10 years or less, are not exceptional in these industrial segments. This is of course no surprise when one takes into account the relatively limited amount of capital that is normally invested in the relatively small production units of fine chemicals and specialties. In terms of R&D, processes for fine chemicals and specialties are mostly far less optimalized in comparison with petrochemicals and bulk chemicals which have often seen a cumulation of process improvements for many years. Thus fine chemicals and specialties are much more prone to major process changes or replacement.

The fact that biotechnology has thus far had its major impact on the pharmaceutical industry is a clear example. How such effects can work out in the fine chemical industry will be explained.

Structure of chemical industry

Very roughly the chemical industry can be divided into petrochemicals, bulk chemicals, fine chemicals, and specialties as shown in Table 18.1. This classification is mainly based on organic chemicals, by far the majority in the chemical industry; both in volume and money terms.

A commercially more useful classification of the chemical industry is given by the quadrant of Kline (Fig. 18.1). Fine chemicals are thus described as small volume/high priced products with well defined and commercially agreed chemical and physical specifications. Very often fine chemicals are mixed up with specialties. With specialties, however, the customer is not interested in chemical or physical specifications as long as the product performs according to his expectations, a feature which can only be determined after the application. A medicine is only then a medicine when it satisfies the patient. With fine chemicals the specifications are agreed beforehand.

In order to understand the potential for product or process substitutions it is also important to notice the difference in cost–price structure between bulk products, fine chemicals, and specialties (Fig. 18.2). The simple fact that more than two-thirds of the cost price of petrochemicals and bulk chemicals is made up of cheap raw materials is already a great hurdle for a change to alternative processes or technologies. For specialties on the other hand raw materials, although sometimes expensive, play a minor role in the end price. Replacement by other new technologies is therefore relatively easy and can lead to, albeit small, decreases in cost price. Very often the small scale and the low investments

Table 18.1 The (organic) chemical industry

	Volume (ton/year)	Price (Dfl/kg)	Added value
Oil and gas	$> 10^9$	0.25	—
Petrochemicals	$> 10^6$	1–2	Very low
Bulk chemicals	$> 10^4$	2–10	Low
Fine chemicals	$> 10^1$	10–1000	High
Specialties	$> 10^0$	> 1000	Very high

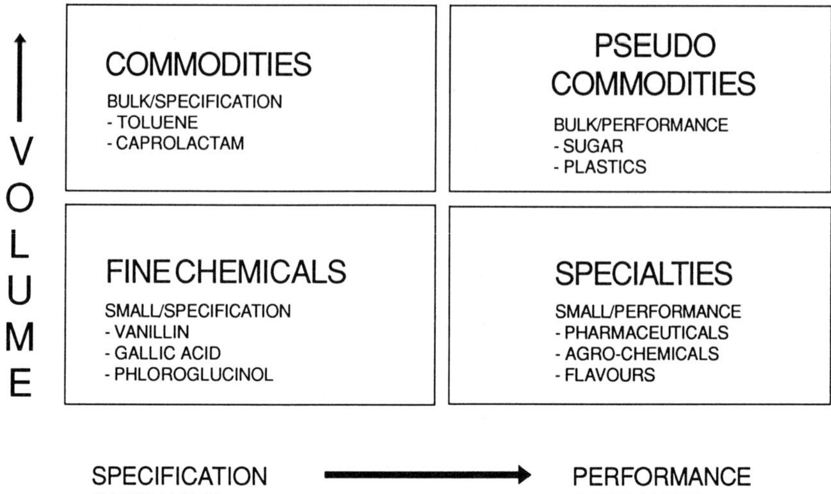

Fig. 18.1. Kline quadrant.

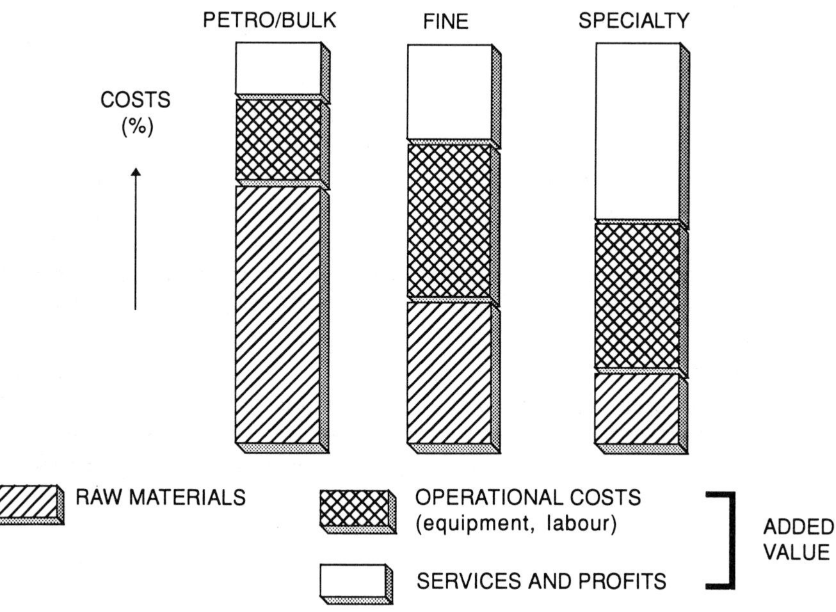

Fig. 18.2. Relative cost structures.

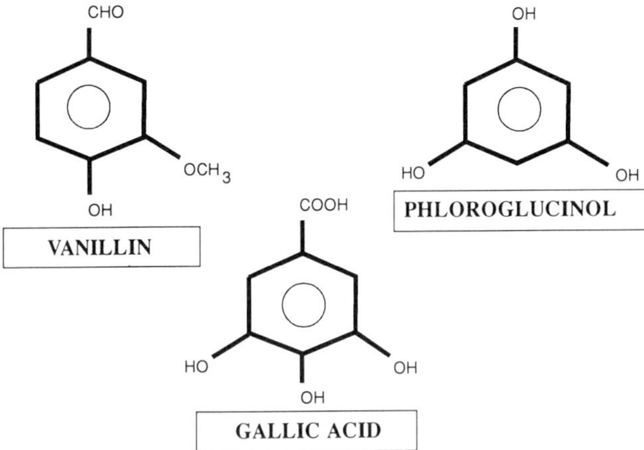

Fig. 18.3. Three simple non-chiral molecules.

in the chemical production processes enhance the chances for substitution (a big part of the operational costs go to labour, quality control, and production overheads).

Fine chemicals

How does this work out for the fine chemical products? From the previous statements about bulk products and specialties an in between position of the fine chemicals becomes clear. The following three examples will clarify that position. We have chosen three very representative examples with rather simple chemical structures (Fig. 18.3). This will focus the attention more on the issue of technology development rather than on the chemistry. It is of interest to note that these examples, important chemicals and in principle stemming from plants, do not necessarily have to be complicated molecules with many chiral centres. Let us now go into some more detail about sources, syntheses, and applications of these molecules and see how competition between chemical synthesis and isolation from natural sources has worked out.

Phloroglucinol

Phloroglucinol (Fig. 18.4) or 1,3,5-trihydroxybenzene, a very simple symmetrical molecule with a world production of around 100 t/year and a

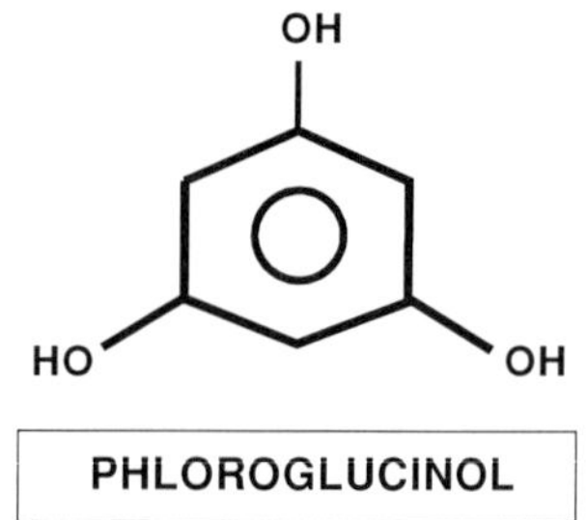

WORLD PRODUCTION	**ca. 100 t/yr**
PRICES	**$ 80 - 100/kg**
APPLICATIONS	**coupler in reprographics** **building block in pharmaceuticals**

Fig. 18.4. Phloroglucinol.

price level of $ 80–100/kg is most representative of fine chemicals. Main applications are as a coupler in diazotype reprographic systems and as a chemical building block in pharmaceuticals such as Buflomedil, an antihypertensive. Phloroglucinol has been on the market for several decades with Andeno as the major supplier.

Despite the fact that phloroglucinol derivatives are found in virtually every plant the industrial manufacture is still by total synthesis. Two routes are practised, both with substantial environmental drawbacks. Traditionally TNT was the preferred starting material, nowadays trihalobenzenes are the preferred starting point (Fig. 18.5). The drawback of these routes is evident from the big loss in molecular weight in going from these intermediate materials to the end product. Together with the usage of reagents and solvents these processes results in the production of 10–20 kg of chemical waste per kg of phloroglucinol. The challenge for new clean routes to phloroglucinol is therefore great.

For two additional reasons the need for clean and modern phloroglucinol manufacture is extremely pressing.

First, phloroglucinol is found as a building block in many plants i.e. in flavonoids. These are secondary plant metabolites which could perhaps be brought to excess production by plant systems. Moreover, the literature already gives both chemical and enzymatic possibilities for the conversion of flavonoids to phloroglucinol or derivatives. The author

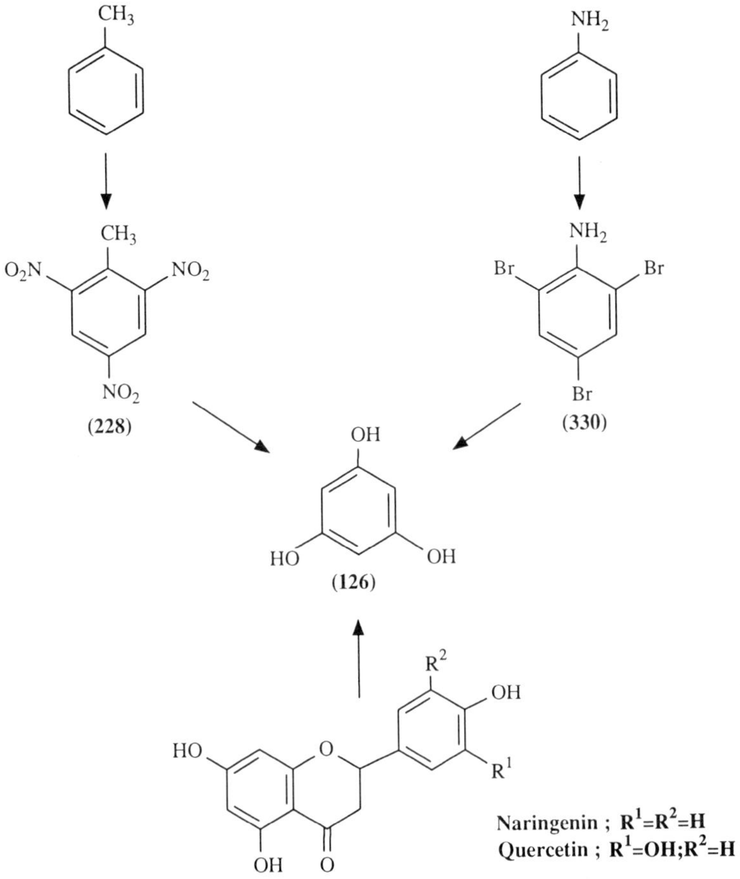

CH_3

CH_3

O_2N NO_2

NO_2

(228)

NH_2

NH_2

Br Br

Br

(330)

OH

HO OH

(126)

R^2

OH

HO O

R^1

OH O

Naringenin ; R^1=R^2=H
Quercetin ; R^1=OH;R^2=H

NARINGIN (M = 580)
QUERCIMERITRIN (M = 464)
and many other Flavones and Flavonoids

Fig. 18.5. Routes to phloroglucinol.

invites you, phytochemists, being experts, to solve these problems and to develop an industrially relevant process for phloroglucinol from plants (Fig. 18.6). I realize, though, that much work remains to be done. Since all flavonoids are high molecular weight compounds we might end up with an elegant isolation process for phloroglucinol but with an even larger amount of waste than in the present chemical processes. There should be more uses, for example in cattle feed, of this waste than with chemical waste.

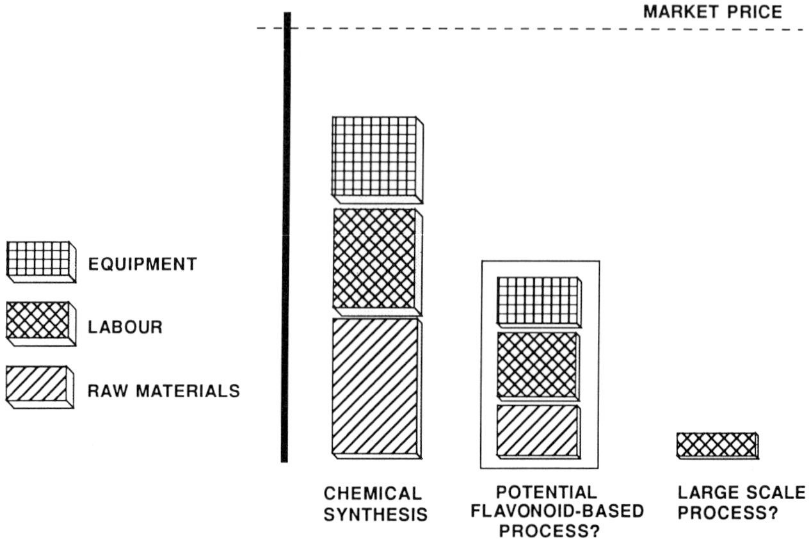

Fig. 18.6. Phloroglucinol: chemical synthesis versus potential plant sources.

Secondly the challenge for new processes should be great because of the large market potential. Phloroglucinol is a proven better alternative for phenol and resorcinol in chip board and other modern construction materials. Also phloroglucinol is a proven growth stimulator for fruit trees bringing about earlier and better fruit bearing. Both these applications require market prices of a few $/kg. So the challenge to develop a new and efficient synthesis is indeed great. In addition to phytochemists, the organic chemists are therefore invited to join in, to investigate, for example, trimerization of acetic acid or ketene directly to phloroglucinol (Fig. 18.7). Nature gives many examples of phloroglucinol skeletons derived from acetic acid. The University of Amsterdam (Prof. Vrieze) has already made some progress in the trimerization of ketene using Ru-catalysis. Trimerization of substituted acetylenes to benzenes is also well known.

In conclusion, phloroglucinol is a very simple molecule with great market potential but outdated manufacturing processes screaming for modern and clean alternatives.

Gallic acid

The second example is gallic acid (Fig. 18.8), 3,4,5-trihydroxybenzoic acid, a somewhat larger member of the fine chemicals family. Annual

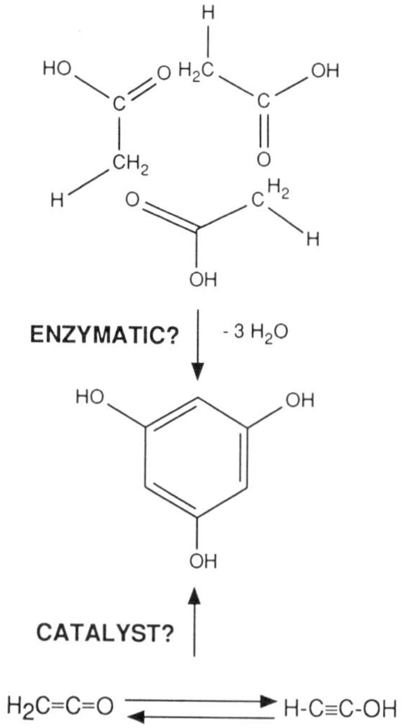

ENZYMATIC? $- 3\,H_2O$

CATALYST?

$H_2C{=}C{=}O \rightleftharpoons H\text{-}C{\equiv}C\text{-}OH$

Fig. 18.7. Challenging processes for phloroglucinol.

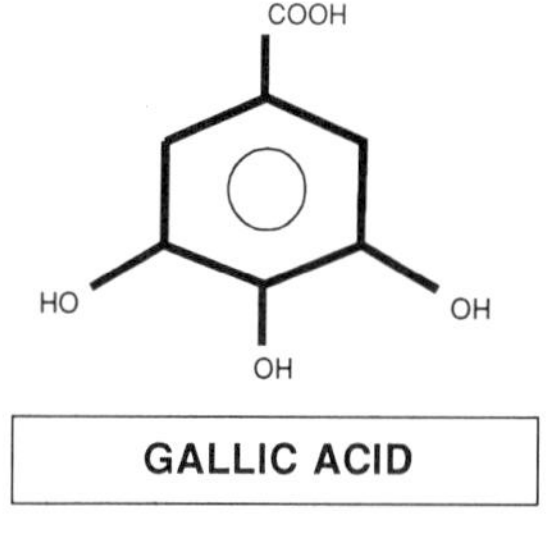

GALLIC ACID

WORLD PRODUCTION	**ca. 2000 t/yr**
PRICES	**ca. $ 10/kg**
APPLICATIONS	**pharmaceutical (as building block) antioxidants dyes, inks, thermographic paper**

Fig. 18.8. Gallic acid.

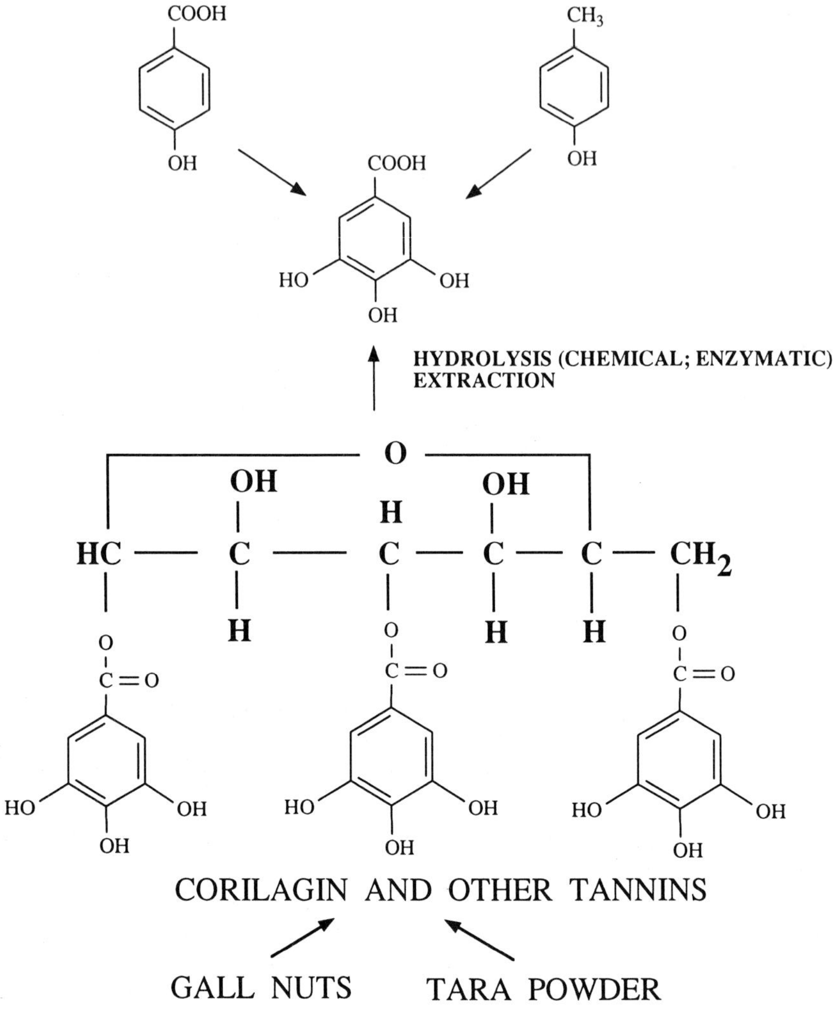

Fig. 18.9. Routes to gallic acid.

production amounts to *c.* 2000 tonnes at a price level of $ 8–10/kg. Like phloroglucinol the major applications are found as a building block for pharmaceuticals (Trimethoprim, to be discussed later) and in various graphical systems.

In contrast to phloroglucinol, gallic acid has always been isolated from plants (Figs 18.9 and 18.10). Chemical routes are feasible, for example starting from cheap chemicals such as *p*-cresol or *p*-hydroxybenzoic acid,

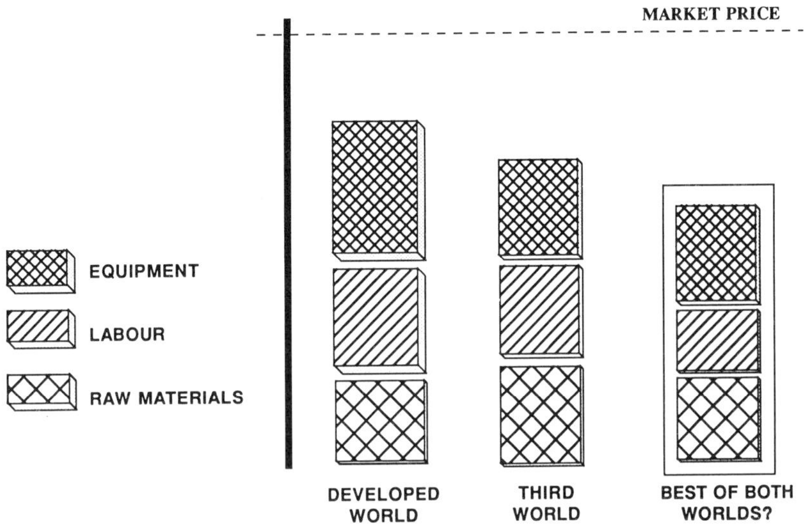

Fig. 18.10. Gallic acid: plant-based processes, developed world versus third world.

but have never reached industrial reality. Modern developments are changing the scene, although not dramatically. Traditionally gallic acid is isolated after chemical hydrolysis of tannins. Tara powder from Latin America is used as a source by European and Indian producers. Gall nuts, are the starting material in China. Chinese and Indian producers are at a slight advantage because of low wages and the easily accessible technology. The best of both worlds has probably been found by a Japanese company called Kikkoman. An enzyme, tannase, derived from *Aspergillus oryzae*, is used to replace the chemical hydrolysis in combination with more simple work-up and extractions. Full chemical synthesis has never made a breakthrough and is unlikely to do so in the future.

Vanillin

At a price of \$ 15–20/kg and with an annual production of close to 10 000 t and applications as a pharmaceutical building block as well as an ingredient for food and flavours, vanillin is on the borderline of bulk and fine chemicals with performance products (Fig. 18.11). It can best be situated in the middle of the Kline quadrant. The main application

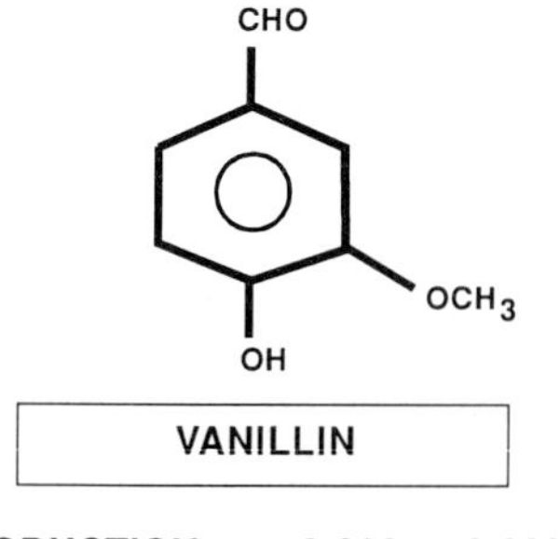

VANILLIN	
WORLD PRODUCTION	8 000 - 10 000 t/yr
PRICES	\$ 15 - \$ 20/kg
APPLICATIONS	food and flavours (as such) pharmaceuticals (as building block)

Fig. 18.11. Vanillin.

in pharmaceuticals is as a starting material for important drugs such as Methyldopa and Trimethoprim.

The industrial synthesis of vanillin is a good example of changing technologies. Traditionally vanillin was produced by a laborious process of chemical oxidation, hydrolysis, and purifications starting from lignin-sulphite mother liquors of paper mills. Isolation from vanilla beans is a negligibly small source but a perfect example of the difference between performance products and specification products. Whereas vanillin from lignin or chemical sources has to be shining white, FDA-approved, and meet strict specifications, the vanillin from the vanilla bean is an ill-defined pleasant smelling mostly dark material highly appreciated in certain flavour compositions. The performance counts and this is reflected in a huge price difference. High grade crystalline vanillin costs only \$ 15–20/kg whereas the bean extracted material reaches prices of up to \$ 2 000/kg.

With growing demand and stagnating development in the lignin based process the way was clear for a chemical synthesis in the 1970s. (Fig. 18.12). Thus Rhône Poulenc developed and commercialized a competitive chemical process based on phenol derived guaiacol on one hand and ethylene-based glyoxylic acid on the other. Compared to the lignin process this process has advantages in labour, transport, investment, and environment (Fig. 18.13). High investments for equipment in the lignin process are the main factor. In raw materials the lignin process still has an advantage which might prove of importance in the future. The traditional producers can absorb a big part of their disadvantages in

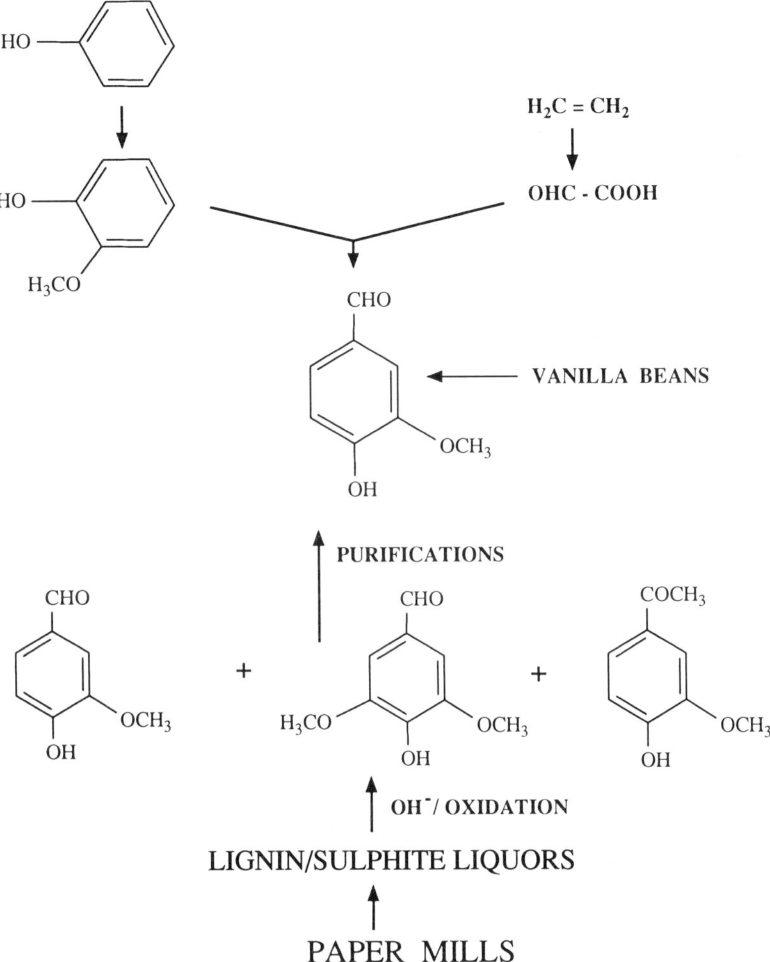

Fig. 18.12. Routes to vanillin.

their fully depreciated production facilities. Therefore the new technology has not fully replaced the old one; the present ratio being 2/1 in favour of the chemical synthesis.

Investments in new plants based on lignin are very unlikely unless major breakthroughs can be found in a more direct release of pure vanillin. Perhaps new enzyme-based techniques could give the lignin process a second life and solve the big pollution problems of paper mills. A lot of research on lignin degradation is already going on including work at

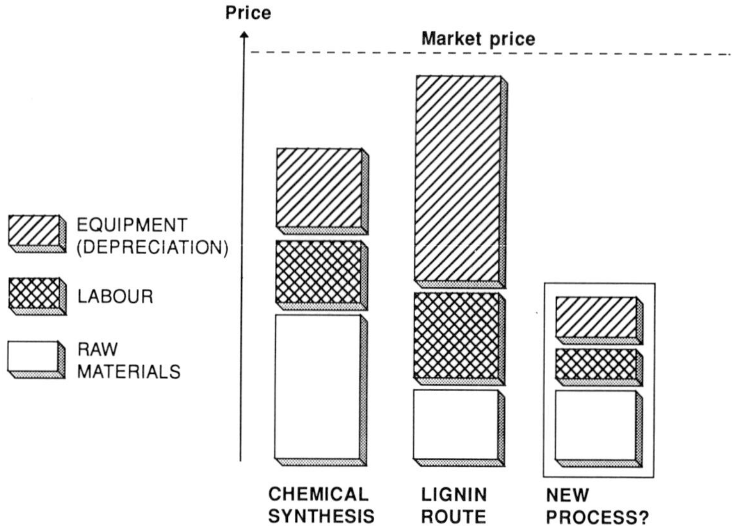

Fig. 18.13. Vanillin: chemical synthesis versus lignin-based route.

DSM-Research. Preservation of the vanillin molecule seems to be a problem.

Genetic engineering and manipulation of the vanilla bean and plants towards increased production or laboratory culture is also a possible new route. Research is going on at the University of Delaware and some small USA-based bioengineering enterprises. It is very unlikely, however, that this will result in a new and competitive process for large scale production. The result will rather be increased profits in the small market for vanilla bean extracts in combination with loss of jobs in the third world countries such as Madagascar. We could better use our biotechnological capabilities to improve lignin-based products and processes.

In summary, unlike phloroglucinol and gallic acid, the vanillin market is big enough to feel the effect of new technologies. The result is a shift from plant-based source, to chemical processes with good chances for a shift back. These effects take place slowly over some decades and are fully economy dictated after an initial technological breakthrough.

Vanillin, gallic acid, and trimethoprim

A pressing question for every research leader in industry is: 'Can technological changes be predicted?'. Is it possible to make the appropriation

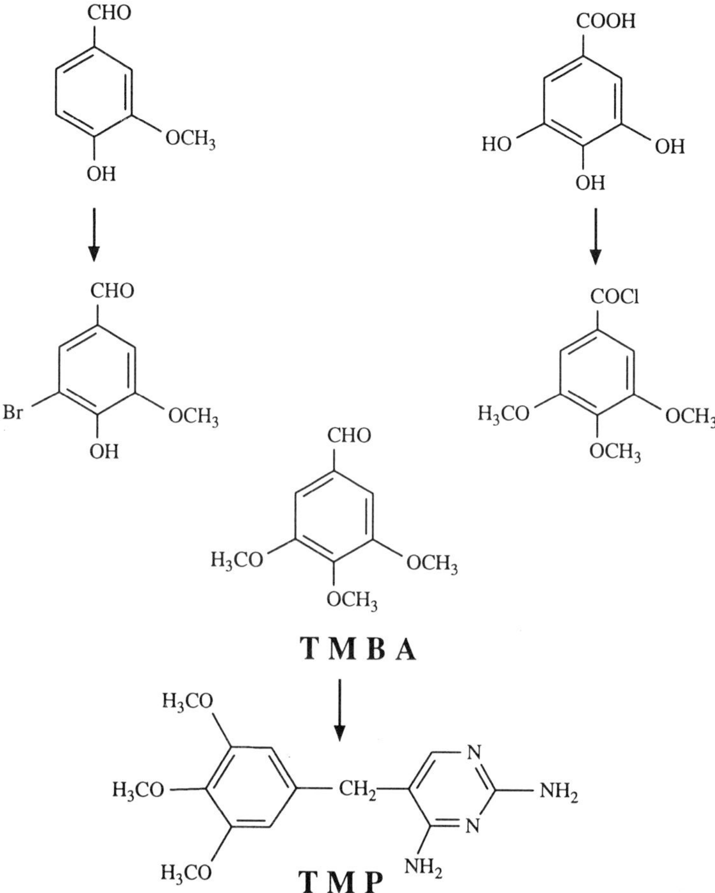

Fig. 18.14. Industrial routes to trimethoprim.

of long-term research money something other than a gamble? The answer to this question is a hesitating yes, but ... A careful combination of technology assessment and competition analysis can give some indications of future changes.

A joint application of gallic acid and vanillin will be used to explain some details of technology forecasting. Trimethoprim, an established and cheap synthetic antibacterial drug, can be produced both from gallic acid and from vanillin (Fig. 18.14). Both go through the common intermediate TMBA, 3,4,5-trimethoxybenzaldehyde. Trimethoprim, TMP, was introduced around 1965 by Hoffmann la Roche and Burroughs Wellcome. The annual consumption has now reached well over 1000 tonnes, and a

1965 - 1975	Synthesis through lignin based vanillin
1970 - 1980	Production via gallic acid (from tara powder) in combinaton with chemical catalysis
1975 - 1985	Synthetic vanillin as main source in combination with lignin based vanillin
1980 - 1990	Manufacture through gallic acid and traditional chemistry in developing countries

Fig. 18.15. Technology changes in TMBA/TMP production.

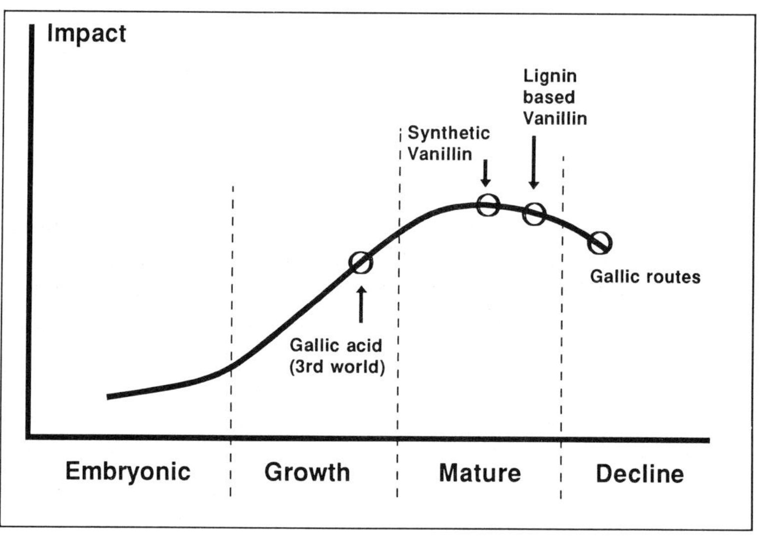

Fig. 18.16. Life-cycle of technologies used in TMBA/TMP synthesis.

significant shift to the developing world of both consumption and production has taken place. This geographic shift has influenced changes in manufacturing technology and vice versa. A technological time table can be drawn (Fig. 18.15).

A more useful economic model is given in the life-cycle graph of Fig. 18.16.

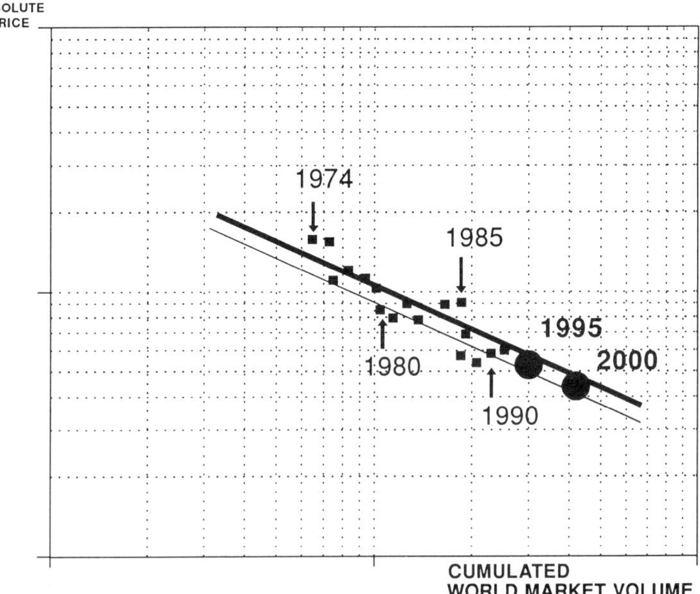

Fig. 18.17. Experience curve. Price versus volume.

What will be the most likely option for the future? Anyone can come up with a range of arguments to defend his preferred process, either new or existing. Detailed process profile analysis, full cost–price calculations, and investment scenarios are needed to make meaningful discrimination in the range of possibilities. It is beyond the scope of this presentation to discuss this in any detail. One helpful factor will be given.

Wouldn't it be nice to know the most likely market price for the future? That would rule out quite a few options. A good analysis of the past will predict the future market price, even in a mathematical way. A plot as given in Fig. 18.17 of the cumulated market volume against the price in constant money, both as the logarithm, does give a straight line. Even large changes in market price are accommodated, such as a dramatic price fall in 1986/1987 caused by large subsidized exports from China. A few years later the market is back to normal due to a combination of factors. China has normalized its export policy to some extent but more important is the development of new and more competitive technology. Based on cheaply available *p*-cresol, fully catalytic chemical processes are introduced which can accomodate the new market conditions in a competitive way (Fig. 18.18). Also the routes via synthetic vanillin are further improved.

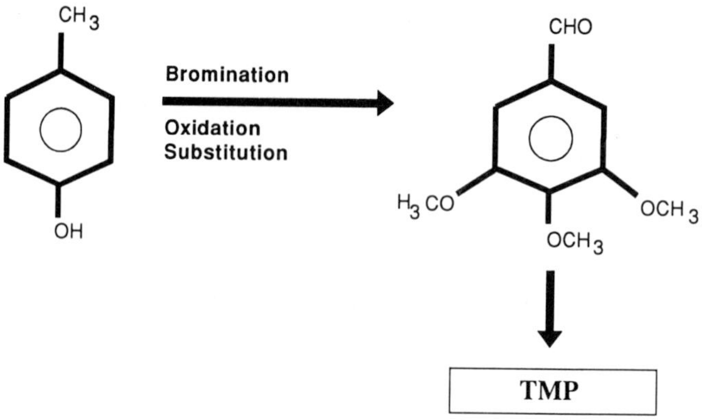

CH3
Bromination
Oxidation
Substitution
OH
CHO
H3 CO
OCH3
OCH3
TMP

Fig. 18.18. Modern process for TMBA/TMP.

PETROCHEMICALS
LIGNIN
DIRECT ROUTES
CHO
OCH3
OH
VANILLIN
Direct
Methoxylation
CHO
H3CO
OCH3
OH
SYRINGALDEHYDE
CHO
Br
OCH3
OH
CHO
H3CO
OCH3
OCH3
TMBA

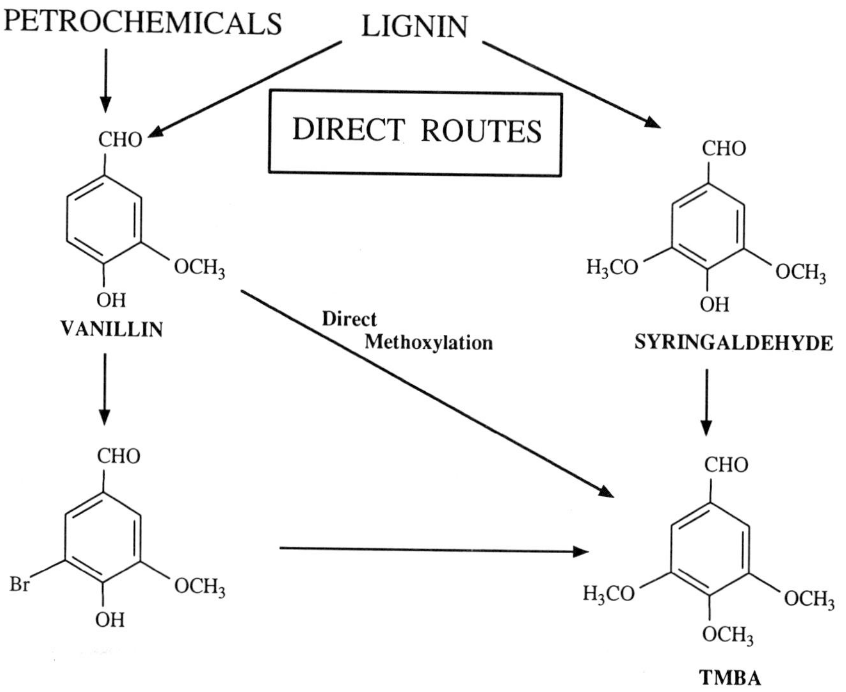

Fig. 18.19. Future industrial routes to TMBA/TMP?

1965 - 1975	Synthesis through lignin based vanillin
1970 - 1980	Production via gallic acid (from tara powder) in combination with chemical catalysis
1975 - 1985	Synthetic vanillin as main source in combination with lignin based vanillin
1980 - 1990	Manufacture through gallic acid and traditional chemistry in developing countries
1985 - 1995?	Fully catalytic chemical processes based on *p*-cresol
1990 - 2000?	Improved routes via gallic acid in developing countries
1995 - 2005?	Revival of lignin based processes; via vanillin or syringaldehyde Direct methoxylation of cheap petrochemicals

Fig. 18.20. Technology changes in TMBA/TMP production.

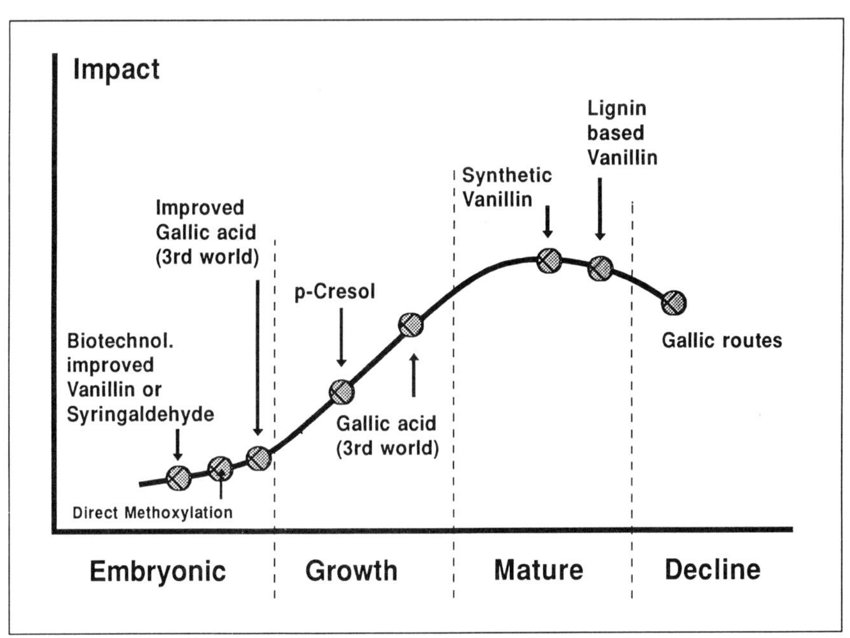

Fig. 18.21. Life-cycle of technologies in TMBA/TMP synthesis.

A good indication of future market developments can be gained by extrapolation. Since the growth in market volume will be roughly known it is easy to estimate where in 1995 or 2000 the market price will be (expressed in 1991 money). In this example a price drop of some 20 per cent is to be expected by 2000. This is accurate enough to determine whether the present technology will be strong enough to sustain these lower prices or new technology will be needed. It is conceivable that new technology will be needed to compete with further improved gallic acid based routes in developing countries. Improved synthesis of vanillin from lignin based on modern biotechnology might be an answer. Cheaper routes to synthetic vanillin via readily available petrochemicals might be an alternative, certainly so in combination with direct methoxylation. Direct isolation of syringaldehyde, also from lignin, might be an attractive alternative (Fig. 18.19). Depending on the source and kind of wood the ratio between vanillin and syringaldehyde is in favour of either one or the other. Thus our timetable might be extended as shown in Fig. 18.20.

Now still embryonic technologies might become reality as shown in the extended life-cycle in Fig. 18.21. But, only the future will tell which forecast is correct.

Conclusions

1. Technology changes in industry occur much more slowly than anticipated from scientific developments.
2. Technology changes can be rationalized and are to a certain extent predictable.
3. Technology changes are always economy driven.
4. There is never a 'winning technology'; every product has at any given time and place its own optimal production process.

Genus and species index

Subject index